网站开发案例课堂

jQuery动态网页设计案例课堂

刘玉红　蒲　娟　编著

清华大学出版社
北　京

内 容 简 介

本书以零基础讲解为宗旨，用实例引导读者深入学习，采取“基础入门→核心技术→移动网页开发→项目实战”的讲解模式，深入浅出地讲解 jQuery 的各项技术及实战技能。

本书第 1 篇“基础入门”主要内容包括：必须了解的 JavaScript 知识、深入学习 JavaScript 对象与数组、jQuery 的基础知识、jQuery 的选择器等；第 2 篇“核心技术”主要内容包括：用 jQuery 控制页面、jQuery 的动画特效、jQuery 的事件处理、jQuery 的功能函数、jQuery 与 Ajax 技术的应用、jQuery 插件开发与使用等；第 3 篇“移动网页开发”主要内容包括：走进 jQuery Mobile、jQuery Mobile UI 组件、jQuery Mobile 事件等；第 4 篇“项目实战”主要内容包括：开发时钟特效系统、开发动态字符演示系统、开发图片堆叠系统、开发商品信息展示系统、开发连锁酒店移动网站。

本书适合任何想学习 jQuery 网页设计的人员，无论您是否从事计算机相关行业，无论您是否接触过 jQuery，通过本书的学习均可快速掌握 jQuery 网页设计的方法和技巧。

图书在版编目(CIP)数据

jQuery 动态网页设计案例课堂/刘玉红，蒲娟编著. —北京：清华大学出版社，2018
(网站开发案例课堂)
ISBN 978-7-302-49193-4

Ⅰ. ①j… Ⅱ. ①刘… ②蒲… Ⅲ. ①JAVA 语言—程序设计 Ⅳ. ①TP312.8

中国版本图书馆 CIP 数据核字(2017)第 330866 号

责任编辑：张彦青
装帧设计：李　坤
责任校对：张彦彬
责任印制：沈　露
出版发行：清华大学出版社
网　　址：http://www.tup.com.cn, http://www.wqbook.com
地　　址：北京清华大学学研大厦 A 座　　邮　　编：100084
社 总 机：010-62770175　　邮　　购：010-62786544
投稿与读者服务：010-62776969, c-service@tup.tsinghua.edu.cn
质量反馈：010-62772015, zhiliang@tup.tsinghua.edu.cn
印 刷 者：清华大学印刷厂
装 订 者：三河市铭诚印务有限公司
经　　销：全国新华书店
开　　本：190mm×260mm　　**印　张**：26.75　　**字　数**：650 千字
版　　次：2018 年 3 月第 1 版　　**印　次**：2018 年 3 月第 1 次印刷
印　　数：1～3000
定　　价：69.00 元

产品编号：073027-01

前　言

“网站开发案例课堂”系列图书是专门为办公技能和网页设计初学者量身定制的一套学习用书。整套书涵盖网页设计、网站开发、数据库设计等方面。整套书具有以下特点。

前沿科技

无论是网站建设、数据库设计还是 HTML 5、CSS 3、JavaScript，我们都精选较为前沿或者用户群最大的领域推进，帮助大家认识和了解最新动态。

权威的作者团队

组织国家重点实验室和资深应用专家联手编著该套图书，融合丰富的教学经验与优秀的管理理念。

学习型案例设计

以技术的实际应用过程为主线，全程采用图解和同步多媒体结合的教学方式，生动、直观、全面地剖析使用过程中的各种应用技能，降低难度，同时提升学习效率。

为什么要写这样一本书

随着网页对用户页面体验要求的提高，JavaScript 再度受到广大技术人员的重视。jQuery 是继 prototype 之后又一个优秀的 JavaScript 框架。本书将全面介绍 jQuery 动态网页设计的知识，主要针对动态网页设计的初学者，让读者能够快速入门和上手，同时本书还介绍比较流行的移动 JavaScript 框架，即 jQuery Mobile。通过本书的项目演练，读者可以很快地掌握流行的工具，提高职业化能力，从而帮助解决公司与求职者的双重需求问题。

本书特色

■ 零基础、入门级的讲解

无论您是否从事计算机相关行业，无论您是否接触过 jQuery 动态网页设计，都能从本书中找到最佳起点。

■ 超多、实用、专业的范例和项目

本书在编排上紧密结合深入学习网页制作技术的先后过程，从 JavaScript 的基本概念开始，带领大家深入地学习各种应用技巧，侧重实战技能，使用简单易懂的实际案例进行分析和操作指导，让读者读起来简明轻松，操作起来有章可循。

■ 随时检测自己的学习成果

内容讲解章节最后的“疑难解惑”板块，均根据本章内容精选而成，从而帮助读者解决自学过程中最常见的疑难问题。

■ 细致入微、贴心提示

本书在讲解过程中，在各章中使用了“注意”“提示”“技巧”等小贴士，使读者在学习过程中更清楚地了解相关操作、理解相关概念，并轻松掌握各种操作技巧。

■ 专业创作团队和技术支持

您在学习过程中遇到任何问题，可加入 QQ 群(案例课堂 VIP)451102631 进行提问，专家人员会在线答疑。

超值资源大放送

■ 全程同步教学录像

涵盖本书所有知识点，详细讲解每个实例及项目的过程及技术关键点。比看书更轻松地掌握书中所有的网页制作和设计知识，而且扩展的讲解部分使您得到比书中更多的收获。

■ 超多容量王牌资源

赠送大量王牌资源，包括实例源代码、教学幻灯片、本书精品教学视频、88 个实用类网页模板、12 部网页开发必备参考手册、jQuery 参考手册、JavaScript 函数速查手册、精选的 JavaScript 实例、CSS 3 属性速查表、CSS+DIV 布局赏析案例、精彩网站配色方案赏析、网页样式与布局案例赏析、Web 前端工程师常见面试题等。读者可以通过 QQ 群(案例课堂 VIP)451102631 获取赠送资源，也可以扫描二维码，下载本书资源。

读者对象

- 没有任何网页设计基础的初学者。
- 有一定的 jQuery 基础，想精通 jQuery 网页设计的人员。
- 有一定的 jQuery 网页设计基础，没有项目经验的人员。
- 正在进行毕业设计的学生。
- 大专院校及培训学校的老师和学生。

创作团队

本书由刘玉红、蒲娟主编，参加编写的人员还有刘玉萍、张金伟、周佳、付红、李园、郭广新、侯永岗、王攀登、刘海松、孙若淞、王月娇、包慧利、陈伟光、胡同夫、王伟、展娜娜、李琪、梁云梁和周浩浩。在编写过程中，我们竭尽所能地将最好的讲解呈现给读者，但也难免有疏漏和不妥之处，敬请不吝指正。若您在学习中遇到困难或疑问，或有任何建议，可写信至信箱 357975357@qq.com。

编　者

目　录

第1篇　基础入门

第1章　必须了解的JavaScript知识..........3

1.1　认识JavaScript..........4

1.1.1　什么是JavaScript..........4

1.1.2　JavaScript的特点..........4

1.2　JavaScript的编写工具..........5

1.2.1　记事本..........5

1.2.2　Dreamweaver CC..........6

1.3　JavaScript在HTML 5中的使用..........6

1.3.1　在HTML 5网页头中嵌入JavaScript代码..........7

1.3.2　在HTML 5网页中嵌入JavaScript代码..........8

1.3.3　在HTML 5中调用已经存在的JavaScript文件..........9

1.4　JavaScript的核心语法..........10

1.4.1　变量的声明和赋值..........10

1.4.2　看透代码中的数据类型..........12

1.4.3　逻辑控制语句..........16

1.5　实战演练——一个简单的JavaScript示例..........19

1.6　疑难解惑..........20

第2章　深入学习JavaScript对象与数组..........21

2.1　了解对象..........22

2.1.1　什么是对象..........22

2.1.2　面向对象编程..........23

2.1.3　JavaScript的内部对象..........24

2.2　对象访问语句..........25

2.2.1　for-in循环语句..........25

2.2.2　with语句..........26

2.3　JavaScript中的数组..........27

2.3.1　结构化数据..........27

2.3.2　创建和访问数组对象..........27

2.3.3　使用for-in语句..........30

2.3.4　Array对象的常用属性和方法..........30

2.4　详解常用的数组对象方法..........39

2.4.1　连接其他数组到当前数组..........39

2.4.2　将数组元素连接为字符串..........40

2.4.3　移除数组中最后一个元素..........41

2.4.4　将指定的数值添加到数组中..........41

2.4.5　反序排列数组中的元素..........42

2.4.6　删除数组中的第一个元素..........43

2.4.7　获取数组中的一部分数据..........44

2.4.8　对数组中的元素进行排序..........45

2.4.9　将数组转换成字符串..........46

2.4.10　将数组转换成本地字符串..........47

2.4.11　在数组开头插入数据..........47

2.5　创建和使用自定义对象..........48

2.5.1　通过定义对象构造函数的方法..........48

2.5.2　通过对象直接初始化的方法..........51

2.5.3　修改和删除对象实例的属性..........52

2.5.4　通过原型为对象添加新属性和新方法..........53

2.5.5　自定义对象的嵌套..........54

2.5.6　内存的分配和释放..........57

2.6　实战演练——利用二维数组创建动态下拉菜单..........57

2.7　疑难解惑..........59

第3章　jQuery的基础知识..........61

3.1　jQuery概述..........62

3.1.1　jQuery能做什么..........62

3.1.2　jQuery的特点..........62

3.1.3　jQuery的技术优势..........63

3.2 下载并配置 jQuery........65
3.2.1 下载 jQuery........66
3.2.2 配置 jQuery........67
3.3 jQuery 的开发工具........67
3.3.1 JavaScript Editor Pro........67
3.3.2 Dreamweaver........68
3.3.3 UltraEdit........69
3.3.4 记事本工具........69
3.4 jQuery 的调试小工具........70
3.4.1 Firebug........70
3.4.2 Blackbird........73
3.4.3 jQueryPad........75
3.5 jQuery 与 CSS 3........75
3.5.1 CSS 3 构造规则........75
3.5.2 浏览器的兼容性........76
3.5.3 jQuery 的引入........77
3.6 实战演练——我的第一个 jQuery 程序........78
3.6.1 开发前的一些准备工作........78
3.6.2 具体的程序开发........79
3.7 疑难解惑........79
第 4 章 jQuery 的选择器........81
4.1 jQuery 的$........82
4.1.1 $符号的应用........82
4.1.2 功能函数的前缀........83
4.1.3 创建 DOM 元素........83
4.2 基本选择器........84
4.2.1 通配符选择器(*)........84
4.2.2 ID 选择器(#id)........85
4.2.3 类名选择器(.class)........87
4.2.4 元素选择器(element)........88
4.2.5 复合选择器........89
4.3 层级选择器........90
4.3.1 祖先后代选择器(ancestor descendant)........90
4.3.2 父子选择器(parent>child)........92
4.3.3 相邻元素选择器(prev+next)........94
4.3.4 兄弟选择器(prev~siblings)........95
4.4 过滤选择器........97
4.4.1 简单过滤选择器........97
4.4.2 内容过滤选择器........104
4.4.3 可见性过滤器........110
4.4.4 表单过滤器........114
4.5 表单选择器........116
4.5.1 :input 选择器........116
4.5.2 :text 选择器........117
4.5.3 :password 选择器........117
4.5.4 :radio 选择器........118
4.5.5 :checkbox 选择器........119
4.5.6 :submit 选择器........121
4.5.7 :reset 选择器........121
4.5.8 :button 选择器........122
4.5.9 :image 选择器........123
4.5.10 :file 选择器........124
4.6 属性选择器........125
4.6.1 [attribute]选择器........125
4.6.2 [attribute=value]选择器........127
4.6.3 [attribute!=value]选择器........128
4.6.4 [attribute$=value]选择器........129
4.7 实战演练——匹配表单中的元素并实现不同的操作........130
4.8 疑难解惑........132

第2篇 核心技术

第 5 章 用 jQuery 控制页面........137
5.1 对页面的内容进行操作........138
5.1.1 对文本内容进行操作........138
5.1.2 对 HTML 内容进行操作........140
5.1.3 移动和复制页面内容........141
5.1.4 删除页面内容........142
5.1.5 克隆页面内容........143
5.2 对标记的属性进行操作........144

5.2.1 获取属性的值............144
5.2.2 设置属性的值............145
5.2.3 删除属性的值............146
5.3 对表单元素进行操作............147
5.3.1 获取表单元素的值............147
5.3.2 设置表单元素的值............148
5.4 对元素的 CSS 样式进行操作............149
5.4.1 添加 CSS 类............149
5.4.2 删除 CSS 类............151
5.4.3 动态切换 CSS 类............153
5.4.4 获取和设置 CSS 样式............154
5.5 实战演练——制作奇偶变色的表格............156
5.6 疑难解惑............158

第 6 章 jQuery 的动画特效............161

6.1 jQuery 的基本动画效果............162
6.1.1 隐藏元素............162
6.1.2 显示元素............165
6.1.3 状态切换............167
6.2 淡入淡出的动画效果............168
6.2.1 淡入隐藏元素............169
6.2.2 淡出可见元素............170
6.2.3 切换淡入淡出元素............171
6.2.4 淡入淡出元素至指定数值............172
6.3 滑动效果............174
6.3.1 滑动显示匹配的元素............174
6.3.2 滑动隐藏匹配的元素............175
6.3.3 通过高度的变化动态切换元素的可见性............176
6.4 自定义的动画效果............178
6.4.1 创建自定义动画............178
6.4.2 停止动画............179
6.5 疑难解惑............181

第 7 章 jQuery 的事件处理............183

7.1 jQuery 的事件机制概述............184
7.1.1 什么是 jQuery 的事件机制............184
7.1.2 事件切换............184
7.1.3 事件冒泡............186
7.2 页面加载响应事件............187
7.3 jQuery 中的事件函数............188
7.3.1 键盘操作事件............188
7.3.2 鼠标操作事件............190
7.3.3 其他的常用事件............193
7.4 事件的基本操作............195
7.4.1 绑定事件............195
7.4.2 触发事件............196
7.4.3 移除事件............197
7.5 实战演练——制作绚丽的多级动画菜单............199
7.6 疑难解惑............204

第 8 章 jQuery 的功能函数............205

8.1 功能函数概述............206
8.2 常用的功能函数............207
8.2.1 操作数组和对象............207
8.2.2 操作字符串............210
8.2.3 序列化操作............212
8.3 调用外部代码............213
8.4 疑难解惑............214

第 9 章 jQuery 与 Ajax 技术的应用............217

9.1 Ajax 快速入门............218
9.1.1 什么是 Ajax............218
9.1.2 Ajax 的关键元素............221
9.1.3 CSS 在 Ajax 应用中的地位............221
9.2 Ajax 的核心技术............222
9.2.1 全面剖析 XMLHttpRequest 对象............222
9.2.2 发出 Ajax 请求............224
9.2.3 处理服务器响应............226
9.3 jQuery 中的 Ajax............227
9.3.1 load()方法............227
9.3.2 $.get()方法和$.post()方法............230
9.3.3 $.getScript()方法和$.getJson()方法............233
9.3.4 $.ajax()方法............236
9.4 疑难解惑............237

第 10 章 jQuery 插件的开发与使用 239

10.1 理解插件 240
10.1.1 什么是插件 240
10.1.2 如何使用插件 240
10.2 流行的插件 241
10.2.1 jQueryUI 插件 242
10.2.2 Form 插件 243
10.2.3 提示信息插件 245
10.2.4 jcarousel 插件 245
10.3 定义自己的插件 246
10.3.1 插件的工作原理 246
10.3.2 自定义一个简单的插件 247
10.4 实战演练——创建拖曳购物车效果 ... 250
10.5 疑难解惑 251

第 3 篇 移动网页开发

第 11 章 走进 jQuery Mobile 255

11.1 认识 jQuery Mobile 256
11.2 跨平台移动设备网页 jQuery Mobile.....256
11.2.1 移动设备模拟器 256
11.2.2 jQuery Mobile 的安装 258
11.2.3 jQuery Mobile 网页的架构 260
11.3 创建多页面的 jQuery Mobile 网页 261
11.4 将页面作为对话框使用 262
11.5 绚丽多彩的页面切换效果 264
11.6 疑难解惑 266

第 12 章 jQuery MobileUI 组件 267

12.1 套用 UI 组件 268
12.1.1 表单组件 268
12.1.2 按钮和按钮组 276
12.1.3 按钮图标 278
12.1.4 弹窗 280
12.2 列表 281
12.2.1 列表视图 281
12.2.2 列表内容 284
12.2.3 列表过滤 286
12.3 面板和可折叠块 287
12.3.1 面板 288
12.3.2 可折叠块 289
12.4 导航条 291
12.5 实战演练——使用 jQuery Mobile 主题 294
12.6 疑难解惑 297

第 13 章 jQuery Mobile 事件 299

13.1 页面事件 300
13.1.1 初始化事件 300
13.1.2 外部页面加载事件 302
13.1.3 页面过渡事件 304
13.2 触摸事件 306
13.2.1 点击事件 306
13.2.2 滑动事件 309
13.3 滚屏事件 311
13.4 定位事件 314
13.5 疑难解惑 316

第 4 篇 项 目 实 战

第 14 章 项目演练 1——开发时钟特效系统 319

14.1 项目需求分析 320
14.2 项目技术分析 321
14.3 系统的代码实现 321
14.3.1 设计首页 322
14.3.2 定义时钟类 323
14.3.3 定义数字时钟的视图类 325

14.3.4 定义圆形指针时钟的视图类....329
14.3.5 合并多个 js 文件.......................331
14.3.6 合并 Clock.js、DigitalView.js 和 CircleView.js 文件................332

第 15 章 项目演练 2——开发动态字符演示系统..343

15.1 项目需求分析..344
15.2 项目技术分析..345
15.3 系统的代码实现......................................345
15.3.1 设计首页..................................345
15.3.2 定义动画的类和执行动画的类..................................348
15.3.3 封装 jQuery 插件.......................358
15.3.4 合并 js 文件和编译 CSS 文件..359
15.3.5 合并 TextAnimate.js 和 jquery.textanimate.js 文件 360

第 16 章 项目演练 3——开发图片堆叠系统..367

16.1 项目需求分析..368
16.2 系统的代码实现......................................369
16.2.1 设计首页..................................369
16.2.2 图片堆叠核心功能...................374
16.2.3 封装 jQuery 插件.......................382
16.2.4 合并 js 文件和编译 CSS 文件.. 383
16.2.5 合并 ImgPile.js 和 jquery.imgpile.js 文件............... 384

第 17 章 项目演练 4——开发商品信息展示系统.. 393

17.1 项目需求分析.. 394
17.2 项目技术分析.. 396
17.3 系统的代码实现.................................... 396
17.3.1 设计首页................................ 396
17.3.2 开发控制器类的文件............... 398
17.3.3 开发数据模型类文件............... 400
17.3.4 开发视图抽象类的文件........... 402
17.3.5 项目中的其他 js 文件说明 405

第 18 章 项目演练 5——开发连锁酒店移动网站.. 407

18.1 连锁酒店订购的需求分析.................. 408
18.2 网站的结构.. 408
18.3 连锁酒店系统的代码实现.................. 409
18.3.1 设计首页................................ 409
18.3.2 订购页面................................ 410
18.3.3 连锁分店页面......................... 415
18.3.4 查看订单页面......................... 417
18.3.5 酒店介绍页面......................... 418

第 I 篇

基础入门

- 第 1 章　必须了解的 JavaScript 知识
- 第 2 章　深入学习 JavaScript 对象与数组
- 第 3 章　jQuery 的基础知识
- 第 4 章　jQuery 的选择器

第1章 必须了解的JavaScript知识

JavaScript 是目前 Web 应用程序开发者使用最为广泛的客户端脚本编程语言。它不仅可以用来开发交互式的 Web 页面，还可以将 HTML、XML 和 Java Applet、Flash 等 Web 对象有机地结合起来，使开发人员能快速生成 Internet 上使用的分布式应用程序。本章主要讲述 JavaScript 的基本入门知识。

1.1 认识 JavaScript

JavaScript 作为一种可以给网页增加交互性的脚本语言，拥有近 20 年的发展历史。它的简单、易学易用特性，使其始终立于不败之地。

1.1.1 什么是 JavaScript

JavaScript 最初由网景公司的 Brendan Eich 设计，是一种动态、弱类型、基于原型的语言，内置支持类。

经过近 20 年的发展，JavaScript 已经成为健壮的、基于对象和事件驱动的、有相对安全性的客户端脚本语言，同时也是一种广泛用于客户端 Web 开发的脚本语言，常用来给 HTML 5 网页添加动态功能，比如响应用户的各种操作。JavaScript 可以弥补 HTML 语言的缺陷，实现 Web 页面客户端动态效果，其主要作用如下。

(1) 动态改变网页内容。

HTML 语言是静态的，一旦编写，内容是无法改变的。JavaScript 可以弥补这种不足，将内容动态地显示在网页中。

(2) 动态改变网页的外观。

JavaScript 通过修改网页元素的 CSS 样式，可以动态地改变网页的外观，如修改文本的颜色、大小等属性，使图片的位置动态地改变。

(3) 验证表单数据。

为了提高网页的运行效率，用户在填写表单时，可以在客户端对数据进行合法性验证，验证成功之后才能提交到服务器上，这样就能减少服务器的负担和降低网络带宽的压力。

(4) 响应事件。

JavaScript 是基于事件的语言，因此可以响应用户或浏览器产生的事件。只有事件产生时才会执行某段 JavaScript 代码，如用户单击“计算”按钮时，程序显示运行结果。

几乎所有浏览器都支持 JavaScript，如 Internet Explorer(IE)、Firefox、Mozilla、Opera 等。

1.1.2 JavaScript 的特点

JavaScript 的特点主要有以下几个方面。

(1) 语法简单，易学易用。

JavaScript 语法简单、结构松散。可以使用任何一种文本编辑器来进行编写。JavaScript 程序运行时不需要编译成二进制代码，只需要支持 JavaScript 的浏览器进行解释。

(2) 解释型语言。

非脚本语言编写的程序通常需要经过“编写→编译→链接→运行”这 4 个步骤，而脚本语言 JavaScript 是解释型语言，只需要经过“编写→运行”这 2 个步骤。

(3) 跨平台。

由于 JavaScript 程序的运行仅依赖于浏览器，所以只要操作系统中安装有支持 JavaScript 的浏览器即可，即 JavaScript 与平台(操作系统)无关。例如，无论是 Windows、UNIX、Linux 操作系统，还是用于手机的 Android、iOS 操作系统，都可以运行 JavaScript。

(4) 基于对象和事件驱动。

JavaScript 把 HTML 页面中的每个元素都当作一个对象来处理，并且这些对象都具有层次关系，像一棵倒立的树，这种关系被称为“文档对象模型(DOM)”。在编写 JavaScript 代码时会接触到大量对象及对象的方法和属性。可以说学习 JavaScript 的过程，就是了解 JavaScript 对象及其方法和属性的过程。因为 JavaScript 基于事件驱动，所以它可以捕捉到用户在浏览器中的操作，可以将原来静态的 HTML 页面变成可以与用户交互的动态页面。

(5) 用于客户端。

尽管 JavaScript 分为服务器端和客户端两种，但目前应用得最多的还是客户端。

1.2 JavaScript 的编写工具

JavaScript 是一种脚本语言，代码不需要编译成二进制形式，而是以文本的形式存在，因此任何文本编辑器都可以作为其开发环境。

通常使用的 JavaScript 编辑器有记事本和 Dreamweaver 等。

1.2.1 记事本

记事本是 Windows 系统自带的文本编辑器，也是最简洁方便的文本编辑器。由于记事本的功能过于单一，所以要求开发者必须熟练掌握 JavaScript 语言的语法、对象、方法和属性等。这对于初学者是个极大的挑战，因此，不建议使用记事本。但是由于记事本简单方便、打开速度快，所以常用来做局部修改。

记事本窗口如图 1-1 所示。

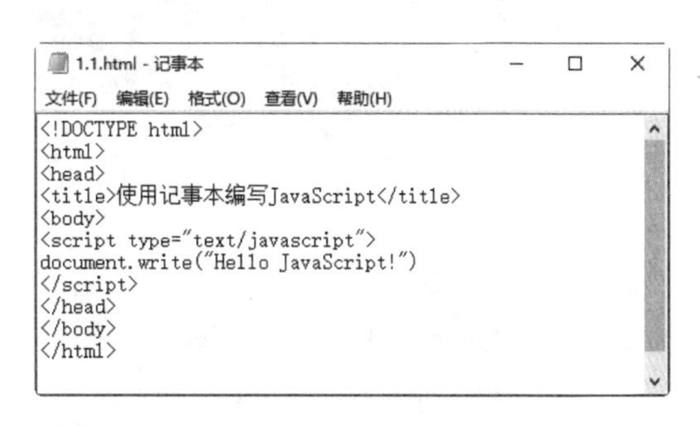

图 1-1 记事本窗口

在记事本中编写 JavaScript 程序的方法很简单，只需要在记事本中打开程序文件，然后在打开的记事本程序窗口中输入相关的 JavaScript 代码即可。

【例 1.1】(示例文件 ch01\1.1.html)

在记事本中编写 JavaScript 脚本。打开记事本文件，在窗口中输入代码(见图 1-1)：

```
<!DOCTYPE html>
<html>
<head>
<title>使用记事本编写 JavaScript</title>
<body>
<script type="text/javascript">
```

```
document.write("Hello JavaScript!")
</script>
</head>
</body>
</html>
```

将记事本文件保存为.html 格式的文件，然后使用 IE 11.0 打开，即可浏览最终的效果，如图 1-2 所示。

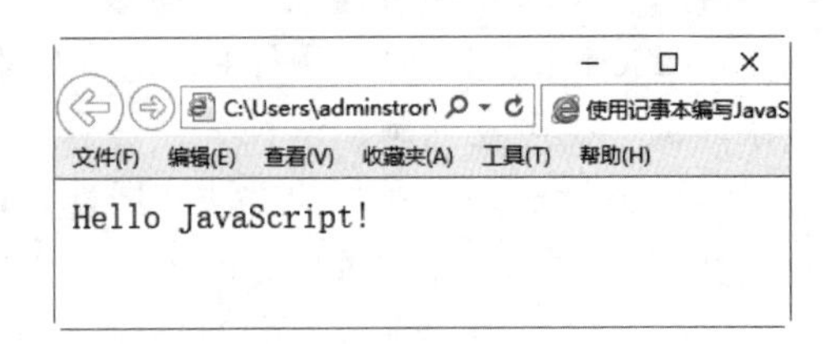

图 1-2　最终效果

1.2.2　Dreamweaver CC

Adobe 公司的 Dreamweaver CC 用户界面非常友好，是一款优秀的网页开发工具，深受广大用户的喜爱。Dreamweaver CC 的主界面如图 1-3 所示。

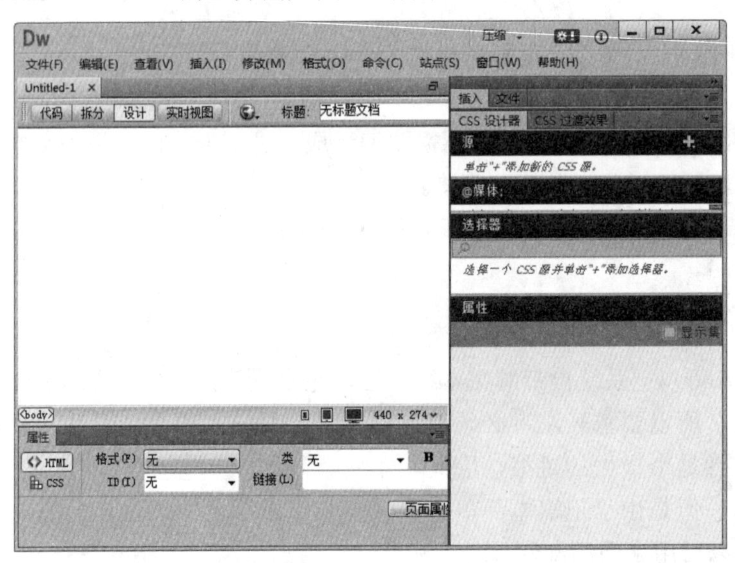

图 1-3　Dreamweaver CC 的主界面

除了上述编辑器外，还有很多种编辑器可以用来编写 JavaScript 程序，如 Aptana、1st JavaScript Editor、JavaScript Menu Master、Platypus JavaScript Editor、SurfMap JavaScript Editor 等。“工欲善其事，必先利其器”，选择一款适合自己的 JavaScript 编辑器，可以让程序员的工作事半功倍。

1.3　JavaScript 在 HTML 5 中的使用

创建好 JavaScript 脚本后，就可以在 HTML 5 中使用 JavaScript 脚本了。把 JavaScript 嵌入 HTML 5 中有多种方式，如在 HTML 5 网页头中嵌入、在 HTML 5 网页中嵌入、在 HTML 5 网页的元素事件中嵌入、在 HTML 5 中调用已经存在的 JavaScript 文件等。

1.3.1 在 HTML 5 网页头中嵌入 JavaScript 代码

如果不是通过 JavaScript 脚本生成 HTML 5 网页的内容，JavaScript 脚本一般放在 HTML 5 网页头部的<head></head>标签对之间。这样，不会因为 JavaScript 影响整个网页的显示结果。

在 HTML 5 网页头部的<head></head>标签对之间嵌入 JavaScript 的代码格式如下：

```
<!DOCTYPE html>
<html>
<head>
<title>在 HTML 5 网页头中嵌入 JavaScript 代码<title>
<script language="JavaScript">
<!--
...
JavaScript 脚本内容
...
//-->
</script>
</head>
<body>
...
</body>
</html>
```

在<script></script>标签对中添加相应的 JavaScript 脚本，这样就可以直接在 HTML 文件中调用 JavaScript 代码，以实现相应的效果。

【例 1.2】(示例文件 ch01\1.2.html)

在 HTML 5 网页头中嵌入 JavaScript 代码：

```
<!DOCTYPE html>
<html>
<head>
    <script language="javascript">
        document.write("欢迎来到 JavaScript 动态世界");
    </script>
</head>
<body>
    <p>学习 JavaScript！！！
</body>
</html>
```

该示例的功能是在 HTML 5 文档里输出一个字符串，即“欢迎来到 JavaScript 动态世界”。在 IE 11.0 中浏览，效果如图 1-4 所示，可以看到网页上输出了两句话，其中第一句就是从 JavaScript 中输出的。

在 JavaScript 的语法中，句末的分号“;”是 JavaScript 程序作为一个语句结束的标识符。

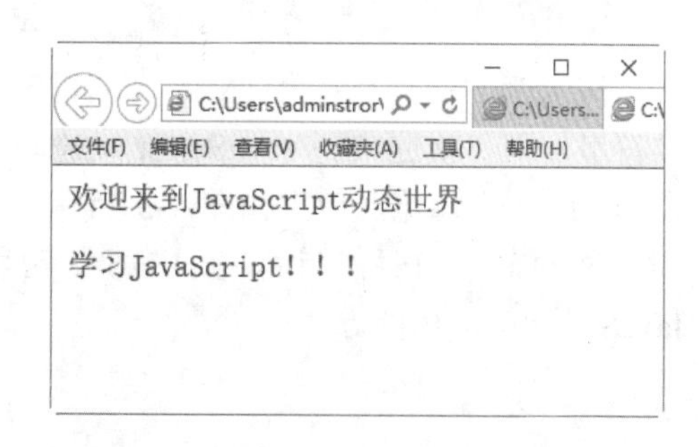

图 1-4 使用 head 中嵌入的 JavaScript 代码

1.3.2　在 HTML 5 网页中嵌入 JavaScript 代码

当需要使用 JavaScript 脚本生成 HTML 5 网页内容时，如某些 JavaScript 实现的动态树，就需要把 JavaScript 放在 HTML 5 网页主体部分的<body></body>标签对中。

具体的代码格式如下：

```
<!DOCTYPE html>
<html>
<head>
<title>在 HTML 5 网页中嵌入 JavaScript 代码<title>
</head>
<body>
<script language="JavaScript">
<!--
...
JavaScript 脚本内容
...
//-->
</script>
</body>
</html>
```

另外，JavaScript 代码可以在同一个 HTML 5 网页的头部与主体部分同时嵌入，并且在同一个网页中可以多次嵌入 JavaScript 代码。

【例 1.3】(示例文件 ch01\1.3.html)

在 HTML 5 网页中嵌入 JavaScript 代码：

```
<!DOCTYPE html>
<html>
<head>
</head>
<body>
    <p>学习 JavaScript！！！ </p>
    <script language="javascript">
       document.write("欢迎来到 JavaScript 动态世界");
    </script>
</body>
</html>
```

该示例的功能是在 HTML 文档里输出一个字符串，即“欢迎来到 JavaScript 动态世界”。在 IE 11.0 中浏览，效果如图 1-5 所示。可以看到，网页输出了两句话，其中第二句就是从 JavaScript 中输出的。

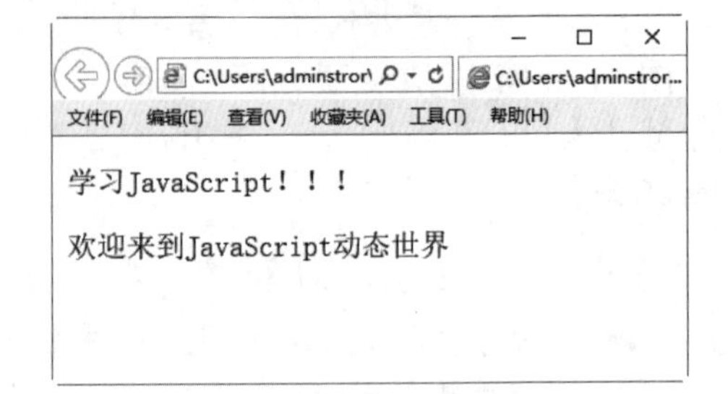

图 1-5　使用 body 中嵌入的 JavaScript 代码

1.3.3 在 HTML 5 中调用已经存在的 JavaScript 文件

如果 JavaScript 的内容较长，或者多个 HTML 5 网页中都调用相同的 JavaScript 程序，可以将较长的 JavaScript 或者通用的 JavaScript 写成独立的.js 文件，直接在 HTML 5 网页中调用。

【例 1.4】(示例文件 ch01\1.4.html)

下面的 HTML 代码就是使用 JavaScript 脚本来调用外部的 JavaScript 文件：

```
<!DOCTYPE html>
<html>
<head>
<title>使用外部文件</title>
<script src = "hello.js"></script>
</head>
<body>
<p>此处引用了一个 javascript 文件
</body>
</html>
```

在 IE 11.0 中浏览，效果如图 1-6 所示。

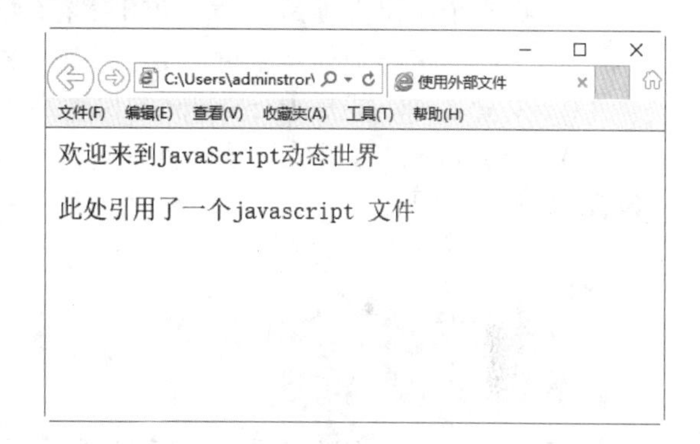

图 1-6 使用导入的 JavaScript 文件

可见通过这种外部引用 JavaScript 文件的方式，也可以实现相应的功能，这种功能具有下面两个优点。

(1) 将脚本程序同现有页面的逻辑结合。通过外部脚本，可以轻易实现多个页面完成同一功能的脚本文件，可以很方便地通过更新一个脚本内容实现批量更新。

(2) 浏览器可以实现对目标脚本文件的高速缓存。这样可以避免引用同样功能的脚本代码而导致下载时间增加。

与 C 语言使用外部头文件(.h 文件等)相似，引入 JavaScript 脚本代码时，使用外部脚本文件的方式符合结构化编程思想，但也有一些缺点，具体表现在以下两个方面。

(1) 并不是所有支持 JavaScript 脚本的浏览器都支持外部脚本。

(2) 外部脚本文件功能过于复杂，或其他原因导致的加载时间过长，则可能导致页面事件得不到处理或得不到正确的处理。程序员必须小心使用并确保脚本加载完成后，其中定义的函数才被页面事件调用，否则浏览器会报错。

综上所述，引入外部 JavaScript 脚本文件的方法是效果与风险并存的。设计人员应该权衡其优缺点，以决定是将脚本代码嵌入到目标 HTML 文件中，还是通过引用外部脚本的方式来实现相同的功能。一般情况下，将实现通用功能的 JavaScript 脚本代码作为外部脚本文件引用，而实现特有功能的 JavaScript 代码则直接嵌入到 HTML 文件中的<head></head>标签对之间载入，使其能及时并正确地响应页面事件。

1.4 JavaScript 的核心语法

JavaScript 的核心语法知识是学习 JavaScript 的必备知识。下面介绍 JavaScript 中最常用的语法知识。

1.4.1 变量的声明和赋值

变量，顾名思义，在程序运行过程中，其值可以改变。变量是存储信息的单元，它对应于某个内存空间。变量用于存储特定数据类型的数据，用变量名代表其存储空间。程序能在变量中存储值和取出值，可以把变量比作超市的货架(内存)，货架上摆放着商品(变量)，可以把商品从货架上取出来(读取)，也可以把商品放入货架(赋值)。

1. 变量的命名

实际上，变量的名称是一个标识符。在 JavaScript 中，用标识符来命名变量和函数，变量的名称可以是任意长度。创建变量名称时，应该遵循以下规则。

(1) 第一个字符必须是一个 ASCII 字符(大小写均可)或一个下划线(_)，但不能是文字。

(2) 后续的字符必须是字母、数字或下划线。

(3) 变量名称不能是 JavaScript 的保留字。

(4) JavaScript 的变量名是严格区分大小写的。例如，变量名称 myCounter 与变量名称 MyCounter 是不同的。

下面给出一些合法的变量命名示例：

```
_pagecount
Part9
Numer
```

下面给出一些错误的变量命名示例：

```
12balloon                   //不能以数字开头
Summary&Went                //“与”符号不能用在变量名称中
```

2. 变量的声明与赋值

JavaScript 是一种弱类型的程序设计语言，变量可以不声明直接使用。所谓声明变量，就是为变量指定一个名称。声明变量后，就可以把它用作存储单元。

JavaScript 中使用关键字 var 来声明变量，在这个关键字之后的字符串将代表一个变量名。声明格式如下：

```
var 标识符;
```

例如，声明变量 username，用来表示用户名，代码如下：

```
var username;
```

另外，一个关键字 var 也可以同时声明多个变量名，多个变量名之间必须用逗号“,”分隔。例如，同时声明变量 username、pwd、age，分别表示用户名、密码和年龄，代码如下：

```
var username,pwd,age;
```

要给变量赋值，可以使用 JavaScript 中的赋值运算符，即等于号(=)。

声明变量名时可以同时赋值，例如，声明变量 username 并赋值为“张三”，代码如下：

```
var username = "张三";
```

声明变量之后，对变量赋值，或者对未声明的变量直接赋值。例如，声明变量 age，然后再为它赋值，以及直接对变量 count 赋值：

```
var age;          //声明变量
age = 18;         //对已声明的变量赋值
count = 4;        //对未声明的变量直接赋值
```

JavaScript 中的变量如果未初始化(赋值)，默认值为 undefined。

3. 变量的作用范围

所谓变量的作用范围，是指可以访问该变量的代码区域。在 JavaScript 中，变量按其作用范围分为全局变量和局部变量。

(1) 全局变量。可以在整个 HTML 文档范围内使用的变量，这种变量通常都是在函数体外定义的变量。

(2) 局部变量。只能在局部范围内使用的变量，这种变量通常都是在函数体内定义的变量，所以只能在函数体中有效。

省略关键字 var 声明的变量，无论是在函数体内还是函数体外，都是全局变量。

【例 1.5】(示例文件 ch01\1.5.html)

创建名为 carname 的变量，并向其赋值 Volvo，然后把它放入 id="demo"的 HTML 段落中。代码如下：

```
<!DOCTYPE html>
<html>
<head>
</head>
<body>
  <p>点击这里来创建变量，并显示结果。</p>
  <button onclick="myFunction()">点击这里</button>
  <p id="demo"></p>
<script type="text/javascript">
function myFunction()
{
  var carname="Volvo";
  document.getElementById("demo").innerHTML=carname;
}
</script>
```

```
</body>
</html>
```

在 IE 11.0 中浏览，效果如图 1-7 所示。单击页面中的“点击这里”按钮，可以看到元素发生了变化，如图 1-8 所示。

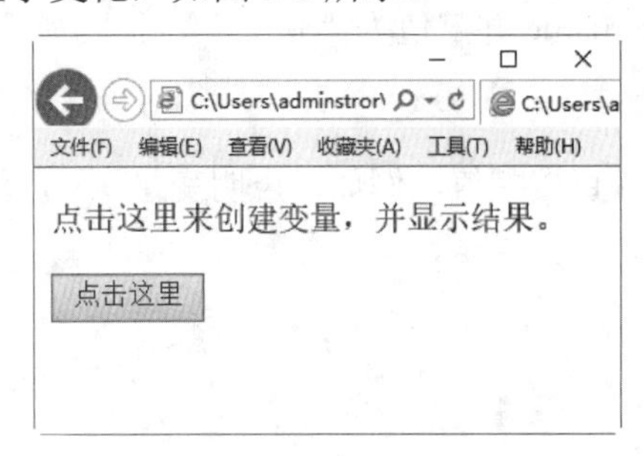

图 1-7　初始效果

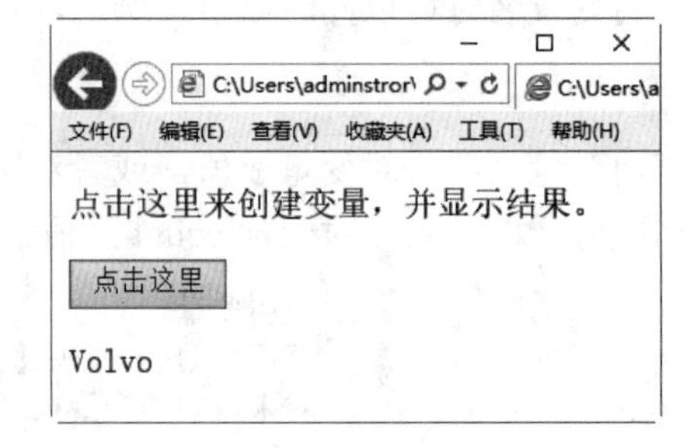

图 1-8　单击按钮后

一个好的编程习惯是，在代码开始处，统一对需要的变量进行声明。

1.4.2　看透代码中的数据类型

每一种计算机语言除了有自己的数据结构外，还具有自己所支持的数据类型。在 JavaScript 脚本语言中，采用的是弱类型方式，即一个变量不必首先做声明，可以在使用或赋值时再确定其数据类型，当然也可以先声明该变量的类型。

1. typeof 运算符

typeof 运算符有一个参数，即要检查的变量或值。例如：

```
var sTemp = "test string";
alert(typeof sTemp);        //输出 string
alert(typeof 86);           //输出 number
```

对变量或值调用 typeof 运算符将返回下列值之一。

- undefined：表示变量是 undefined 类型的。
- boolean：表示变量是 boolean 类型的。
- number：表示变量是 number 类型的。
- string：表示变量是 string 类型的。
- object：表示变量是一种引用类型或 null 类型的。

【例 1.6】(示例文件 ch01\1.6.html)

typeof 运算符的使用：

```
<!DOCTYPE html>
<html>
<head>
</head>
```

```
<body>
<script type="text/javascript">
  typeof(1);
  typeof(NaN);
  typeof(Number.MIN_VALUE);
  typeof(Infinity);
  typeof("123");
  typeof(true);
  typeof(window);
  typeof(document);
  typeof(null);
  typeof(eval);
  typeof(Date);
  typeof(sss);
  typeof(undefined);
  document.write("typeof(1): "+typeof(1)+"<br>");
  document.write("typeof(NaN): "+typeof(NaN)+"<br>");
  document.write("typeof(Number.MIN_VALUE): "
    + typeof(Number.MIN_VALUE)+"<br>")
  document.write("typeof(Infinity): "+typeof(Infinity)+"<br>")
  document.write("typeof(\"123\"): "+typeof("123")+"<br>")
  document.write("typeof(true): "+typeof(true)+"<br>")
  document.write("typeof(window): "+typeof(window)+"<br>")
  document.write("typeof(document): "+typeof(document)+"<br>")
  document.write("typeof(null): "+typeof(null)+"<br>")
  document.write("typeof(eval): "+typeof(eval)+"<br>")
  document.write("typeof(Date): "+typeof(Date)+"<br>")
  document.write("typeof(sss): "+typeof(sss)+"<br>")
  document.write("typeof(undefined): "+typeof(undefined)+"<br>")
</script>
</body>
</html>
```

在 IE 11.0 中浏览，效果如图 1-9 所示。

图 1-9　使用 typeof 运算符

2. undefined 类型

undefined 是未定义类型的变量，表示变量还没有赋值，如“var a;”，或者赋予一个不存在的属性值，例如“var a = String.notProperty;”。

此外，JavaScript 中有一种特殊类型的常量 NaN，表示“非数字”。当在程序中由于某种原因发生计算错误后，将产生一个没有意义的值，此时 JavaScript 返回的就是 NaN。

【例 1.7】(示例文件 ch01\1.7.html)

使用 undefined：

```
<!DOCTYPE html>
<html>
<head>
</head>
<body>
<script type="text/javascript">
  var person;
  document.write(person + "<br />");
</script>
</body>
</html>
```

在 IE 11.0 中浏览，效果如图 1-10 所示。

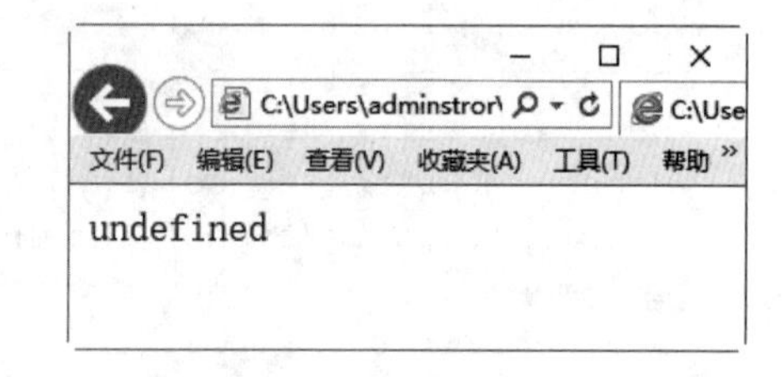

图 1-10 使用 undefined 的变量

3. null 类型

JavaScript 中的关键字 null 是一个特殊的值，表示空值，用于定义空的或不存在的引用。不过，null 不等同于空的字符串或 0。由此可见，null 与 undefined 的区别是：null 表示一个变量被赋予了一个空值，而 undefined 则表示该变量还未被赋值。

【例 1.8】(示例文件 ch01\1.8.html)

使用 null：

```
<!DOCTYPE html>
<html>
<head>
</head>
<body>
<script type="text/javascript">
  var person;
  document.write(person + "<br />");
  var car = null;
  document.write(car + "<br />");
</script>
</body>
</html>
```

在 IE 11.0 中浏览，效果如图 1-11 所示。

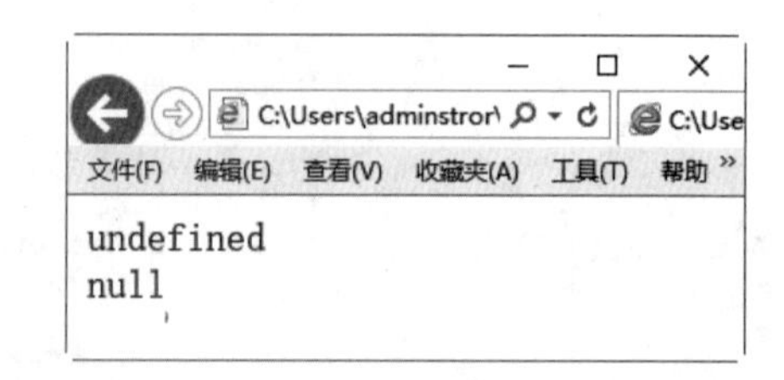

图 1-11 使用被赋予了一个空值的变量

4. boolean 类型

布尔类型 boolean 表示一个逻辑数值，用于表示两种可能的情况。逻辑真，用 true 表示；逻辑假，用 false 表示。通常，我们使用 1 表示真，0 表示假。

【例 1.9】(示例文件 ch01\1.9.html)

使用 boolean 类型：

```
<!DOCTYPE html>
<html>
<head>
</head>
<body>
<script type="text/javascript">
  var b1 = Boolean(""); //返回 false，空字符串
  var b2 = Boolean("s"); //返回 true，非空字符串
  var b3 = Boolean(0); //返回 false，数字 0
  var b4 = Boolean(1); //返回 true，非 0 数字
  var b5 = Boolean(-1); //返回 true，非 0 数字
  var b6 = Boolean(null); //返回 false
  var b7 = Boolean(undefined); //返回 false
  var b8 = Boolean(new Object()); //返回 true，对象
  document.write(b1 + "<br>")
  document.write(b2 + "<br>")
  document.write(b3 + "<br>")
  document.write(b4 + "<br>")
  document.write(b5 + "<br>")
  document.write(b6 + "<br>")
  document.write(b7 + "<br>")
  document.write(b8 + "<br>")
</script>
</body>
</html>
```

在 IE 11.0 中浏览，效果如图 1-12 所示。

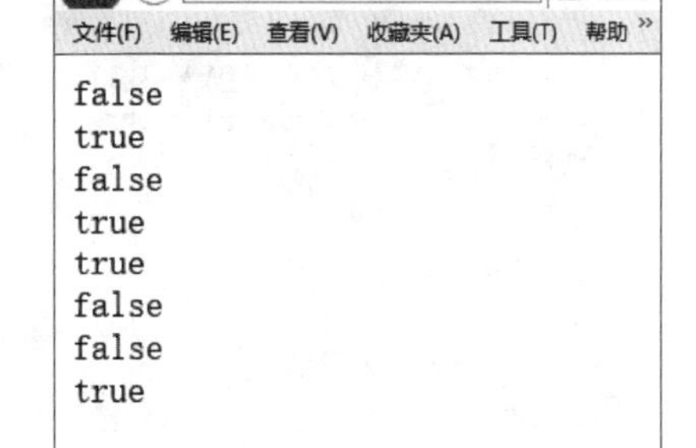

图 1-12　使用 boolean 类型

5. number 类型

JavaScript 的数值类型可以分为 4 类，即整数、浮点数、内部常量和特殊值。整数可以为正数、0 或者负数；浮点数可以包含小数点，也可以包含一个 e(大小写均可，在科学记数法中表示“10 的幂”)，或者同时包含这两项。整数可以以 10(十进制)、8(八进制)和 16(十六进制)作为基数来表示。

【例 1.10】(示例文件 ch01\1.10.html)

输出数值：

```
<!DOCTYPE html>
<html><head></head>
<body>
```

```
<script type="text/javascript">
  var x1 = 36.00;
  var x2 = 36;
  var y = 123e5;
  var z = 123e-5;
  document.write(x1 + "<br />")
  document.write(x2 + "<br />")
  document.write(y + "<br />")
  document.write(z + "<br />")
</script>
</body>
</html>
```

在 IE 11.0 中浏览，效果如图 1-13 所示。

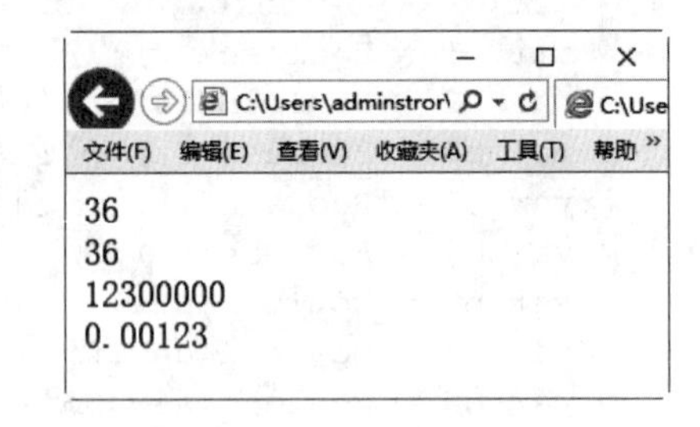

图 1-13 输出数值

6. string 类型

字符串是用一对单引号('')或双引号("")和引号中的内容构成的。字符串也是 JavaScript 中的一种对象，有专门的属性。引号中间的部分可以是任意多的字符，如果没有，则是一个空字符串。

如果要在字符串中使用双引号，则应该将其包含在使用单引号的字符串中，使用单引号时则反之。

【例 1.11】(示例文件 ch01\1.11.html)

输出字符串：

```
<!DOCTYPE html>
<html><head></head>
<body>
<script type="text/javascript">
  var string1 = "Bill Gates";
  var string2 = 'Bill Gates';
  var string3 = "Nice to meet you!";
  var string4 = "He is called 'Bill'";
  var string5 = 'He is called "Bill"';
  document.write(string1 + "<br>")
  document.write(string2 + "<br>")
  document.write(string3 + "<br>")
  document.write(string4 + "<br>")
  document.write(string5 + "<br>")
</script>
</body>
</html>
```

在 IE 11.0 中浏览，效果如图 1-14 所示。

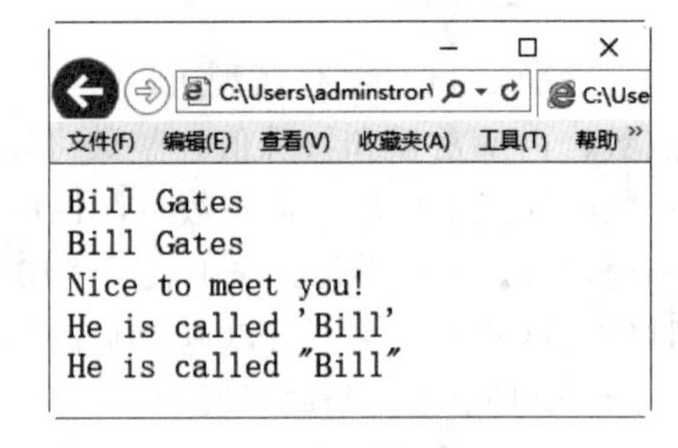

图 1-14 输出字符串

1.4.3 逻辑控制语句

JavaScript 编程中对程序流程的控制主要是通过条件判断、循环控制语句及 continue、

break 来完成的，其中条件判断按预先设定的条件执行程序，它包括 if 语句和 switch 语句；而循环控制语句则可以重复完成任务，它包括 while 语句、do-while 语句及 for 语句。

1. 条件判断语句

条件判断语句就是对语句中条件的值进行判断，进而根据不同的值执行不同的语句。条件判断语句主要包括两大类，分别是 if 判断语句和 switch 多分支语句。

(1) if 语句是使用得最为普遍的条件选择语句。每一种编程语言都有一种或多种形式的 if 语句，在编程中它是经常被用到的。

if 语句的语法格式如下：

```
if(条件语句)
{
    执行语句;
}
```

其中的“条件语句”可以是任何一种逻辑表达式。如果“条件语句”的返回结果为 true，则程序先执行后面大括号{}中的“执行语句”，然后接着执行它后面的其他语句。如果“条件语句”的返回结果为 false，则程序跳过“条件语句”后面的“执行语句”，直接去执行程序后面的其他语句。大括号的作用就是将多条语句组合成一个复合语句，作为一个整体来处理。如果大括号中只有一条语句，这对大括号就可以省略。

(2) if-else 语句通常用于一个条件需要两个程序分支来执行的情况。if-else 语句的语法格式如下：

```
if (条件)
{
    当条件为 true 时执行的代码
}
else
{
    当条件不为 true 时执行的代码
}
```

这种格式在 if 从句的后面添加了一个 else 从句，这样当条件语句的返回结果为 false 时，执行 else 后面部分的代码。

(3) 使用 if-else-if 语句来选择多个代码块之一来执行。if-else-if 语句的语法格式如下：

```
if (条件 1)
{
    当条件 1 为 true 时执行的代码
}
else if (条件 2)
{
    当条件 2 为 true 时执行的代码
}
else
{
    当条件 1 和 条件 2 都不为 true 时执行的代码
}
```

(4) switch 选择语句用于将一个表达式的结果与多个值进行比较，并根据比较结果选择执行语句。

switch 语句的语法格式如下：

```
switch (表达式)
{
    case 取值 1:
       语句块 1; break;
    case 取值 2:
       语句块 2; break;
    ...
    case 取值 n;
       语句块 n; break;
    default:
       语句块 n+1;
}
```

case 语句只是相当于定义一个标记位置，程序根据 switch 条件表达式的结果，直接跳转到第一个匹配的标记位置处，开始顺序执行后面的所有程序代码，包括后面的其他 case 语句下的代码，直至遇到 break 语句或函数返回语句为止。default 语句是可选的，它匹配上面所有的 case 语句定义的值以外的其他值，也就是当前面所有取值都不满足时，就执行 default 后面的语句块。

2. 循环控制语句

循环控制语句，顾名思义，主要就是在满足条件的情况下反复执行某一个操作，循环控制语句主要包括 while 语句、do-while 语句和 for 语句。

(1) while 语句是循环语句，也是条件判断语句。while 语句的语法格式如下：

```
while(条件表达式语句)
{
    执行语句块
}
```

当“条件表达式语句”的返回值为 true 时，则执行大括号{}中的语句块，然后再次检测条件表达式的返回值，如果返回值仍然是 true，则重复执行大括号{}中的语句块，直到返回值为 false 时，结束整个循环过程，接着往下执行 while 代码段后面的程序代码。

(2) do-while 语句的功能和 while 语句差不多，只不过它是在执行完第一次循环之后才检测条件表达式的值，这意味着包含在大括号中的代码块至少要被执行一次。另外，do-while 语句结尾处的 while 条件语句的括号后有一个分号“;”，该分号一定不能省略。

do-while 语句的语法格式如下：

```
do
{
    执行语句块
} while(条件表达式语句);
```

(3) for 语句通常由两部分组成：一是条件控制部分；二是循环部分。for 语句的语法格式如下：

```
for(初始化表达式; 循环条件表达式; 循环后的操作表达式)
{
    执行语句块
}
```

在使用 for 循环前，要先设定一个计数器变量，可以在 for 循环之前预先定义，也可以在使用时直接进行定义。在上述语法格式中，“初始化表达式”表示计数器变量的初始值；“循环条件表达式”是一个计数器变量的表达式，决定了计数器的最大值；“循环后的操作表达式”表示循环的步长，也就是每循环一次，计数器变量值的变化，该变化可以是增大的，也可以是减小的，或进行其他运算。for 循环是可以嵌套的，也就是在一个循环里还可以有另一个循环。

1.5　实战演练——一个简单的 JavaScript 示例

本例是一个简单的 JavaScript 程序，主要用来说明如何编写 JavaScript 程序以及在 HTML 中如何使用。本例主要实现的功能为：页面打开时显示“尊敬的客户，欢迎您光临本网站”对话框，关闭页面时弹出“欢迎下次光临！”对话框，效果如图 1-15 和图 1-16 所示。

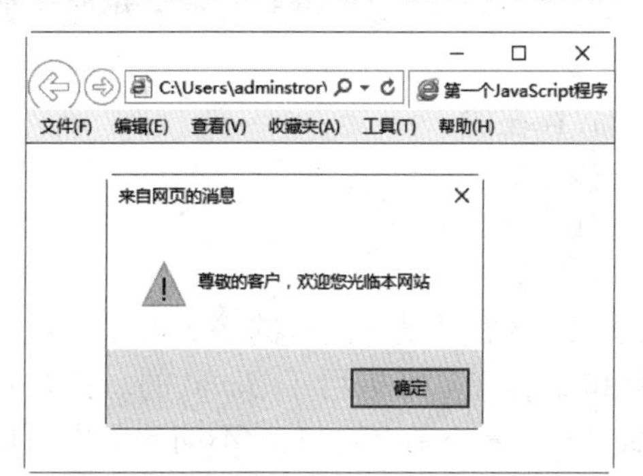

图 1-15　页面加载时的效果

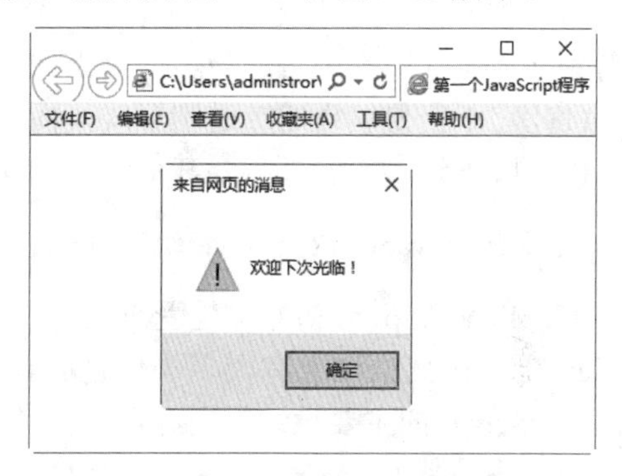

图 1-16　页面关闭时的效果

具体操作步骤如下。

step 01 新建 HTML 5 文档，输入以下代码：

```
<!DOCTYPE html>
<html>
<head>
<title>第一个 JavaScript 程序</title>
</head>
<body>
</body>
</html>
```

step 02 保存 HTML 5 文件，选择相应的保存位置，文件名为 welcome.html。

step 03 在 HTML 文档的 head 部分输入如下代码：

```
<script>
    //页面加载时执行的函数
    function showEnter(){
        alert("尊敬的客户，欢迎您光临本网站");
    }
    //页面关闭时执行的函数
    function showLeave(){
        alert("欢迎下次光临！");
    }
    //页面加载事件触发时调用函数
    window.onload=showEnter;
    //页面关闭事件触发时调用函数
    window.onbeforeunload=showLeave;
</script>
```

step 04 保存网页，浏览最终效果。

1.6 疑难解惑

疑问 1：什么是脚本语言？

答：脚本语言是由传统编程语言简化而来的语言，它与传统编程语言有很多相似之处，也有不同之处。脚本语言的最显著特点如下。

(1) 它不需要编译成二进制形式，而是以文本形式存在的。

(2) 脚本语言一般都需要其他语言的调用，不能独立运行。

疑问 2：JavaScript 是 Java 的变种吗？

答：JavaScript 最初的确是受 Java 启发而开始设计的，而且设计的目的之一就是“看上去像 Java”，因此语法上有很多类似之处，许多名称和命名规范也借用了 Java 的。但是实际上，JavaScript 的主要设计原则源自 Self 和 Scheme，它与 Java 在本质上是不同的。之所以与 Java 在名称上近似，是因为当时网景为了营销上的考虑与 Sun 公司达成协议的结果。其实从本质上讲，JavaScript 更像是一门函数式编程语言，而非面向对象的语言，它使用一些智能的语法和语义来仿真高度复杂的行为。其对象模型极为灵活、开放和强大，具有全部的反射性。

疑问 3：JavaScript 与 JScript 相同吗？

答：为了取得技术优势，微软推出了 JScript 来迎战 JavaScript 脚本语言。为了加强互用性，ECMA 国际协会(前身为欧洲计算机制造商协会)建立了 ECMA-262 标准(ECMAScript)。现在 JavaScript 与 JScript 两者都属于 ECMAScript 的实现。

疑问 4：JavaScript 是一门简单的语言吗？

答：尽管 JavaScript 是作为给非编程人员的脚本语言，而不是作为给编程人员的编程语言来推广和宣传的，但是，JavaScript 是一门具有丰富特性的语言，它有着与其他编程语言一样的复杂性，甚至更加复杂。实际上，我们必须对 JavaScript 有扎实的理解，才能用它来编写比较复杂的程序。

第 2 章 深入学习 JavaScript 对象与数组

对象是 JavaScript 最基本的数据类型之一，是一种复合的数据类型，它将多种数据类型集中在一个数据单元中，同时允许通过对象名来存取这些数据的值。数组是 JavaScript 中唯一用来存储和操作有序数据集的数据结构。本章主要介绍对象与数组的基本概念和基础知识。

2.1 了解对象

在 JavaScript 中，对象包括内置对象、自定义对象等多种类型，使用这些对象，可大大简化 JavaScript 程序的设计，并提供直观、模块化的方式进行脚本程序开发。

2.1.1 什么是对象

对象(Object)是一件事、一个实体、一个名词，是可以获得的东西，是可以想象有自己标识的任何东西。对象是类的实例化。有些对象是活的，有些对象不是。以自然人为例，我们来构造一个对象，其中 Attribute 表示对象属性，Method 表示对象行为，如图 2-1 所示。

在计算机语言中也存在对象，可以定义为相关变量和方法的软件集。对象主要由下面两个部分组成。

(1) 一组包含各种类型数据的属性。

(2) 允许对属性中的数据进行的操作，即相关方法。

以 HTML 文档中的 document 对象为例，其中包含各种属性和方法，如图 2-2 所示。

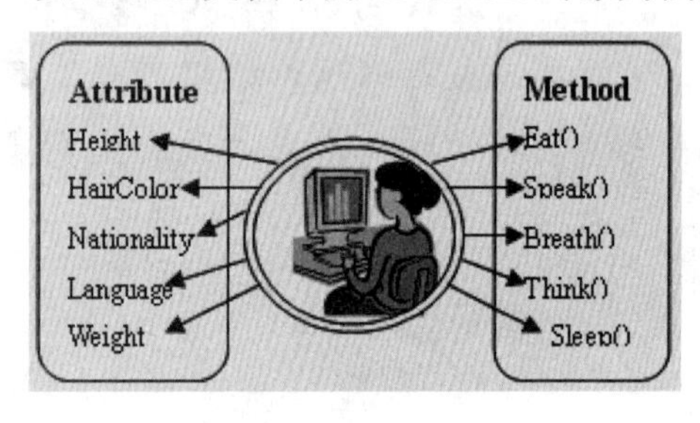

图 2-1 对象的属性和行为

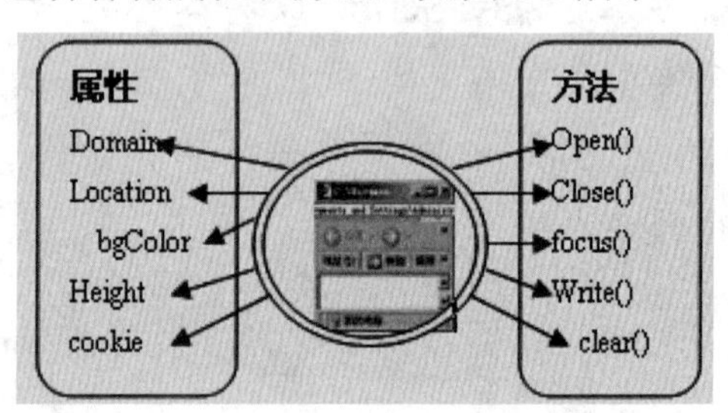

图 2-2 以 HTML 文档中的 document 为例构造的对象

凡是能够提取一定的度量数据，并能通过某种方式对度量数据实施操作的客观存在都可以构成一个对象。可以用属性来描述对象的状态，使用方法和事件来处理对象的各种行为。

(1) 属性。用来描述对象的状态，通过定义属性值来定义对象的状态。在图 2-1 中，定义了字符串 Nationality 来表示人的国籍，所以 Nationality 成为人的某个属性。

(2) 方法。针对对象行为的复杂性，对象的某些行为可以用通用的代码来处理，这些代码就是方法。在图 2-2 中，定义了 Open()方法来处理文件的打开情况。

(3) 事件。由于对象行为的复杂性，对象的某些行为不能使用通用的代码来处理，需要用户根据实际情况来编写处理该行为的代码，该代码称为事件。

JavaScript 是基于对象的编程语言，除循环和关系运算符等语言构造之外，其所有的特征几乎都是按照对象的方法进行处理的。

JavaScript 支持的对象主要包括以下 4 种。

(1) JavaScript 核心对象。包括基本数据类型的相关对象(如 String、Boolean、Number)、允许创建用户自定义和组合类型的对象(如 Object、Array)和其他能简化 JavaScript 操作的对象

(如 Math、Date、RegExp、Function)。

(2) 浏览器对象。包括不属于 JavaScript 语言本身但被绝大多数浏览器所支持的对象，如控制浏览器窗口和用户交互界面的 Window 对象、提供客户端浏览器配置信息的 Navigator 对象。

(3) 用户自定义对象。Web 应用程序开发者用于完成特定任务而创建的自定义对象，可自由设计对象的属性、方法和事件处理程序，编程灵活性较大。

(4) 文本对象。由文本域构成的对象，在 DOM 中定义，同时赋予很多特定的处理方法，如 insertData()、appendData()等。

2.1.2 面向对象编程

面向对象程序设计(Object-Oriented Programming，OOP)是一种起源于 20 世纪 60 年代的 Simula 语言，其自身理论已经十分完善，并被多种面向对象程序设计语言实现。面向对象编程的基本原则是：计算机程序由单个能够起到子程序作用的单元或对象组合而成。面向对象编程具有 3 个最基本的特点：重用性、灵活性和扩展性。这种方法将软件程序中的每一个元素作为一个对象看待，同时定义对象的类型、属性和描述对象的方法。为了实现整体操作，每个对象都应该能够接收信息、处理数据和向其他对象发送信息。

面向对象编程主要包含如下 3 个重要的概念。

1. 继承

继承性是子类自动共享父类数据结构和方法的机制，这是类之间的一种关系。在定义和实现一个类的时候，可以在一个已经存在的类的基础之上来进行，把这个已经存在的类所定义的内容作为自己的内容，并加入若干新的内容。继承性是面向对象程序设计语言不同于其他语言的最重要的特点，是其他语言所没有的。继承主要分为以下两种类型。

(1) 在类层次中，子类只继承一个父类的数据结构和方法，则称为单重继承。

(2) 在类层次中，子类继承多个父类的数据结构和方法，则称为多重继承。

在软件开发中，类的继承性使所建立的软件具有开放性、可扩充性，这是信息组织与分类的行之有效的方法，简化了对象、类的创建工作量，增加了代码重用性。

继承性提供了类规范的等级结构。通过类的继承关系，使公共的特性能够共享，提高了软件的重用性。

2. 封装

封装的作用是将对象的实现过程通过函数等方式封装起来，使用户只能通过对象提供的属性、方法和事件等接口去访问对象，而不需要知道对象的具体实现过程。封装的目的是增强安全性和简化编程，使用者不必了解具体的实现细节，而只是要通过外部接口——特定的访问权限来使用类的成员。

封装允许对象运行的代码相对于调用者来说是完全独立的，调用者通过对象及相关接口参数来访问此接口。只要对象的接口不变，即使对象的内部结构或实现方法发生了改变，程序的其他部分也不用做任何处理。

3. 多态

多态性是指相同的操作或函数、过程可作用于多种类型的对象上并获得不同的结果。不同的对象收到同一消息可以产生不同的结果，这种现象称为多态性。多态性允许每个对象以适合自身的方式去响应共同的消息。多态性增强了软件的灵活性和重用性。

需要说明的是：JavaScript 脚本是基于对象的脚本编程语言，而不是面向对象的编程语言。其原因在于：JavaScript 是以 DOM 和 BOM 中定义的对象模型及操作方法为基础的，但又不具备面向对象编程语言所必须具备的显著特征，如分类、继承、封装、多态、重载等。另外，JavaScript 还支持 DOM 和 BOM 提供的对象模型，用于根据其对象模型层次结构来访问目标对象的属性并对对象施加相应的操作。

在 JavaScript 语言中，之所以任何类型的对象都可以赋予任意类型的数值，是因为 JavaScript 为弱类型的脚本语言，即变量在使用前无须任何声明，在浏览器解释运行其代码时，才检查目标变量的数据类型。

2.1.3 JavaScript 的内部对象

JavaScript 的内部对象按照使用方式可以分为静态对象和动态对象两种。在引用动态对象的属性和方法时，必须使用 new 关键字来创建一个对象实例，然后才能使用“对象实例名.成员”的方式来访问其属性和方法；在引用静态对象属性和方法时，不需要使用 new 关键字来创建对象实例，直接使用“对象名.成员”的方式来访问其属性和方法即可。

JavaScript 中常见的内部对象如表 2-1 所示。

表 2-1 JavaScript 中常见的内部对象

对 象 名	功 能	静态/动态
Object	使用该对象可以在程序运行时为 JavaScript 对象随意添加属性	动态对象
String	用于处理或格式化文本字符串以及确定和定位字符串中的子字符串	动态对象
Date	使用 Date 对象执行各种日期和时间的操作	动态对象
Event	用来表示 JavaScript 的事件	静态对象
FileSystemObject	主要用于实现文件操作功能	动态对象
Drive	主要用于收集系统中的物理或逻辑驱动器资源中的内容	动态对象
File	用于获取服务器端指定文件的相关属性	静态对象
Folder	用于获取服务器端指定文件夹的相关属性	静态对象

2.2 对象访问语句

在 JavaScript 中，用于对象访问的语句有两种，分别是 for-in 循环语句和 with 语句。下面详细介绍这两种语句的用法。

2.2.1 for-in 循环语句

for-in 循环语句与 for 语句十分相似，该语句用来遍历对象的每一个属性，每次都会将属性名作为字符串保存在变量中。

for-in 语句的语法格式如下：

```
for(variable in object){
   statement
}
```

其中各项说明如下。

(1) variable：变量名，声明一个变量的 var 语句、数组的一个元素或者对象的一个属性。

(2) object：对象名，或者是计算结果为对象的表达式。

(3) statement：通常是一个原始语句或者语句块，由它构建循环的主体。

【例 2.1】(示例文件 ch02\2.1.html)

for-in 语句的使用：

```
<!DOCTYPE html>
<html>
<head>
<title>使用 for in 语句</title>
</head>
<body>
<script type="text/javascript">
var myarray = new Array()
myarray[0] = "星期一"
myarray[1] = "星期二"
myarray[2] = "星期三"
myarray[3] = "星期四"
myarray[4] = "星期五"
myarray[5] = "星期六"
myarray[6] = "星期日"
for (var i in myarray)
{
   document.write(myarray[i] + "<br />")
}
</script>
</body>
</html>
```

在 IE 11.0 中浏览，效果如图 2-3 所示。

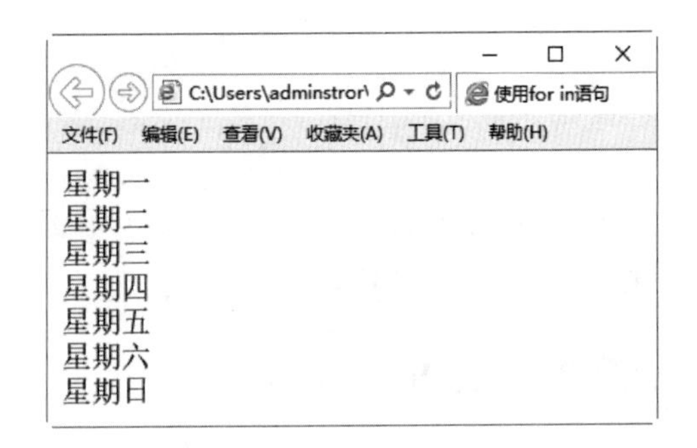

图 2-3　使用 for-in 语句

2.2.2　with 语句

有了 with 语句，在存取对象属性和方法时就不用重复指定参考对象了，在 with 语句块中，凡是 JavaScript 不识别的属性和方法都和该语句块指定的对象有关。

with 语句的语法格式如下：

```
with object {
   statements
}
```

对象指明了当语句组中对象缺省时的参考对象，这里我们用较为熟悉的 document 对象对 with 语句举例。例如，当使用与 document 对象有关的 write()或 writeln()方法时，往往使用如下形式：

```
document.writeln("Hello!");
```

当需要显示大量数据时，就会多次使用同样的 document.writeln()语句，这时就可以像下面的程序那样，把所有以 document 对象为参考对象的语句放到 with 语句块中，从而达到减少语句量的目的。

【例 2.2】(示例文件 ch02\2.2.html)

with 语句的使用：

```
<!DOCTYPE html>
<html>
<head>
<title>with 语句的使用</title>
</head>
<body>
<script type="text/javascript">
var date_time = new Date();
with(date_time){
    var a = getMonth() + 1;
    alert(getFullYear() + "年" + a + "月" + getDate() + "日"
      + getHours() + ":" + getMinutes() + ":" + getSeconds());
}
var date_time = new Date();
alert(date_time.getFullYear() + "年" + date_time.getMonth() + 1 + "月"
  + date_time.getDate() + "日" + date_time.getHours() + ":"
  + date_time.getMinutes() + ":" + date_time.getSeconds());
</script>
</body>
</html>
```

在 IE 11.0 中浏览，效果如图 2-4 所示。

图 2-4　with 语句的使用

2.3　JavaScript 中的数组

数组是有序数据的集合，JavaScript 中的数组元素允许属于不同数据类型。用数组名和下标可以唯一地确定数组中的元素。

2.3.1　结构化数据

在 JavaScript 程序中，Array(数组)被定义为有序的数据集。最好将数组想象成一个表，与电子数据表很类似。在 JavaScript 中，数组被限制为一个只有一列数据的表，但却有许多行，用来容纳所有数据。JavaScript 浏览器为 HTML 文档和浏览器属性中的对象创建了许多内部数组。例如，如果文档中含有 5 个链接，则浏览器就保留一张链接的表。

可以通过数组语法中的编号(0 是第一个)访问它们：数组名后紧跟着的是方括号中的索引数。例如 Document.links[0]，代表着文档中的第一个链接。在许多 JavaScript 应用程序中，可以靠与表单元素的交互作用，用数组来组织网页使用者所访问的数据。

对于许多 JavaScript 应用程序，可以将数组作为有组织的数据仓库来使用，这些数据是页面的浏览者基于他们与表单元素的交互而访问的数据。数组是 JavaScript 增强页面重新创建服务器端复制 CGI 程序行为的一种方式。当嵌在脚本中的数据集与典型的.gif 文件一样大的时候，用户在载入页面时不会感觉有很长的时间延迟，而且还有充分的权利对小数据库集进行即时查询，而不需要对服务器进行任何回调。这种面向数据库的数组是 JavaScript 的一个重要应用，称为 serverlessCGIs(无服务器 CGI)。

当设计一个应用程序时，要寻找潜在应用程序数组的线索。如果有许多对象或者数据点使用同样的方式与脚本进行交互，那么就可以使用数组结构。例如，除 Internet Explorer 3 外，在每一个浏览器中，可以在一个订货表单中为每列的文本域指定类似的名称，这里，类似名称的对象可以作为数组元素处理。为了重复处理订货表单的行计算，脚本可以在很少的 JavaScript 语句中使用数组语法来完成，而不是对每个域都用代码编写许多语句。

还可以创建类似 Java 哈希表的数组：哈希表是一个查找表，如果知道与表目有关联的名称，就能立刻找到想要的数据点。如果认为数据是一个表的形式，就可以使用数组。

2.3.2　创建和访问数组对象

数组是具有相同数据类型的变量集合，这些变量都可以通过索引进行访问。数组中的变量称为数组的元素，数组能够容纳元素的数量称为数组的长度。数组中的每个元素都具有唯一的索引(或称为下标)与其相对应，在 JavaScript 中，数组的索引从零开始。

Array 对象是常用的内置动作脚本对象，它将数据存储在已编号的属性中，而不是已命名的属性中。数组元素的名称做索引。数组用于存储和检索特定类型的信息，如学生列表或游戏中的一系列移动。Array 对象类似 String 和 Date 对象，需要使用 new 关键字和构造函数来创建。

可以在创建一个 Array 对象时初始化它：

```
myArray = new Array()
myArray = new Array([size])
myArray = new Array([element0[, element1[, ...[, elementN]]]])
```

其中各项的含义如下。

- size：可选，指定一个整数，表示数组的大小。
- element0,...,elementN：可选，为要放到数组中的元素。创建数组后，能够用[]符号访问数组单个元素。

由此可知，创建数组对象有 3 种方法。

(1) 新建一个长度为零的数组：

```
var 数组名 = new Array();
```

例如，声明数组为 myArr1，长度为 0，代码如下：

```
var myArr1 = new Array();
```

(2) 新建一个长度为 n 的数组：

```
var 数组名 = new Array(n);
```

例如，声明数组为 myArr2，长度为 6，代码如下：

```
var myArr2 = new Array(6);
```

(3) 新建一个指定长度的数组，并赋值：

```
var 数组名 = new Array(元素 1,元素 2,元素 3,...);
```

例如，声明数组为 myArr3，并且分别赋值为 1、2、3、4，代码如下：

```
var myArr3 = new Array(1,2,3,4);
```

上面这行代码创建一个数组 myArr3，并且包含 4 个元素：myArr3[0]、myArr3[1]、myArr3[2]、myArr3[3]，这 4 个元素的值分别为 1、2、3、4。

【例 2.3】(示例文件 ch02\2.3.html)

下列代码是构造一个长度为 5 的数组，为其添加元素后，使用 for 循环语句枚举其元素：

```
<!DOCTYPE html>
<html>
<head>
<script language=JavaScript>
myArray = new Array(5);
myArray[0] = "a";
```

```
myArray[1] = "b";
myArray[2] = "c";
myArray[3] = "d";
myArray[4] = "e";
for (i=0; i<5; i++){
   document.write(myArray[i] + "<br>");
}
</script>
<META content="MSHTML 6.00.2900.5726" name=GENERATOR>
</head>
<body>
</body>
</html>
```

在 IE 11.0 中浏览，效果如图 2-5 所示。

图 2-5　显示构造的数组

只要构造了一个数组，就可以使用中括号[]通过索引位置(也是基于 0 的数值)来访问它的元素。每个数组对象实体也可以看作是一个对象，因为每个数组都是由它所包含的若干个数组元素组成的，每个数组元素都可以看作是这个数组对象的一个属性。可以用表示数组元素位置的数来标识。也就是说，数组对象使用数组元素的下标来进行区分，数组元素的下标从 0 开始索引，第一个下标为 0，后面依次加 1。访问数据的语法格式如下：

```
document.write(mycars[0])
```

【例 2.4】(示例文件 ch02\2.4.html)

使用方括号直接构造并访问数组：

```
<!DOCTYPE html>
<html>
<head>
<META http-equiv=Content-Type content="text/html; charset=gb2312">
<script language=JavaScript>
  myArray = [["a1","b1","c1"],["a2","b2","c2"],["a3","b3","c3"]];
  for (var i=0; i<=2; i++){
     document.write(myArray[i])
     document.write("<br>");
  }
  document.write("<hr>");
  for (i=0; i<3; i++){
     for (j=0; j<3; j++){
        document.write(myArray[i][j] + "  ");
     }
     document.write("<br>");
  }
</script>
```

```
<META content="MSHTML 6.00.2900.5726" name=GENERATOR>
</head>
<body>
</body>
</html>
```

在 IE 11.0 中浏览，效果如图 2-6 所示。

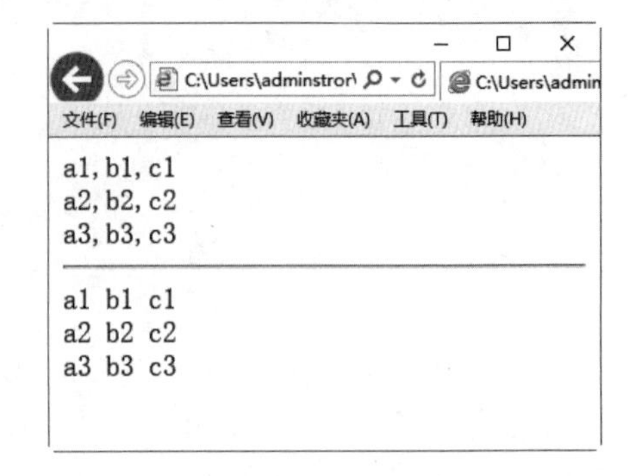

图 2-6　访问构造的数组

2.3.3　使用 for-in 语句

在 JavaScript 中，可以使用 for-in 语句来控制循环输出数组中的元素，而不需要事先知道对象属性的个数。具体的语法格式为 for (key in myArray)，其中 myArray 表示数组名。

【例 2.5】(示例文件 ch02\2.5.html)

for-in 语句的具体用法：

```
<!DOCTYPE html>
<html>
 <head>
 <META http-equiv=Content-Type content="text/html; charset=gb2312">
 <script language=JavaScript>
    myArray = new Array(5);
    myArray[0] = "a";
    myArray[1] = "b";
    myArray[2] = "c";
    myArray[3] = "d";
    myArray[4] = "e";
    for (key in myArray){
        document.write(myArray[key] + "<br>");
    }
 </script>
 <META content="MSHTML 6.00.2900.5726" name=GENERATOR>
 </head>
 <body>
 </body>
</html>
```

在 IE 11.0 中浏览，效果如图 2-7 所示。

图 2-7　循环输出数组中的元素

2.3.4　Array 对象的常用属性和方法

JavaScript 提供了一个 Array 内部对象来创建数组，通过调用 Array 对象的各种方法，可以方便地对数组进行排序、删除、合并等操作。

1. Array 对象常用的属性

Array 对象的属性主要有 2 个，分别是 length

属性和 prototype 属性。下面详细介绍这两个属性。

1) length

该属性的作用是指定数组中元素数量的非从零开始的整数。当将新元素添加到数组时，此属性会自动更新。其语法格式为 my_array.length。

【例 2.6】(示例文件 ch02\2.6.html)

下面的示例是解释 length 属性是如何更新的：

```
<!DOCTYPE html>
<html>
 <head>
 <META http-equiv=Content-Type content="text/html; charset=gb2312">
 <script language=JavaScript>
     my_array = new Array();
     document.write(my_array.length + "<br>"); //初始长度为 0
     my_array[0] = 'a';
     document.write(my_array.length + "<br>"); //将长度更新为1
     my_array[1] = 'b';
     document.write(my_array.length + "<br>"); //将长度更新为2
     my_array[9] = 'c';
     document.write(my_array.length + "<br>"); //将长度更新为10
 </script>
 </head>
 <body>
 </body>
</html>
```

在 IE 11.0 中浏览，效果如图 2-8 所示。

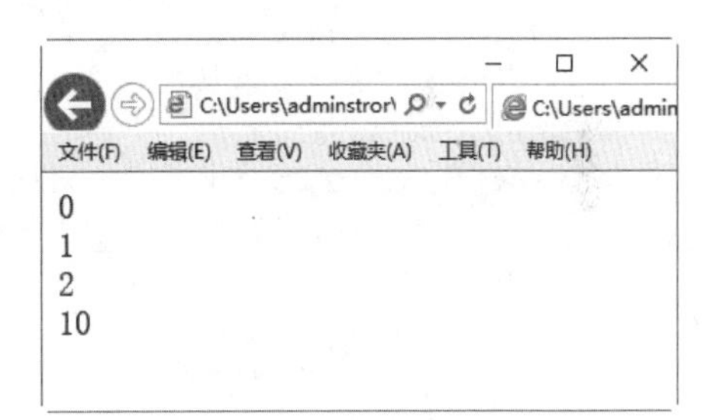

图 2-8　给数组指定相应的整数

2) prototype

该属性是所有 JavaScript 对象所共有的属性，其作用是将新定义的属性或方法添加到 Array 对象中，这样该对象的实例就可以调用该属性或方法了。其语法格式如下：

```
Array.prototype.methodName = functionName;
```

其中各项的作用说明如下。

- methodName：必选项，新增方法的名称。
- functionName：必选项，要添加到对象中的函数名称。

【例 2.7】(示例文件 ch02\2.7.html)

下面为 Array 对象添加返回数组中最大元素值。必须声明该函数，并将它加入 Array.prototype，且使用它。代码如下：

```
<!DOCTYPE html>
<html>
 <head>
 <META http-equiv=Content-Type content="text/html; charset=gb2312">
 <script>
```

```
  //添加一个属性，用于统计删除的元素个数
  Array.prototype.removed = 0;
  //添加一个方法，用于删除指定索引的元素
  Array.prototype.removeAt=function(index)
  {
    if(isNaN(index) || index<0)
    {return false;}
    if(index>=this.length)
    {index=this.length-1}
    for(var i=index; i<this.length; i++)
    {
       this[i] = this[i+1];
    }
    this.length -= 1
    this.removed++;
  }
  //添加一个方法，输出数组中的全部数据
  Array.prototype.outPut=function(sp)
  {
      for(var i=0; i<this.length; i++)
      {
         document.write(this[i]);
         document.write(sp);
      }
      document.write("<br>");
  }
  //定义数组
  var arr = new Array(1,2,3,4,5,6,7,8,9);
  //测试添加的方法和属性
  arr.outPut(" ");
  document.write("删除一个数据<br>");
  arr.removeAt(2);
  arr.outPut(" ");
  arr.removeAt(4);
  document.write("删除一个数据<br>");
  arr.outPut(" ")
  document.write("一共删除了" + arr.removed + "个数据");
</script>
</head>
<body>
</body>
</html>
```

在 IE 11.0 中浏览，效果如图 2-9 所示。

这段代码利用 prototype 属性分别向 Array 对象中添加了 2 个方法和 1 个属性，分别实现了删除指定索引处的元素、输出数组中的所有元素和统计删除元素个数的功能。

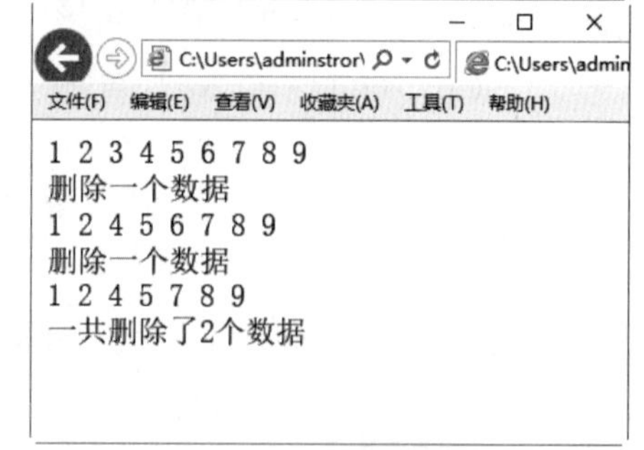

图 2-9　删除数组中的数据

2. Array对象常用的方法

Array 对象常用的方法有连接方法 concat、分隔方法 join、追加方法 push、反序方法 reverse、排序方法 sort 等。

1) concat

该方法的作用是把当前数组与指定的数组相连接，返回一个新的数组，该数组中含有前面两个数组的全部元素，其长度为两个数组的长度之和。其基本的语法格式为 array1.concat(array2)，对其中参数说明如下。

- array1：必选项，数组名称。
- array2：必选项，数组名称，该数组中的元素将被添加到数组 array1 中。

【例 2.8】(示例文件 ch02\2.8.html)

定义两个数组 array1 和 array2，然后把这两个数组连接起来，并将值赋给数组 array：

```
<!DOCTYPE html>
<html>
 <head>
 <META http-equiv=Content-Type content="text/html; charset=gb2312">
 <script language=JavaScript>
     var array1 = new Array(1,2,3,4,5,6);
        var array2 = new Array(7,8,9,10);
        var array = array1.concat(array2);
        //自定义函数，输出数组中所有数据
        function writeArr(arrname,sp)
        {
          for(var i=0; i<arrname.length; i++)
          {
             document.write(arrname[i]);
             document.write(sp);
          }
          document.write("<br>");
        }
        document.write("数组1: ");
        writeArr(array1,",");
        document.write("数组2: ");
        writeArr(array2,",");
        document.write("数组3: ");
        writeArr(array,",");
     </script>
 </head>
 <body>
 </body>
</html>
```

在 IE 11.0 中浏览，效果如图 2-10 所示。

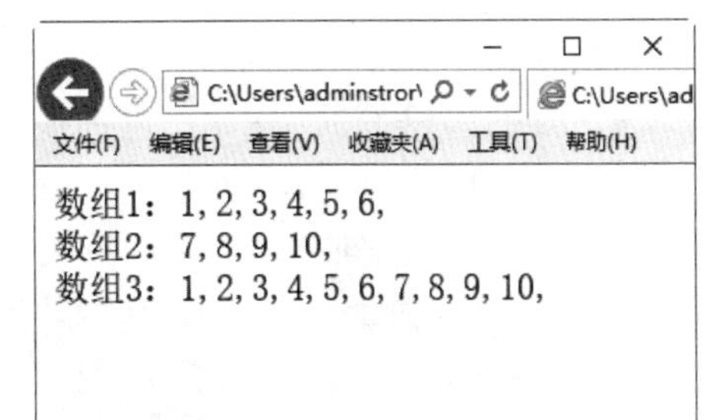

图 2-10 连接数组

2) join

该方法与 String 对象 split 方法的作用相反，是将数组中的所有元素连接为一个字符串。如果数组中的元素不是字符串，则该元素将首先被转化为字符串，各个元素之间可以以指定的分隔符进行连接。其语法格式为 array.join(separator)，其中 array 必选项为数组的名称，而 separator 必选项是连接各个元素之间的分隔符。

【例 2.9】(示例文件 ch02\2.9.html)

对比 split 方法和 join 方法：

```
<!DOCTYPE html>
<html>
 <head>
 <META http-equiv=Content-Type content="text/html; charset=gb2312">
 <script language=JavaScript>
    var str1 = "this ia a test";
    var arr = str1.split(" ");
    var str2 = arr.join(",");
    with(document){
       write(str1);
       write("<br>分割为数组，数组长度" + arr.length + ",重新连接如下：<br>");
       write(str2);
    }
 </script>
 </head>
 <body>
 </body>
</html>
```

在 IE 11.0 中浏览，效果如图 2-11 所示。

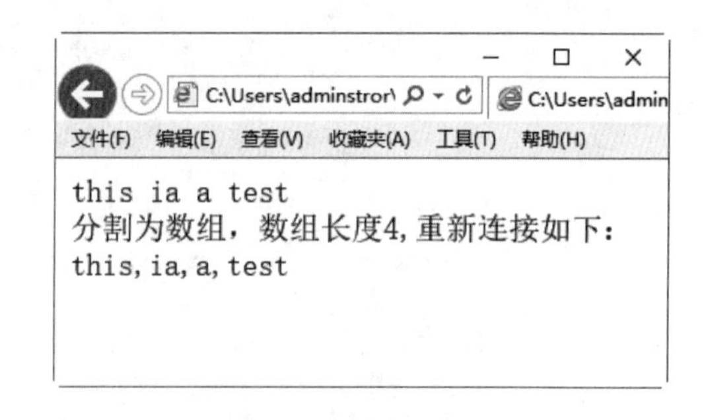

图 2-11　将数组中所有元素连接为一个字符串

上述代码首先使用 split 方法以空格为分隔符将字符串分隔存储到数组中，再调用 join 方法以“,”(逗号)为分隔符，将数组中的各个元素重新连接为一个新字符串。

3)　push

该方法可以将所指定的一个或多个数据添加到数组中，该方法的返回值为添加新数据后数组的长度。其语法格式为 array.push([data1[,data2[,...[,datan]]]])，其中参数的作用如下。

- array：必选项，数组名称。
- data1、data2、datan：可选参数，将被添加到数组中的数据。

【例 2.10】(示例文件 ch02\2.10.html)

利用 push 方法向数组中添加新数据：

```
<!DOCTYPE html>
<html>
 <head>
 <META http-equiv=Content-Type content="text/html; charset=gb2312">
```

```
<script language=JavaScript>
var arr = new Array();
document.write("向数组中写入数据：");  //单个数据写入数组
for (var i=1; i<=5; i++)
{
   var data = arr.push(Math.ceil(Math.random()*10));
   document.write(data);
   document.write("个,");
}
document.write("<br>");  //批量写入数组
var data = arr.push("a",4.15,"hello");
document.write("批量写入，数组长度已为" + data + "<br>");
var newarr = new Array(1,2,3,4,5);
document.write("向数组中写入另一个数组<br>");  //写入新数组
arr.push(newarr);
document.write("全部数据如下:<br>");
document.write(arr.join(","));
</script>
</head>
<body>
</body>
</html>
```

在 IE 11.0 中浏览，效果如图 2-12 所示。上述代码使用 push 方法，向数组中逐个和批量添加数据。

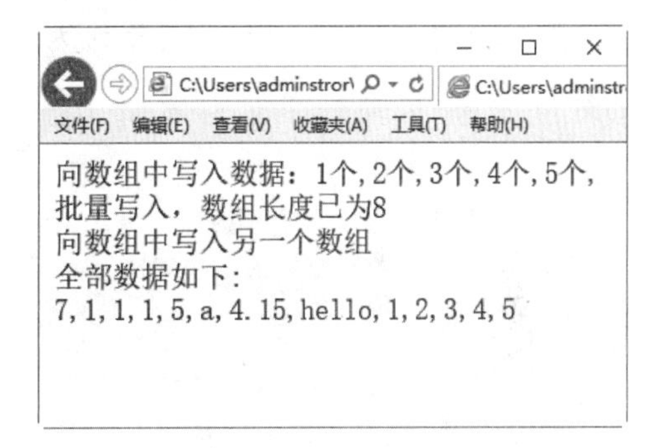

图 2-12　使用 push 方法向数组中添加数据

4)　reverse

该方法可以将数组中的元素反序排列，数组中所包含的内容和数组的长度不会改变。其语法格式为 array.reverse()，其中 array 为数组的名称。

【例 2.11】(示例文件 ch02\2.11.html)

将数组中的元素反序排列：

```
<!DOCTYPE html>
<html>
<head>
<META http-equiv=Content-Type content="text/html; charset=gb2312">
<script>
   var arr = new Array(1,2,3,4,5,6);
   with (document)
   {
     write("数组为:");
     write(arr.join(","));
     arr.reverse();
     write("<br>反序后的数组为:")
     write(arr.join("-"));
   }
```

```
</script>
</head>
<body>
</body>
</html>
```

在 IE 11.0 中浏览，效果如图 2-13 所示。

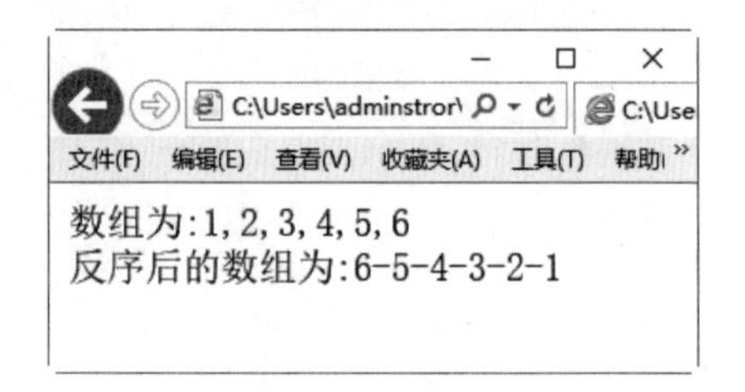

图 2-13 将数组中的元素反序排列

5) slice

该方法将提取数组中的一个片段或子字符串，并将其作为新数组返回，而不修改原始数组。返回的数组包括 start 元素到 end 元素(但不包括该元素)的所有元素。

其语法格式如下：

```
my_array.slice([start[, end]])
```

其中各项的含义如下。

- start：指定片段起始点索引的数值。
- end：指定片段终点索引的数值。如果省略此参数，则片段包括数组中从开头 start 到结尾的所有元素。

【例 2.12】(示例文件 ch02\2.12.html)

将数组中的一个片段或子字符串作为新数组返回，而不修改原始数组：

```
<!DOCTYPE html>
<html>
<head>
<META http-equiv=Content-Type content="text/html; charset=gb2312">
<Script language="JavaScript">
   var myArray = [1, 2, 3, 4, 5, 6,7];
   newArray = myArray.slice(1, 6);
   document.write(newArray);
   document.write("<br>");
   newArray = myArray.slice(1);
   document.write(newArray);
</Script>
</head>
<body>
</body>
</html>
```

在 IE 11.0 中浏览，效果如图 2-14 所示。

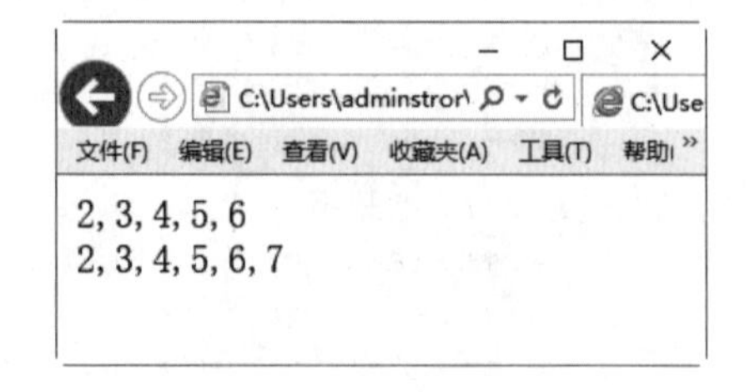

图 2-14 作为新数组返回

6) sort

该方法对数组中的所有元素按 Unicode 编码进行排序，并返回经过排序后的数组。sort 方法默认按升序进行排列，但也可以通过指定对比函数来实现特殊的排序要求，对比函数的格式为 comparefunction(arg1,arg2)。其中，comparefunction

为排序函数的名称，该函数必须包含两个参数：arg1 和 arg2，分别代表了两个将要进行对比的字符。该函数的返回值决定了如何对 arg1 和 arg2 进行排序。如果返回值为负，则 arg2 将排在 arg1 的后面；如果返回值为 0，arg1、arg2 视为相等；如果返回值为正，则 arg2 将排在 arg1 的前面。

sort 方法的语法格式如下：

```
array.sort([cmpfun(arg1,arg2)])
```

各项说明如下。

- array：必选项，数组名称。
- cmpfun：可选项，比较函数。
- arg1、arg2：可选项，比较函数的两个参数。

【例 2.13】(示例文件 ch02\2.13.html)

使用 sort 方法对数组中的数据进行排序：

```
<!DOCTYPE html>
<html>
<head>
<META http-equiv=Content-Type content="text/html; charset=gb2312">
<Script language="JavaScript">
  var arr = new Array(1,6,3,40,1,"a","b","A","B");
  writeArr("排序前",arr);
  writeArr("升序排列",arr.sort());
  writeArr("降序排列,字母不分大小写",arr.sort(desc));
  writeArr("严格降序排列",arr.sort(desc1));
  //自定义函数输出提示信息和数组元素
  function writeArr(str,array)
  {
    document.write(str + ":");
    document.write(array.join(","));
    document.write("<br>");
  }
  //按降序排列，字母不区分大小写
  function desc(a,b)
  {
    var a = new String(a);
    var b = new String(b);
    //如果 a 大于 b，则返回-1，所以 a 排在前 b 排在后
    return -1*a.localeCompare(b);
  }
  //严格降序
  function desc1(a,b)
  {
    var stra = new String(a);
    var strb = new String(b);
    var ai = stra.charCodeAt(0);
    var bi = strb.charCodeAt(0);
    if(ai>bi)
```

```
        return -1;
      else
        return 1;
    }
</script>
</head>
<body>
</body>
</html>
```

在 IE 11.0 中浏览，效果如图 2-15 所示。这段代码中定义了两个对比函数，其中 desc 进行降序排列，但字母不区分大小写；desc1 进行严格降序排列。

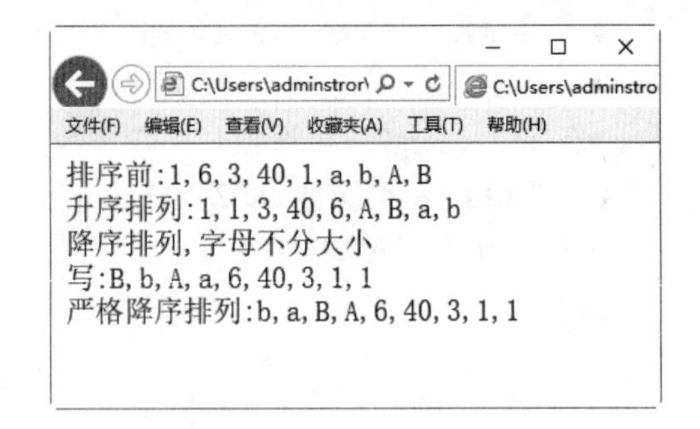

图 2-15 对数组进行排序

7) splice

该方法可以通过指定起始索引和数据个数的方式，删除或替换数组中的部分数据。该方法的返回值为被删除或替换掉的数据。其语法格式如下：

```
array.splice(start,count[,data1[,
data2,[,...[,datacount]]]])
```

其中，array 必选项是数组名称；start 必选项为整数起始索引；count 必选项为要删除或替换的数组的个数；data 为可选项，是用于替换指定数据的新数据。

如果没有指定 data 参数，则该指定的数据将被删除；如果指定了 data 参数，则数组中的数据将被替换。

【例 2.14】(示例文件 ch02\2.14.html)

splice 方法的具体使用过程：

```
<!DOCTYPE html>
<html>
<head>
<META http-equiv=Content-Type content="text/html; charset=gb2312">
<Script language="JavaScript">
   var arr = new Array(0,1,2,3,4,5,6,7,8,9,10);
   var rewith = new Array("a","b","c");
   var tmp1 = arr.splice(2,5,rewith);
   with(document)
   {
      writeArr("替换了 5 个数据",tmp1);
      writeArr("替换为: ",rewith);
      writeArr("替换后",arr);
      var tmp2=arr.splice(5,2);
      writeArr("删除 2 个数据",tmp2);
      writeArr("替换后",arr);
   }
```

```
    //自定义函数输出提示信息和数组元素
    function writeArr(str,array)
    {
        document.write(str + ":");
        document.write(array.join(","));
        document.write("<br>");
    }
</script>
</head>
<body>
</body>
</html>
```

在 IE 11.0 中浏览，效果如图 2-16 所示。上述代码分别演示了如何使用 splice 方法替换和删除数组中指定数目的数据。

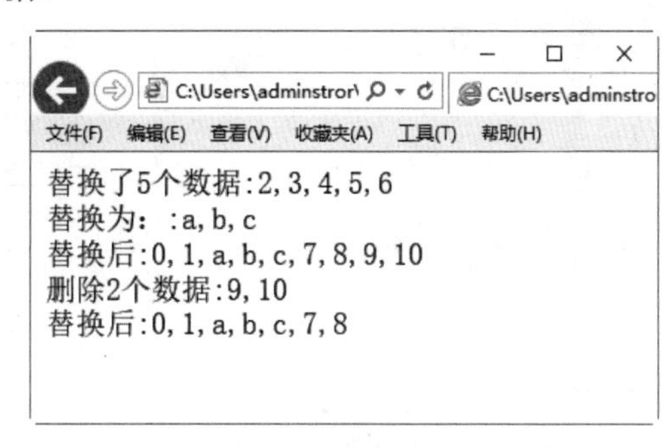

图 2-16　替换和删除数组中指定数目的数据

2.4　详解常用的数组对象方法

在 JavaScript 中，数组对象的方法有 11 种。下面详细介绍常用的数组对象方法的使用。

2.4.1　连接其他数组到当前数组

使用 concat()方法可以连接两个或多个数组。该方法不会改变现有的数组，而仅仅会返回被连接数组的一个副本。

其语法格式如下：

```
arrayObject.concat(array1,array2,...,arrayN)
```

其中 arrayN 是必选项，该参数可以是具体的值，也可以是数组对象，还可以是任意多个。

【例 2.15】(示例文件 ch02\2.15.html)

使用 concat()方法连接 3 个数组：

```
<!DOCTYPE html>
<html>
<body>
<script type="text/javascript">
```

```
    var arr = new Array(3)
    arr[0] = "北京"
    arr[1] = "上海"
    arr[2] = "广州"
    var arr2 = new Array(3)
    arr2[0] = "西安"
    arr2[1] = "天津"
    arr2[2] = "杭州"
    var arr3 = new Array(2)
    arr3[0] = "长沙"
    arr3[1] = "温州"
    document.write(arr.concat(arr2,arr3))
</script>
</body>
</html>
```

在 IE 11.0 中浏览，效果如图 2-17 所示。

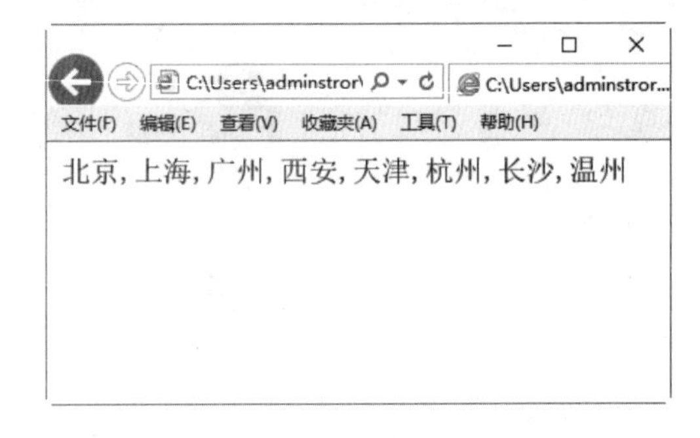

图 2-17 使用 concat()方法连接 3 个数组

2.4.2 将数组元素连接为字符串

使用 join()方法可以把数组中的所有元素放入一个字符串。其语法格式如下：

```
arrayObject.join(separator)
```

其中 separator 是可选项，用于指定要使用的分隔符，如果省略该参数，则使用逗号作为分隔符。

【例 2.16】(示例文件 ch02\2.16.html)

使用 join()方法将数组元素连接为字符串：

```
<!DOCTYPE html>
<html>
<body>
<script type="text/javascript">
    var arr = new Array(3);
    arr[0] = "河北"
    arr[1] = "石家庄"
    arr[2] = "廊坊"
    document.write(arr.join());
    document.write("<br />");
    document.write(arr.join("."));
</script>
</body>
</html>
```

在 IE 11.0 中浏览，效果如图 2-18 所示。

2.4.3 移除数组中最后一个元素

使用 pop()方法可以移除并返回数组中最后一个元素。其语法格式如下：

```
arrayObject.pop()
```

pop()方法将移除 arrayObject 的最后一个元素，把数组长度减 1，并且返回它移除的元素的值。如果数组已经为空，则 pop()不改变数组，并返回 undefined。

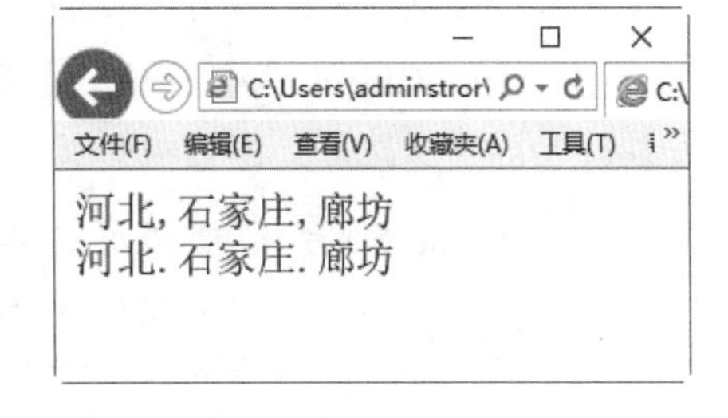

图 2-18 使用 join()方法将数组元素连接为字符串

【例 2.17】(示例文件 ch02\2.17.html)

使用 pop()方法移除数组最后一个元素：

```
<!DOCTYPE html>
<html>
<body>
<script type="text/javascript">
    var arr = new Array(3)
    arr[0] = "河南"
    arr[1] = "郑州"
    arr[2] = "洛阳"
    document.write("数组中原有元素：" + arr)
    document.write("<br />")
    document.write("被移除的元素：" + arr.pop())
    document.write("<br />")
    document.write("移除元素后的数组元素：" + arr)
</script>
</body>
</html>
```

在 IE 11.0 中浏览，效果如图 2-19 所示。

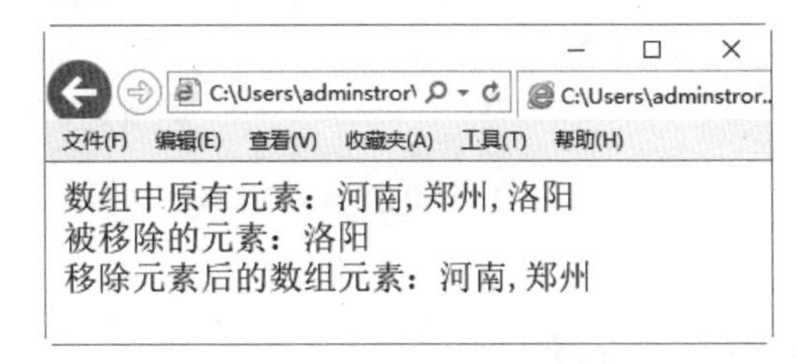

图 2-19 使用 pop()方法移除数组最后一个元素

2.4.4 将指定的数值添加到数组中

使用 push()方法可以向数组的末尾添加一个或多个元素，并返回新的长度。

语法格式如下：

```
arrayObject.push(newelement1,newelement2,...,newelementN)
```

其中，arrayObject 为必选项，是数组对象。newelementN 为可选项，表示添加到数组中的元素。

push()方法可以把其参数顺序添加到 arrayObject 的尾部。它直接修改 arrayObject，而不是创建一个新的数组。

push()方法和 pop()方法使用数组提供的先进后出的栈功能。

【例 2.18】(示例文件 ch02\2.18.html)

使用 push()方法将指定数值添加到数组中：

```
<!DOCTYPE html>
<html>
<body>
<script type="text/javascript">
    var arr = new Array(3)
    arr[0] = "河南"
    arr[1] = "河北"
    arr[2] = "江苏"
    document.write("原有的数组元素：" + arr)
    document.write("<br />")
    document.write("添加元素后数组的长度：" + arr.push("吉林"))
    document.write("<br />")
    document.write("添加数值后的数组：" + arr)
</script>
</body>
</html>
```

在 IE 11.0 中浏览，效果如图 2-20 所示。

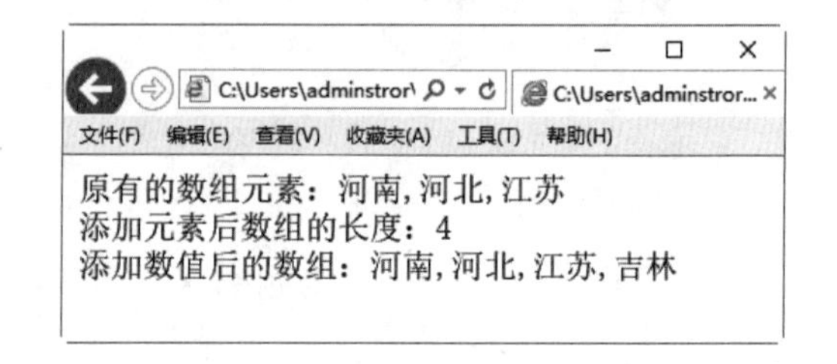

图 2-20 使用 push()方法将指定数值添加到数组中

2.4.5 反序排列数组中的元素

使用 reverse()方法可以颠倒数组中元素的顺序。其语法格式如下：

```
arrayObject.reverse()
```

该方法会改变原来的数组，而不会创建新的数组。

【例 2.19】(示例文件 ch02\2.19.html)

使用 reverse()方法颠倒数组中的元素顺序：

```
<!DOCTYPE html>
<html>
<body>
<script type="text/javascript">
    var arr = new Array(3);
```

```
    arr[0] = "张三";
    arr[1] = "李四";
    arr[2] = "王五";
    document.write(arr + "<br />");
    document.write(arr.reverse());
</script>
</body>
</html>
```

在 IE 11.0 中浏览，效果如图 2-21 所示。

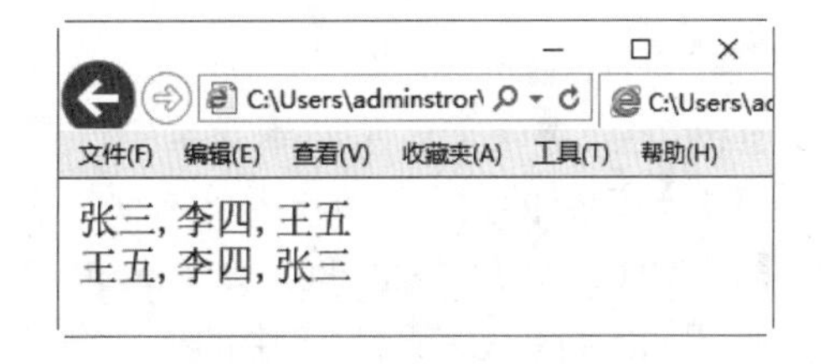

图 2-21 使用 reverse()方法颠倒数组中的元素顺序

2.4.6 删除数组中的第一个元素

使用 shift()方法可以把数组中的第一个元素删除，并返回第一个元素的值。其语法格式如下：

```
arrayObject.shift()
```

其中，arrayObject 为必选项，是数组对象。

如果数组是空的，那么 shift()方法将不进行任何操作，返回 undefined。请注意，该方法不创建新数组，而是直接修改原有的 arrayObject。

【例 2.20】(示例文件 ch02\2.20.html)

使用 shift()方法删除数组中的第一个元素：

```
<!DOCTYPE html>
<html>
<body>
<script type="text/javascript">
    var arr = new Array(4)
    arr[0] = "北京"
    arr[1] = "上海"
    arr[2] = "广州"
    arr[3] = "天津"
    document.write("原有数组元素为：" + arr)
    document.write("<br />")
    document.write("删除数组中的第一个元素为：" + arr.shift())
    document.write("<br />")
    document.write("删除元素后的数组为：" + arr)
</script>
</body>
</html>
```

在 IE 11.0 中浏览，效果如图 2-22 所示。

2.4.7 获取数组中的一部分数据

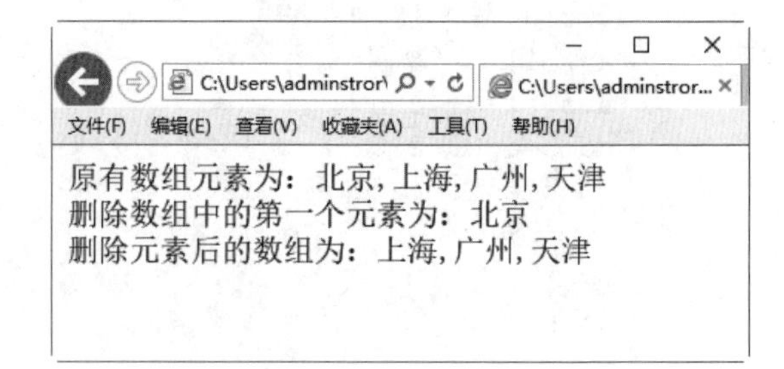

图 2-22 使用 shift()方法删除数组中的第一个元素

使用 slice()方法可从已有的数组中返回选定的元素。

其语法格式如下：

```
arrayObject.slice(start,end)
```

其中，arrayObject 为必选项，是数组对象。start 为必选项，表示开始元素的位置，是从 0 开始计算的索引。end 为可选项，表示结束元素的位置，也是从 0 开始计算的索引。

【例 2.21】(示例文件 ch02\2.21.html)

使用 slice()方法获取数组中的一部分数据：

```
<!DOCTYPE html>
<html>
<body>
<script type="text/javascript">
    var arr = new Array(6)
    arr[0] = "黑龙江"
    arr[1] = "吉林"
    arr[2] = "辽宁"
    arr[3] = "内蒙古"
    arr[4] = "河北"
    arr[5] = "山东"
    document.write("原有数组元素：" + arr)
    document.write("<br />")
    document.write("获取的部分数组元素：" + arr.slice(2,3))
    document.write("<br />")
    document.write("获取部分元素后的数据：" + arr)
</script>
</body>
</html>
```

在 IE 11.0 中浏览，效果如图 2-23 所示，可以看出，获取部分数组元素后的数组其前后是不变的。

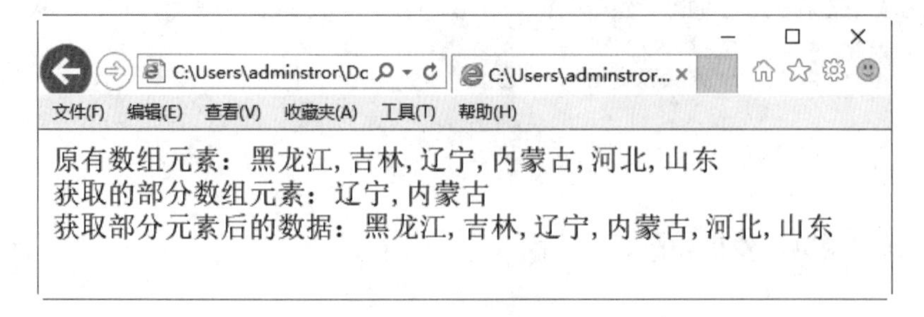

图 2-23 使用 slice()方法获取数组中的一部分数据

2.4.8 对数组中的元素进行排序

使用sort()方法可以对数组的元素进行排序。其语法格式如下：

```
arrayObject.sort(sortby)
```

其中，arrayObject为必选项，是数组对象。sortby为可选项，用来确定元素顺序的函数的名称，如果这个参数被省略，那么元素将按照ASCII字符顺序进行升序排序。

【例2.22】(示例文件ch02\2.22.html)

新建数组x，并赋值1、20、8、12、6、7，使用sort()方法排序数组，并输出x数组到页面：

```
<!DOCTYPE html>
<html>
<head>
<title>数组排序</title>
<script type="text/javascript">
  var x = new Array(1,20,8,12,6,7);   //创建数组
  document.write("排序前数组:" + x.join(",") + "<p>"); //输出数组元素
  x.sort();   //按字符升序排列数组
  document.write(
    "没有使用比较函数排序后数组:" + x.join(",") + "<p>");   //输出排序后的数组
  x.sort(asc);  //有比较函数的升序排列
  /*升序比较函数*/
  function asc(a,b)
  {
       return a-b;
  }
  document.write("排序升序后数组:" + x.join(",") + "<p>"); //输出排序后的数组
  x.sort(des); //有比较函数的降序排列
  /*降序比较函数*/
  function des(a,b)
  {
       return b-a;
  }
  document.write("排序降序后数组:" + x.join(",")); //输出排序后的数组
</script>
</head>
<body>
</body>
</html>
```

在IE 11.0中浏览，效果如图2-24所示。

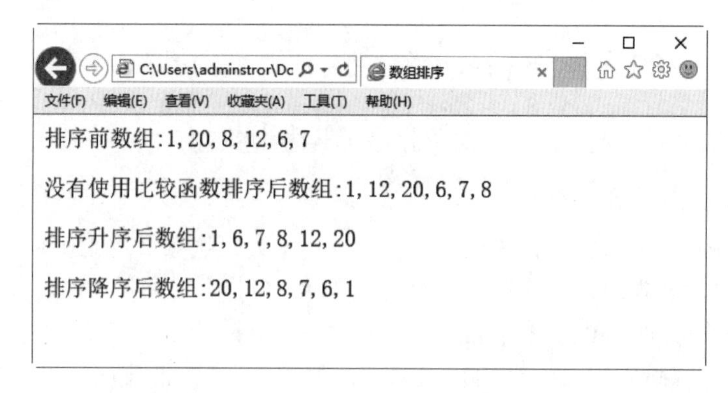

图 2-24 使用 sort()方法排序数组

在没有使用比较函数进行排序时，sort()方法是按字符的 ASCII 值排序，先从第一个字符比较，如果第 1 个字符相等，再比较第 2 个字符，依此类推。

对于数值型数据，如果按字符比较，若得到的结果并不是用户所需要的，需要借助于比较函数。比较函数有两个参数，分别代表每次排序时的两个数组项。sort()排序时，每次比较两个数组项都会执行这个函数，并把两个比较的数组项作为参数传递给这个函数。当函数返回值大于 0 的时候，就交换两个数组的顺序，否则就不交换。即函数返回值小于 0 表示升序排列，函数返回值大于 0 表示降序排列。

2.4.9 将数组转换成字符串

使用 toString()方法可以把数组转换为字符串，并返回结果。其语法格式如下：

```
arrayObject.toString()
```

【例 2.23】(示例文件 ch02\2.23.html)

将数组转换成字符串：

```
<!DOCTYPE html>
<html>
<body>
<script type="text/javascript">
   var arr = new Array(3);
   arr[0] = "北京";
   arr[1] = "上海";
   arr[2] = "广州";
   document.write(arr.toString());
</script>
</body>
</html>
```

在 IE 11.0 中浏览，效果如图 2-25 所示。可以看出，数组中的元素之间用逗号分隔。

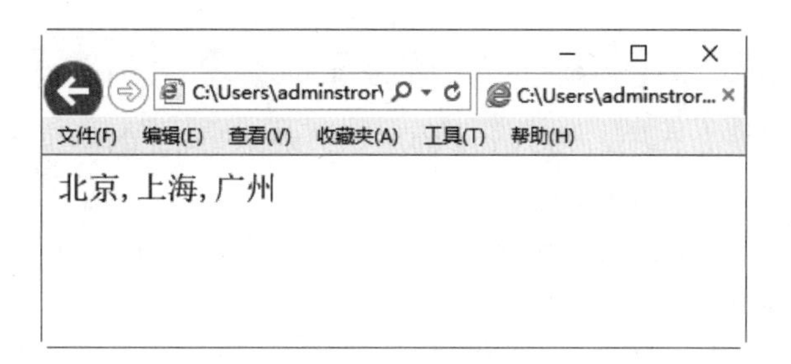

图 2-25　将数组转换成字符串

2.4.10　将数组转换成本地字符串

使用 toLocaleString()方法可以把数组转换为本地字符串。其语法格式如下：

```
arrayObject.toLocaleString()
```

该转换首先调用每个数组元素的 toLocaleString()方法，然后使用地区特定的分隔符把生成的字符串连接起来，形成一个字符串。

【例 2.24】(示例文件 ch02\2.24.html)

将数组转换成本地字符串：

```
<!DOCTYPE html>
<html>
<body>
<script type="text/javascript">
  var arr = new Array(3);
  arr[0] = "北京";
  arr[1] = "上海";
  arr[2] = "广州";
  document.write(arr.toLocaleString());
</script>
</body>
</html>
```

在 IE 11.0 中浏览，效果如图 2-26 所示。可以看出，数组中的元素之间用全角逗号分隔。

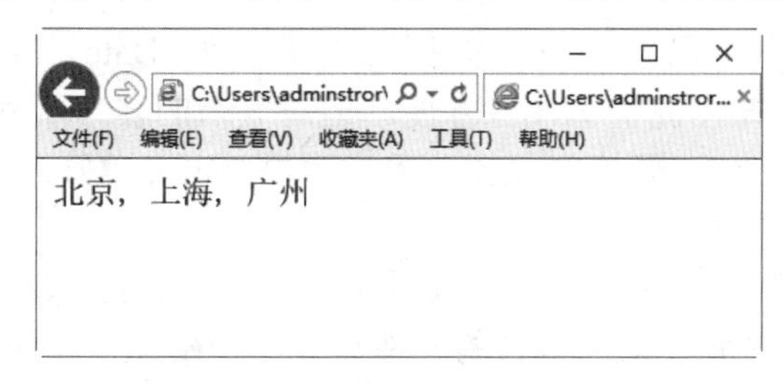

图 2-26　将数组转换成本地字符串

2.4.11　在数组开头插入数据

使用 unshift()方法可以将指定的元素插入数组开始位置并返回该数组。其语法格式如下：

```
arrayObject.unshift(newelement1,newelement2,...,newelementN)
```

其中，arrayObject 是必选项，为 Array 的对象；newelementN 是可选项，为要添加到该数组对象的新元素。

【例 2.25】(示例文件 ch02\2.25.html)

在数组开头插入数据：

```
<!DOCTYPE html>
<html>
<body>
<script type="text/javascript">
  var arr = new Array();
  arr[0] = "北京";
  arr[1] = "上海";
  arr[2] = "广州";
  document.write(arr + "<br />");
  document.write(arr.unshift("天津") + "<br />");
  document.write(arr);
</script>
</body>
</html>
```

在 IE 11.0 中浏览，效果如图 2-27 所示。

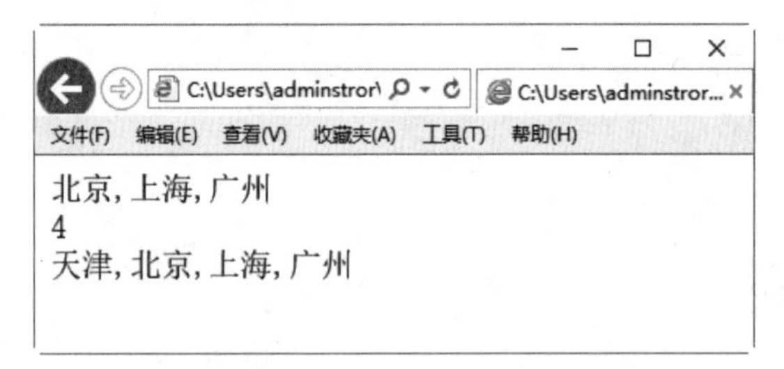

图 2-27　在数组开头插入数据

2.5　创建和使用自定义对象

目前在 JavaScript 中，已经存在一些标准的类，如 Date、Array、RegExp、String、Math、Number 等，这为编程提供了许多方便。但对复杂的客户端程序而言，这些还远远不够。在 JavaScript 脚本语言中，还有浏览器对象、用户自定义对象、文本对象等，其中用户自定义对象占据举足轻重的地位。

JavaScript 作为基于对象的编程语言，其对象实例通过构造函数来创建。每一个构造函数包括一个对象原型，定义了每个对象包含的属性和方法。在 JavaScript 脚本中创建自定义对象的方法主要有两种：通过定义对象构造函数的方法和通过对象直接初始化的方法。

2.5.1　通过定义对象构造函数的方法

在实际使用中，可以首先定义对象的构造函数，然后使用 new 操作符来生成该对象的实例，从而创建自定义对象。

【例 2.26】(示例文件 ch02\2.26.html)

通过定义对象构造函数的方法创建自定义对象：

```
<!DOCTYPE html>
<html>
<head>
<meta http-equiv="Content-Type" content="text/html; charset=gb2312">
<title>自定义对象</title>
<script language="JavaScript" type="text/javascript">
<!--
//对象的构造函数
function Student(iName,iAddress,iGrade,iScore)
{
  this.name = iName;
  this.address = iAddress;
  this.grade = iGrade;
  this.Score = iScore;
  this.information = showInformation;
}
//定义对象的方法
function showInformation()
{
  var msg = "";
  msg = "学生信息：\n"
  msg += "\n 学生姓名 : " + this.name + " \n";
  msg += "家庭地址 : " + this.address + "\n";
  msg += "班级 : " + this.grade + " \n";
  msg += "分数 : " + this.Score;
  window.alert(msg);
}
//生成对象的实例
var ZJDX = new Student("刘明明","新疆乌鲁木齐 100 号","401","99");
-->
</script>
</head>
<body>
<br>
<center>
<form>
 <input type="button" value="查看" onclick="ZJDX.information()">
</form>
</center>
</body>
</html>
```

在 IE 11.0 中浏览，效果如图 2-28 所示。单击【查看】按钮，即可看到含有学生信息的提示框，如图 2-29 所示。

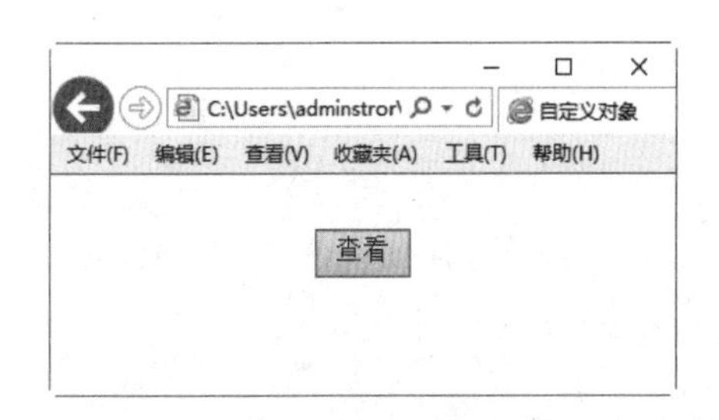

图 2-28　显示初始结果

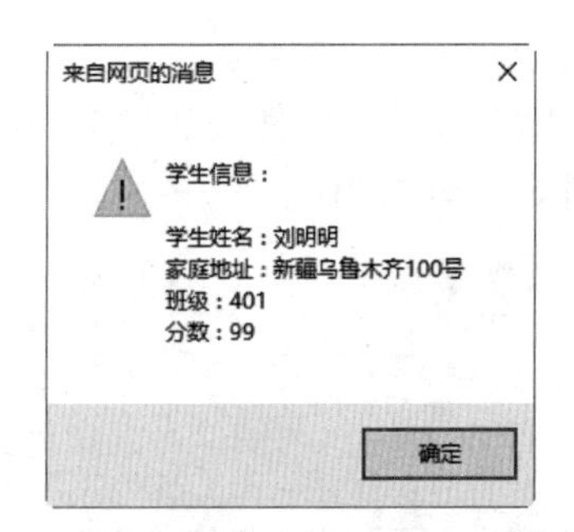

图 2-29　含有学生信息的提示框

在该示例中，用户需要先定义一个对象的构造函数，再通过 new 关键字来创建该对象的实例。定义对象的构造函数如下：

```
function Student(iName,iAddress,iGrade,iScore)
{
  this.name = iName;
  this.address = iAddress;
  this.grade = iGrade;
  this.score = iScore;
  this.information = showInformation;
}
```

当调用该构造函数时，浏览器给新的对象分配内存，并将该对象传递给函数。this 操作符是指向新对象引用的关键词，用于操作这个新对象。语句“this.name=iName;”使用作为函数参数传递过来的 iName 值在构造函数中给该对象的 name 属性赋值，该属性属于所有 Student 对象，而不仅仅属于 Student 对象的某个实例，如上面的 ZJDX。对象实例的 name 属性被定义和赋值后，可以通过“var str=ZJDX.name;”语句来访问该实例的该属性。

使用同样的方法继续添加 address、grade、score 等其他属性，但 information 不是对象的属性，而是对象的方法：

```
this.information = showInformation;
```

方法 information 指向的外部函数 showInformation 的结构如下：

```
function showInformation()
{
  var msg = "";
  msg = "学生信息：\n"
  msg += "\n学生姓名 : " + this.name + " \n";
  msg += "家庭地址 : " + this.address + "\n";
  msg += "班级 : " + this.grade + " \n";
  msg += "分数 : " + this.Score;
  window.alert(msg);
}
```

同样，由于被定义为对象的方法，在外部函数中也可以使用 this 操作符指向当前的对象，并通过 this.name 等方式访问它的某个属性。在构建对象的某个方法时，如果代码比较简单，也可以使用非外部函数的做法，改写 Student 对象的构造函数：

```
function Student(iName,iAddress,iGrade,iScore)
{
  this.name = iName;
  this.address = iAddress;
  this.grade = iGrade;
  this.score = iScore;
  this.information = function()
                  {
                    var msg = " ";
                    msg = "学生信息\n"
                    msg += "\n 学生姓名 : " + this.name + " \n";
                    msg += "家庭地址 : " + this.address + "\n";
                    msg += "班级 : " + this.grade + " \n";
                    msg += "分数 : " + this.Score;
                    window.alert(msg);
                  };
}
```

2.5.2　通过对象直接初始化的方法

此方法通过直接初始化对象来创建自定义对象，与定义对象的构造函数方法不同的是，该方法无须生成此对象的实例。将上面 HTML 文件中的 JavaScript 脚本部分做如下修改：

```
<script language="JavaScript" type="text/javascript">
<!--
//直接初始化对象
var ZJDX = {
              name:"刘明明",
              address:" 新疆乌鲁木齐 100 号",
              grade:" 401",
              score:"99",
              information:showInformation
           };
//定义对象的方法
function showInformation()
{
  var msg = "";
  msg = "学生信息: \n"
  msg += "\n 学生姓名 : " + this.name + " \n";
  msg += "家庭地址 : " + this.address + "\n";
  msg += "班级 : " + this.grade + " \n";
  msg += "分数 : " + this.Score;
  window.alert(msg);
}
-->
</script>
```

在 IE 中浏览修改后的 HTML 文档，可以看到与前面相同的结果。

该方法适合只需要生成某个应用对象并进行相关操作的情况下使用，代码紧凑，编程效率高。但若要生成若干个对象的实例，就必须为生成每个实例重复相同的代码结构，而只是参数不同而已，代码的重用性比较差，不符合面向对象的编程思路。因此，应尽量避免使用该方法创建自定义对象。

2.5.3 修改和删除对象实例的属性

JavaScript 脚本可以动态添加对象实例的属性，同时，也可以动态修改、删除某个对象实例的属性。将上面 HTML 文件中的 function showInformation()部分做如下修改：

```
function showInformation()
{
  var msg = "";
  msg = "自定义对象实例：\n\n"
  msg += " 学生姓名 : " + this.name + " \n";
  msg += " 家庭地址 : " + this.address + "\n";
  msg += " 班级 : " + this.grade + " \n";
  msg += " 分数 : " + this.score + " \n\n";
  //修改对象实例的 score 属性
  this.score = 88;
  msg += "修改对象实例的属性：\n\n"
  msg += " 学生姓名 : " + this.name + " \n";
  msg += " 所在地址 : " + this.address + "\n";
  msg += " 班级 : " + this.grade + " \n";
  msg += " 分数 : " + this.score + " \n\n";
  //删除对象实例的 score 属性
  delete this.score;
  msg += "删除对象实例的属性：\n\n"
  msg += " 学生姓名 : " + this.name + " \n";
  msg += " 家庭地址 : " + this.address + "\n";
  msg += " 班级 : " + this.grade + " \n";
  msg += " 分数 : " + this.score + " \n\n";
  window.alert(msg);
}
```

保存更改，程序运行后，在原始页面中单击【查看】按钮，弹出信息框，如图 2-30 所示。

在执行“this.score=88;”语句后，对象实例的 score 属性值更改为 88；而执行 delete this.score 语句后，对象实例的 score 属性变为 undefined，同任何不存在的对象属性一样为未定义类型，但并不能删除对象实例本身，否则将返回错误。

可见，JavaScript 动态添加、修改、删除对象实例的属性过程十分简单。之所以称为对象实例的属性而不是对象的属性，是因为该属性只在对象的特定实例中才存在，而不能通过某种方法将某个属性赋予特定对象的所有实例。

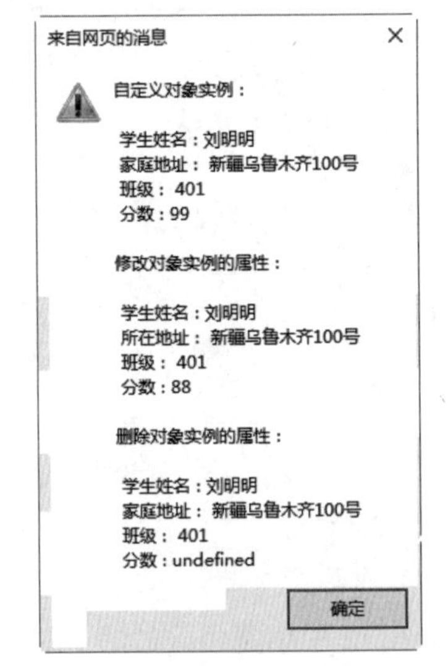

图 2-30 修改和删除对象实例的属性

JavaScript 脚本中的 delete 运算符用于删除对象实例的属性，而在 C++中，delete 运算符不能删除对象的实例。

2.5.4 通过原型为对象添加新属性和新方法

JavaScript 中，对象的 prototype 属性是用来返回对象类型原型引用的。使用 prototype 属性，能提供对象的类的一组基本功能，并且对象的新实例会“继承”赋予该对象原型的操作。所有 JavaScript 内部对象都有只读的 prototype 属性。可以向其原型中动态添加功能(属性和方法)，但该功能不能被赋予不同的原型，而用户定义的功能可以被赋予新的原型。

【例 2.27】(示例文件 ch02\2.27.html)

给已存在的对象添加新属性和新方法：

```
<!DOCTYPE html>
<html>
<head>
<title>自定义对象</title>
<script language="JavaScript" type="text/javascript">
<!--
//对象的构造函数
function Student(iName,iAddress,iGrade,iScore)
{
  this.name = iName;
  this.address = iAddress;
  this.grade = iGrade;
  this.score = iScore;
  this.information = showInformation;
}
//定义对象的方法
function showInformation()
{
  var msg = "";
  msg = "通过原型给对象添加新属性和新方法：\n\n"
  msg += "原始属性:\n";
  msg += "学生姓名：" + this.name + " \n";
  msg += "家庭住址：" + this.address + "\n";
  msg += "班级：" + this.grade + " \n";
  msg += "分数：" + this.score + " \n\n";
  msg += "新属性:\n";
  msg += "性别：" + this.addAttributeOfSex + " \n";
  msg += "新方法:\n";
  msg += "方法返回 : " + this.addMethod + "\n";
  window.alert(msg);
}
function MyMethod()
{
  var AddMsg = "New Method Of Object!";
  return AddMsg;
}
```

```
//生成对象的实例
var ZJDX = new Student("刘明明","新疆乌鲁木齐 100 号","401","88");
Student.prototype.addAttributeOfSex = "男";
Student.prototype.addMethod = MyMethod();
-->
</script>
</head>
<body>
<br>
<center>
<form>
  <input type="button" value="查看" onclick="ZJDX.information()">
</form>
</center>
</body>
</html>
```

将上述代码保存为 HTML 文件，再在 IE 中打开该网页。在打开的网页中单击【查看】按钮，即可看到含有新添加性别信息的提示框，如图 2-31 所示。

图 2-31　通过原型给对象添加新属性和新方法

在上面的程序中，是通过调用对象的 prototype 属性给对象添加新属性和新方法的：

```
Student.prototype.addAttributeOfSex = "男";
Student.prototype.addMethod = MyMethod();
```

原型属性为对象的所有实例所共享，用户利用原型添加对象的新属性和新方法后，可以通过对象引用的方法来修改。

2.5.5　自定义对象的嵌套

与面向对象编程方法相同的是，JavaScript 允许对象的嵌套使用，可以将对象的某个实例作为另外一个对象的属性来看待，如下面的程序：

```
<!DOCTYPE html>
<html>
<head>
<meta http-equiv="Content-Type" content="text/html; charset=gb2312">
<title>自定义对象嵌套</title>

<script language="JavaScript" type="text/javascript">
<!--
//对象的构造函数
//构造嵌套的对象
var StudentData={
             age:"26",
             Tel:"1810000000",
             teacher:"张老师"
            };
//构造被嵌入的对象
var ZJDX={
          name:"刘明明",
          address:"新疆乌鲁木齐100号",
          grade:"401",
          score:"86",
          //嵌套对象 StudentData
          data:StudentData,
          information:showInformation
         };
//定义对象的方法
function showInformation()
{
  var msg = "";
  msg = "对象嵌套实例：\n\n";
  msg += "被嵌套对象直接属性值:\n"
  msg += "学生姓名：" + this.name+"\n";
  msg += "家庭地址：" + this.address + "\n";
  msg += "年级：" + this.grade + "\n";
  msg += "分数：" + this.number + "\n\n";
  msg += "访问嵌套对象直接属性值:\n"
  msg += "年龄：" + this.data.age + "\n";
  msg += "联系电话：" + this.data.Tel + " \n";
  msg += "班主任：" + this.data.teacher + " \n";
  window.alert(msg);
}
-->
</script>

</head>
<body>
<br>
<center>
<form>
  <input type="button" value="查看" onclick="ZJDX.information()">
</form>
```

```
</center>
</body>
</html>
```

在上述 JavaScript 代码中，先构造对象 StudentData，包含学生的相关联系信息，代码如下：

```
var StudentData={
                  age:"26",
                  Tel:"1810000000",
                  teacher:"张老师""
                 };
```

然后构建 ZJDX 对象，同时嵌入 StudentData 对象，代码如下：

```
var ZJDX={
              name:"刘明明",
              address:"新疆乌鲁木齐 100 号",
              grade:"401",
              score:"86",
              //嵌套对象 StudentData
              data:StudentData,
              information:showInformation
             };
```

可以看出，在构建 ZJDX 对象时，StudentData 对象作为其自身属性 data 的取值而引入，并可通过如下的代码进行访问：

```
this.data.age
this.data.Tel
this.data.teacher
```

程序运行后，在打开的网页中单击【查看】按钮，即可弹出信息框，如图 2-32 所示。

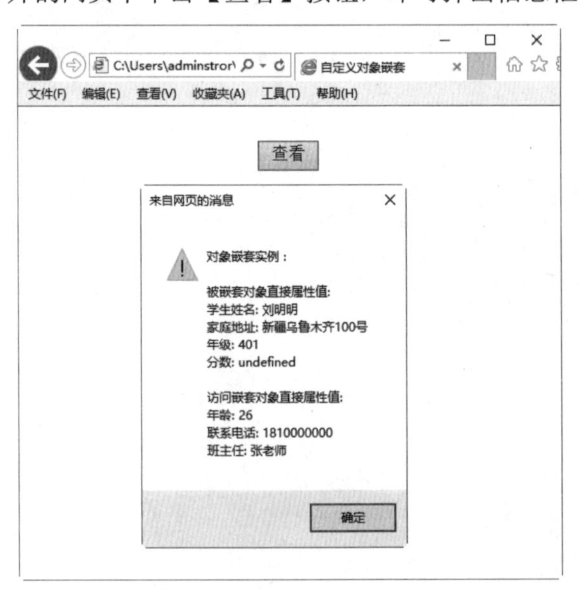

图 2-32　自定义对象的嵌套

2.5.6　内存的分配和释放

在创建对象时，浏览器自动为其分配内存空间，并在关闭当前页面时释放。下面介绍对象创建过程中内存的分配和释放问题。

JavaScript 是基于对象的编程语言，而不是面向对象的编程语言，因此缺少指针的概念。面向对象的编程语言在动态分配和释放内存等方面起着非常重要的作用，那么 JavaScript 中的内存如何管理呢？在创建对象的同时，浏览器自动为创建的对象分配内存空间，JavaScript 将新对象的引用传递给调用的构造函数；而在对象清除时，其占据的内存将被自动回收，其实整个过程都是浏览器的功劳，JavaScript 只是创建该对象。

浏览器中的这种内存管理机制称为“内存回收”，它动态分析程序中每个占据内存空间的数据(变量、对象等)。如果该数据被程序标记为不可再用时，浏览器将调用内部函数将其占据的内存空间释放，实现内存的动态管理。在自定义的对象使用过后，可以通过给其赋空值的方法来标记对象占据的空间可予以释放，如“ZJDX=null;”。浏览器将根据此标记动态释放其占据的内存，否则将保存该对象，直至当前程序再次使用它为止。

2.6　实战演练——利用二维数组创建动态下拉菜单

二维数组又称为矩阵。行列数相等的矩阵称为方阵。对称矩阵 $a_{ij}=a_{ji}$。对角矩阵是 n 阶方阵的所有非零元素都集中在主对角线上。

许多编程语言中都提供定义和使用二维或多维数组的功能。JavaScript 通过 Array 对象创建的数组都是一维的，但是可以通过在数组元素中使用数组来实现二维数组。

下面的 HTML 5 文档就是通过使用一个二维数组来改变下拉菜单内容的：

```
<!DOCTYPE html>
<HTML>
<HEAD>
<TITLE>动态改变下拉菜单内容</TITLE>
</HEAD>

<SCRIPT LANGUAGE=javascript>
   //定义一个二维数组 aArray，用于存放城市名称
   var aCity = new Array();
   aCity[0] = new Array();
   aCity[1] = new Array();
   aCity[2] = new Array();
   aCity[3] = new Array();
   //赋值，每个省份的城市存放于数组的一行
   aCity[0][0] = "--请选择--";
   aCity[1][0] = "--请选择--";
   aCity[1][1] = "广州市";
   aCity[1][2] = "深圳市";
   aCity[1][3] = "珠海市";
   aCity[1][4] = "汕头市";
```

```
    aCity[1][5] = "佛山市";
    aCity[2][0] = "--请选择--";
    aCity[2][1] = "长沙市";
    aCity[2][2] = "株洲市";
    aCity[2][3] = "湘潭市";
    aCity[3][0] = "--请选择--";
    aCity[3][1] = "杭州市";
    aCity[3][2] = "台州市";
    aCity[3][3] = "温州市";
    function ChangeCity()
    {
        var i,iProvinceIndex;
        iProvinceIndex = document.frm.optProvince.selectedIndex;
        iCityCount = 0;
        while (aCity[iProvinceIndex][iCityCount] != null)
            iCityCount++;
        //计算选定省份的城市个数
        document.frm.optCity.length = iCityCount; //改变下拉菜单的选项数
        for (i=0; i<=iCityCount-1; i++) //改变下拉菜单的内容
            document.frm.optCity[i] = new Option(aCity[iProvinceIndex][i]);
        document.frm.optCity.focus();
    }
</SCRIPT>

<BODY ONfocus=ChangeCity()>
  <H3>选择省份及城市</H3>
  <FORM NAME="frm">
   <P>省份:
    <SELECT NAME="optProvince" SIZE="1" ONCHANGE=ChangeCity()>
     <OPTION>--请选择--</OPTION>
     <OPTION>广东省</OPTION>
     <OPTION>湖南省</OPTION>
     <OPTION>浙江省</OPTION>
    </SELECT>
 </P>
 <P>城市:
  <SELECT NAME="optCity" SIZE="1">
   <OPTION>--请选择--</OPTION>
  </SELECT>
 </P>
</FORM>
</BODY>
</HTML>
```

在 IE 中打开上面的 HTML 文档，其显示结果如图 2-33 所示。在第一个下拉列表中选择一个省份，然后在第二个下拉列表中即可看到相应的城市，如图 2-34 所示。

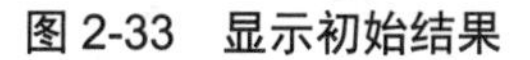
图 2-33　显示初始结果

图 2-34　选择省份对应的城市

2.7　疑 难 解 惑

疑问 1：JavaScript 支持的对象主要包括哪些？

答：JavaScript 主要支持下列对象。

(1)　JavaScript 核心对象。

包括同基本数据类型相关的对象(如 String、Boolean、Number)、允许创建用户自定义和组合类型的对象(如 Object、Array)和其他能简化 JavaScript 操作的对象(如 Math、Date、RegExp、Function)。

(2)　浏览器对象。

包括不属于 JavaScript 语言本身但被绝大多数浏览器所支持的对象，如控制浏览器窗口和用户交互界面的 Window 对象、提供客户端浏览器配置信息的 Navigator 对象。

(3)　用户自定义对象。

Web 应用程序开发者用于完成特定任务而创建的自定义对象，可自由设计对象的属性、方法和事件处理程序，编程灵活性较大。

(4)　文本对象。

对象由文本域构成，在 DOM 中定义，同时赋予很多特定的处理方法，如 insertData()、appendData()等。

疑问 2：如何获取数组的长度？

答：获取数组长度的代码如下：

```
var arr = new Array();
var len = arr.length;
```

第3章 jQuery的基础知识

当今，随着互联网的快速发展，程序员开始越来越重视程序功能上的封装与开发，进而可以从烦琐的 JavaScript 中解脱出来，以便后人在遇到相同问题时可以直接使用，提高项目的开发效率，其中 jQuery 就是一个优秀的 JavaScript 脚本库。

3.1 jQuery 概述

jQuery 是一个兼容多浏览器的 JavaScript 框架，它的核心理念是“写得更少，做得更多”。jQuery 在 2006 年 1 月由美国人 John Resig 在纽约的 Barcamp 发布，吸引了来自世界各地众多 JavaScript 高手的加入。如今，jQuery 已经成为最流行的 JavaScript 框架之一。

3.1.1 jQuery 能做什么

最开始时，jQuery 所提供的功能非常有限，仅仅能增强 CSS 的选择器功能，而如今 jQuery 已经发展到集 JavaScript、CSS、DOM 和 Ajax 于一体的优秀框架，其模块化的使用方式使开发者可以很轻松地开发出功能强大的静态或动态网页。目前，很多网站的动态效果就是利用 jQuery 脚本库制作出来的，如中国网络电视台、CCTV、京东商城等。

下面介绍京东商城应用的 jQuery 效果。访问京东商城的首页时，在右侧有一个话费、旅行、彩票、游戏栏目，这里应用 jQuery 实现了选项卡的效果，将鼠标指针移动到【话费】栏目上，选项卡中将显示手机话费充值的相关内容，如图 3-1 所示；将鼠标指针移动到【游戏】栏目上，选项卡中将显示游戏充值的相关内容，如图 3-2 所示。

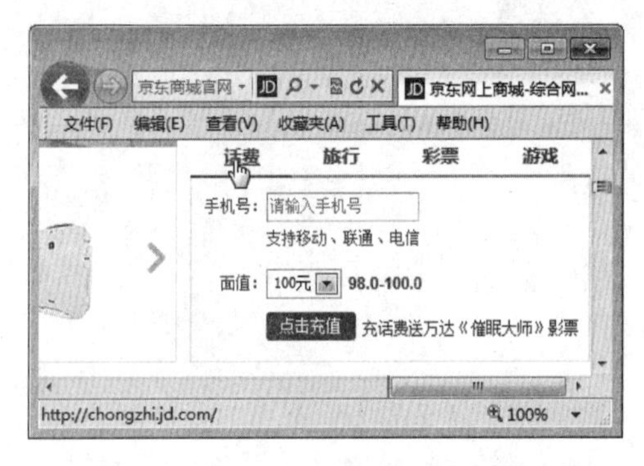

图 3-1　显示手机话费充值的相关内容

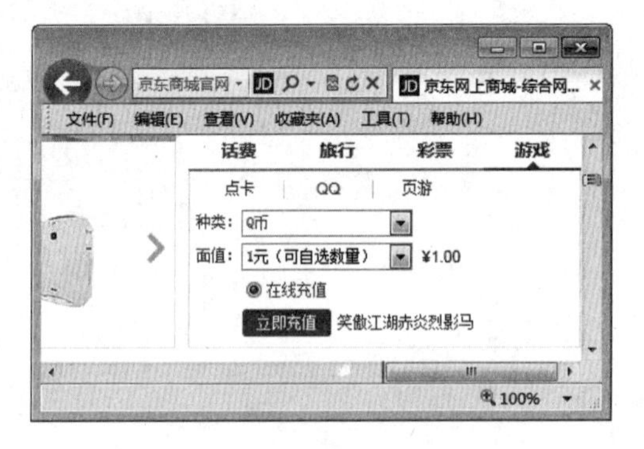

图 3-2　显示游戏充值的相关内容

3.1.2 jQuery 的特点

jQuery 是一个简洁快速的 JavaScript 脚本库，其独特的选择器、链式的 DOM 操作方式、事件绑定机制、封装完善的 Ajax 都是其他 JavaScript 库望尘莫及的。

jQuery 的主要特点如下。

(1) 代码短小精悍。jQuery 是一个轻量级的 JavaScript 脚本库，其代码非常短小，采用 Dean Edwards 的 Packer 压缩后，只有不到 30KB 的大小；如果服务器端启用 gzip 压缩，甚至只有 16KB 的大小。

(2) 强大的选择器支持。jQuery 可以让操作者使用从 CSS 1 到 CSS 3 几乎所有的选择器，以及 jQuery 独创的高级而复杂的选择器。

(3) 出色的 DOM 操作封装。jQuery 封装了大量常用 DOM 操作，使用户编写 DOM 操作相关程序的时候能够得心应手，顺畅地完成各种原本非常复杂的操作，让 JavaScript 新手也能写出出色的程序。

(4) 可靠的事件处理机制。jQuery 的事件处理机制汲取了 JavaScript 专家 Dean Edwards 编写的事件处理函数的精华，使得 jQuery 处理事件绑定的时候相当可靠。在预留退路方面，jQuery 也做得非常不错。

(5) 完善的 Ajax。jQuery 将所有的 Ajax 操作封装到一个$.ajax 函数中，使得用户处理 Ajax 的时候能够专心处理业务逻辑，而无须关心复杂的浏览器兼容性和 XML Http Request 对象的创建和使用的问题。

(6) 出色的浏览器兼容性。作为一个流行的 JavaScript 库，浏览器的兼容性自然是必须具备的条件之一，jQuery 能够在 IE 6.0+、Firefox 2+、Safari 2.0+和 Opera 9.0+下正常运行。同时修复了一些浏览器之间的差异，使用户不用在开展项目前因为忙于建立一个浏览器兼容库而焦头烂额。

(7) 丰富的插件支持。任何事物的壮大，如果没有很多人的支持，是永远发展不起来的。jQuery 的易扩展性，吸引了来自全球的开发者来共同编写 jQuery 的扩展插件。目前已经有几百种官方插件支持。

(8) 开源特点。jQuery 是一个开源的产品，任何人都可以自由地使用。

3.1.3 jQuery 的技术优势

jQuery 最大的技术优势就是简洁实用，能够使用短小的代码来实现复杂的网页预览效果。下面通过例子来介绍 jQuery 的技术优势。

在日常生活中，经常会遇到各种各样以表格形式出现的数据，当数据量很大或者表格格式过于一致时，会使人感觉混乱，所以工作人员常常通过奇偶行异色来实现使数据一目了然的效果。如果利用 JavaScript 来实现隔行变色的效果，需要用 for 循环遍历所有行，当行数为偶数的时候，添加不同类别即可。

【例 3.1】(示例文件 ch03\3.1.html)

JavaScript 实现表格奇偶行异色：

```
<!DOCTYPE html>
<html>
<head>
<title>JavaScript 表格奇偶行异色</title>
<style>
<!--
.datalist{
   border: 1px solid #007108;        /* 表格边框 */
   font-family: Arial;
   border-collapse: collapse;        /* 边框重叠 */
   background-color: #d999dc;        /* 表格背景色: 紫色 */
   font-size: 14px;
}
.datalist th{
   border: 1px solid #007108;        /* 行名称边框 */
```

```
    background-color: #000000;      /* 行名称背景色：黑色*/
    color: #FFFFFF;                 /* 行名称颜色：白色 */
    font-weight: bold;
    padding-top: 4px; padding-bottom: 4px;
    padding-left: 12px; padding-right: 12px;
    text-align: center;
}
.datalist td{
    border: 1px solid #007108;      /* 单元格边框 */
    text-align: left;
    padding-top: 4px; padding-bottom: 4px;
    padding-left: 10px; padding-right: 10px;
}
.datalist tr.altrow{
    background-color: #a5e5ff;      /* 隔行变色：蓝色 */
}
-->
</style>
<script language="javascript">
window.onload = function(){
    var oTable = document.getElementById("Table");
    for(var i=0; i<Table.rows.length; i++){
        if(i%2==0)        //偶数行时
            Table.rows[i].className = "altrow";
    }
}
</script>
</head>
<body>
<table class="datalist" summary="list of members in EE Study" id="Table">
    <tr>
        <th scope="col">姓名</th>
        <th scope="col">性别</th>
        <th scope="col">出生日期</th>
        <th scope="col">移动电话</th>
    </tr>
    <tr>
        <td>张三</td>
        <td>女</td>
        <td>8 月 10 日</td>
        <td>13012345678</td>
    </tr>
    <tr>
        <td>李四</td>
        <td>男</td>
        <td>5 月 25 日</td>
        <td>13112345678</td>
    </tr>
    <tr>
        <td>王五</td>
```

```
            <td>男</td>
            <td>7月3日</td>
            <td>13312345678</td>
        </tr>
        <tr>
            <td>赵六</td>
            <td>男</td>
            <td>10月2日</td>
            <td>13212345678</td>
        </tr>
</table>
</body>
</html>
```

运行结果如图 3-3 所示。

下面使用 jQuery 来实现表格奇偶行异色。当引入 jQuery 时，jQuery 的选择器会自动选择奇偶行。具体的实现代码如下。

【例 3.2】 (示例文件 ch03\3.2.html)

jQuery 实现表格奇偶行异色：

```
<script language="javascript" src="jquery.min.js"></script>
<script language="javascript">
$(function(){
    $("table.datalist tr:nth-child(odd)").addClass("altrow");
});
</script>
```

运行结果与 JavaScript 的结果完全一样，如图 3-4 所示，但是代码量减少，一行代码就轻松实现，语法也十分简单。

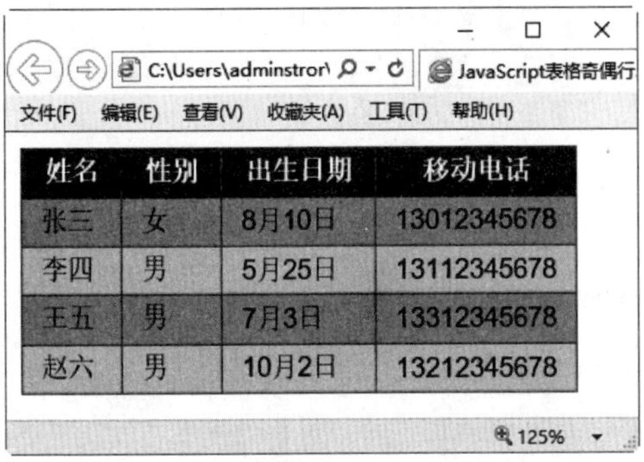

姓名	性别	出生日期	移动电话
张三	女	8月10日	13012345678
李四	男	5月25日	13112345678
王五	男	7月3日	13312345678
赵六	男	10月2日	13212345678

图 3-3　JavaScript 实现表格奇偶行异色

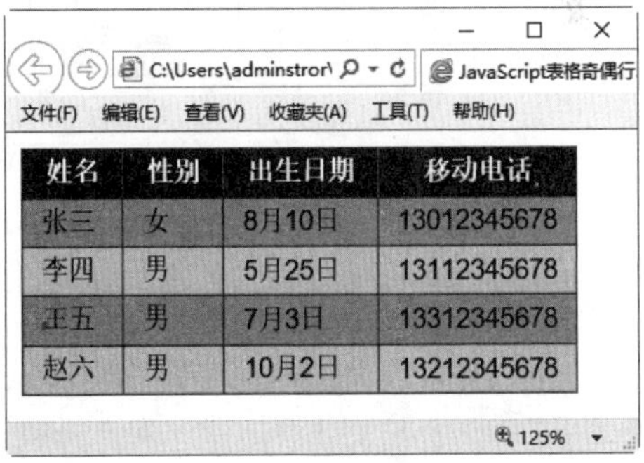

姓名	性别	出生日期	移动电话
张三	女	8月10日	13012345678
李四	男	5月25日	13112345678
王五	男	7月3日	13312345678
赵六	男	10月2日	13212345678

图 3-4　jQuery 实现表格奇偶行异色

3.2　下载并配置 jQuery

要想在开发网站的过程中应用 jQuery 库，需要下载并配置它。下面介绍如何下载与配置 jQuery。

3.2.1　下载 jQuery

jQuery 是一个开源的脚本库，可以从其官方网站(http://jquery.com)下载，下载 jQuery 库的具体操作步骤如下。

step 01 打开 IE 浏览器，在地址栏中输入 http://jquery.com，按 Enter 键，即可进入 jQuery 官方网站的首页，如图 3-5 所示。

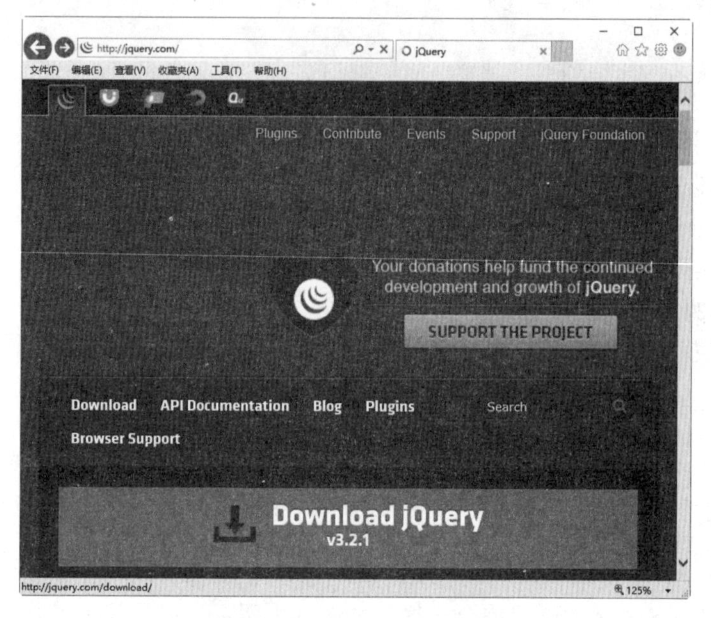

图 3-5　jQuery 官方网站的首页

step 02 在 jQuery 官方网站的首页中单击 jQuery 库的下载链接，如图 3-6 所示。

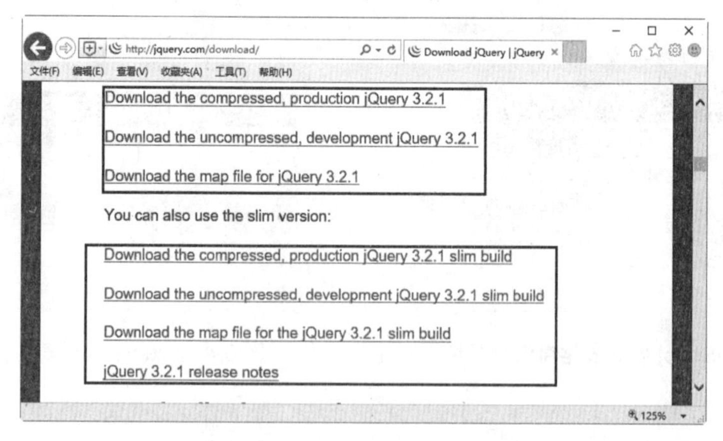

图 3-6　单击 jQuery 库的下载链接

step 03 这样即可打开迅雷下载对话框，在其中设置下载文件保存的位置，单击【立即下载】按钮，即可下载 jQuery 库，如图 3-7 所示。

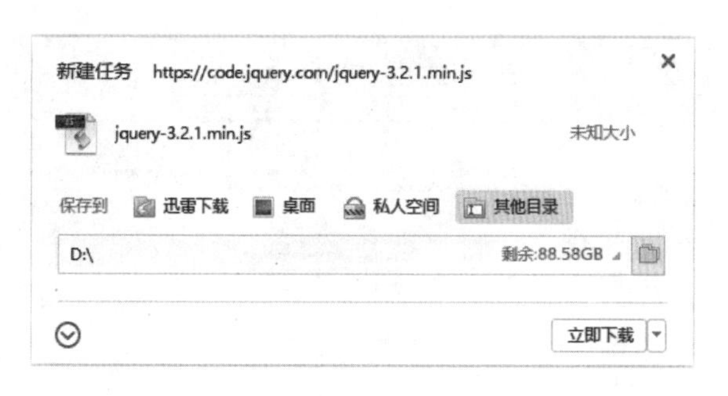

图 3-7　下载 jQuery 库

3.2.2　配置 jQuery

将 jQuery 库下载到本地计算机后，还需要在项目中配置 jQuery 库，即把下载的后缀名为.js 的文件放置到项目的指定文件夹中，通常放置在 JS 文件夹中，然后根据需要应用到 jQuery 的页面中。使用下面的语句，将其引用到文件中：

```
<script src="jquery.min.js" type="text/javascript"></script>
<!--或者-->
<script Language="javascript" src="jquery.min.js"></script>
```

引用 jQuery 的<script>标签必须放在所有的自定义脚本的<script>之前，否则在自定义的脚本代码中应用不到 jQuery 脚本库。

3.3　jQuery 的开发工具

适合开发 jQuery 的工具很多，常用的有 JavaScript Editor Pro、Dreamweaver、文本编辑器 UltraEdit 等。其中，最普通的文本编辑器就可以用来作为 jQuery 的开发工具。

3.3.1　JavaScript Editor Pro

JavaScript Editor Pro 是一款专业的 JavaScript 脚本编辑器，支持多种网页脚本语言编辑(JavaScript、HTML、CSS、VBScript、PHP 和 ASP 语法标注等)和内嵌的预览功能，还提供了大量的 HTML 标签、属性、事件和 JavaScript 事件、功能、属性、语句、动作等代码库，同时有着贴心的代码自动补全功能，可轻松插入到网页中。

JavaScript Editor 编辑器可以使用内置的“函数和变量”导航工具帮助用户浏览代码时提供智能提示，以简化代码编写过程，有效地减少语法等错误。

软件发布者提供了免费版的下载。免费版的软件名称叫 Free JavaScript Editor。需要注意的是，该免费版提供了 21 天的试用期限，下载地址是 http://www.yaldex.com/Free_JavaScript_Editor.htm。在 IE 浏览器中输入该下载地址，然后按 Enter 键，即可进入下载页面，如图 3-8 所示。

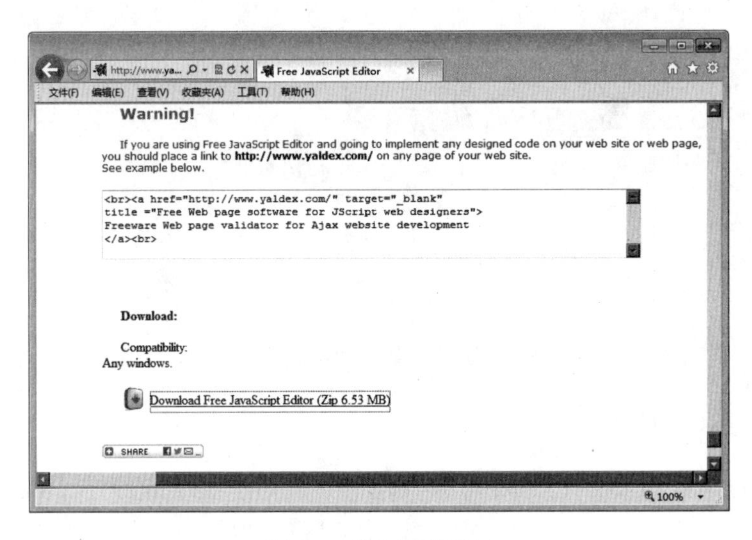

图 3-8　进入下载页面

下载完毕后，双击下载的安装程序，按照软件安装提示，即可将 Free JavaScript Editor 安装到自己的电脑中。最后双击桌面上的快捷图标，即可打开 Free JavaScript Editor 工作界面，如图 3-9 所示。

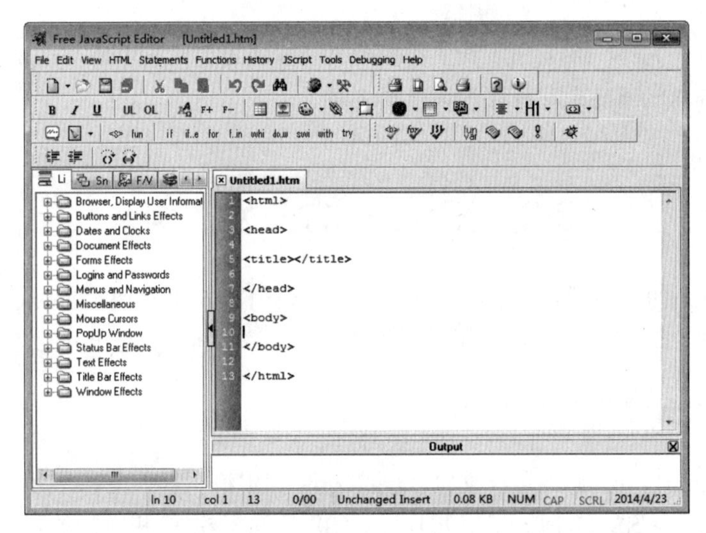

图 3-9　Free JavaScript Editor 工作界面

3.3.2　Dreamweaver

Dreamweaver 是由 Macromedia 公司所开发的著名网站开发工具，它使用所见即所得的接口，是一个非常优秀并深受广大用户喜爱的开发工具，同时还具有 HTML 编辑功能。目前，该工具有 Mac 和 Windows 系统的版本。其中，最新的版本已经更新到 CC，Dreamweaver CC 的主界面，如图 3-10 所示。

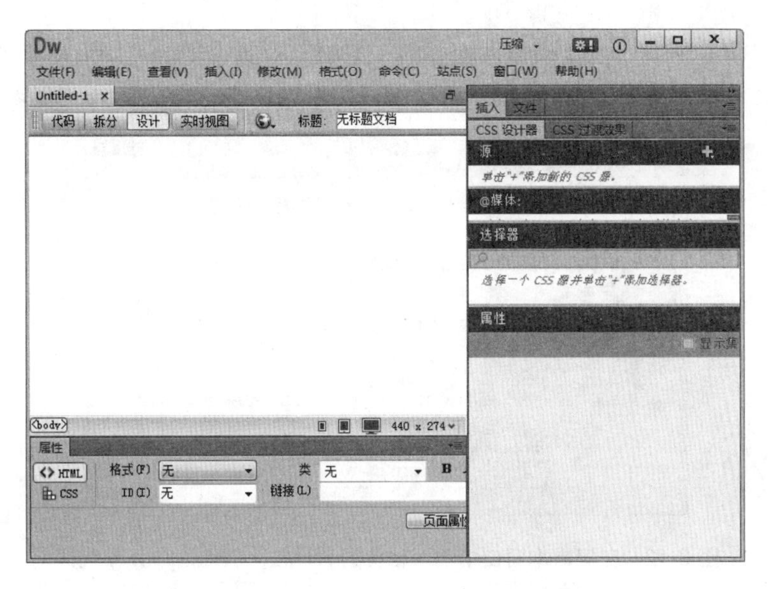

图 3-10 Dreamweaver CC 的主界面

3.3.3 UltraEdit

UltraEdit 是一套功能强大的文本编辑器，可以编辑文本、HTML、十六进制、ASCII 码，完全可以取代记事本文件，内建英文单词检查、C++及 VB 指令突显，可同时编辑多个文件，而且即使开启很大的文件，速度也不会慢。同时，它也是高级 PHP、Perl、Java 和 JavaScript 程序编辑器。软件附有 HTML 标签颜色显示、搜寻替换以及无限制还原功能。目前最新的版本为 UltraEdit v18.00。如图 3-11 所示为 UltraEdit 的工作界面。该软件的特点有：打开文件速度快、列操作功能强大、有代码折叠功能、可以进行十六进制编辑。

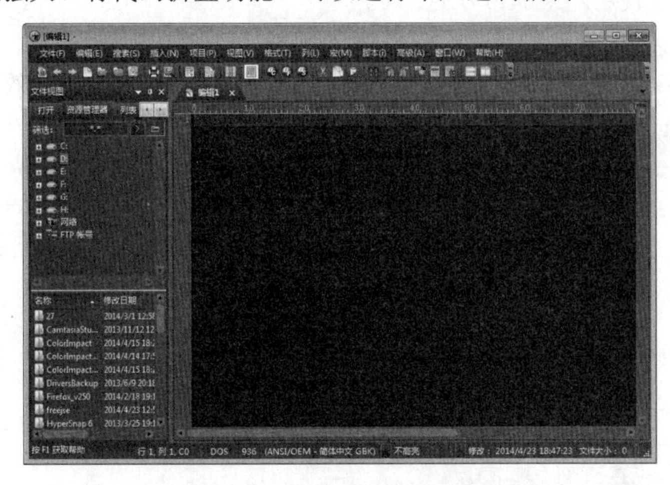

图 3-11 UltraEdit 的工作界面

3.3.4 记事本工具

单击 Windows 桌面上的【开始】按钮，选择【所有程序】→【附件】→【记事本】命令，打开一个记事本窗口，在其中输入相关的 HTML、CSS、jQuery 代码，如图 3-12 所示。然后将记事本文件以扩展名.html 或.htm 进行保存，可以在浏览器中打开文档以查看效果。

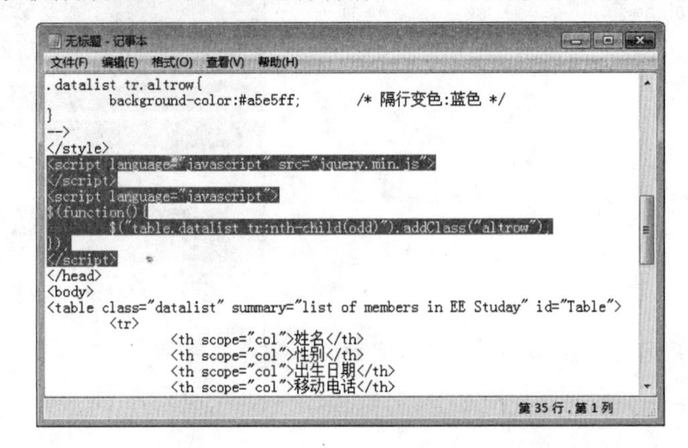

图 3-12　在记事本窗口中输入相关的 HTML、CSS、jQuery 代码

3.4　jQuery 的调试小工具

常用的 jQuery 的调试工具主要有 Firebug、Blackbird、jQueryPad 等。下面就来介绍 jQuery 调试工具的使用方法。

3.4.1　Firebug

Firebug 是火狐(Firefox)浏览器的一个插件。该插件可以调试所有网站语言，如 HTML、CSS、JavaScript 等，使用起来非常方便。

使用 Firebug 调试 jQuery 的具体操作步骤如下。

step 01 双击桌面上的 Firefox 快捷图标，打开火狐浏览器的工作界面，效果如图 3-13 所示。

图 3-13　火狐浏览器的工作界面

step 02 选择【工具】→【附加组件】命令，如图 3-14 所示。

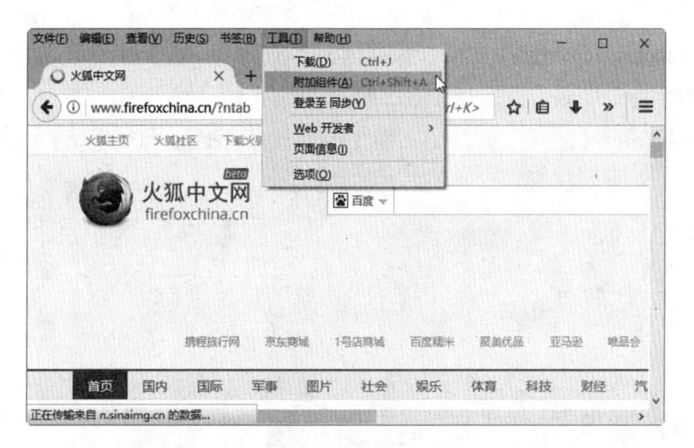

图 3-14 选择【工具】→【附加组件】命令

step 03 随即进入火狐浏览器的【附加组件管理器】工作界面中，如图 3-15 所示。

图 3-15 火狐浏览器的【附加组件管理器】工作界面

step 04 在【附加组件管理器】工作界面中的搜索文本框中输入 fireBug，然后单击右侧的搜索按钮，即可在界面中显示有关的插件信息，如图 3-16 所示。

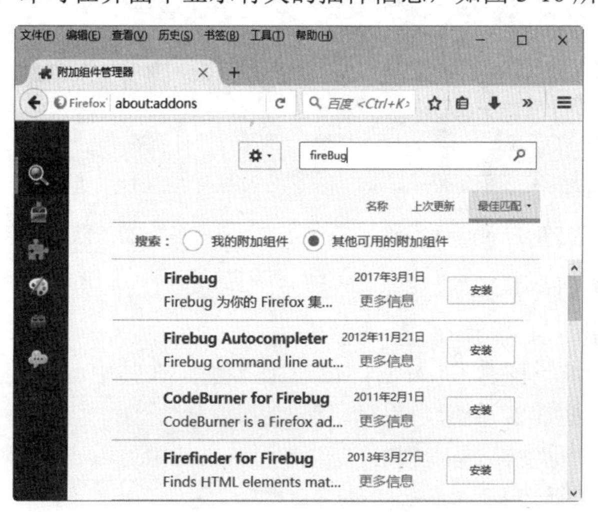

图 3-16 显示有关的插件信息

step 05 单击 Firebug 插件后面的【更多信息】超级链接，即可在打开的界面中查看有关 Firebug 的相关说明性信息，如图 3-17 所示。

step 06 单击 Firebug 页面下方的【安装】按钮，Firefox 开始自动下载并安装 Firebug 插件，如图 3-18 所示。

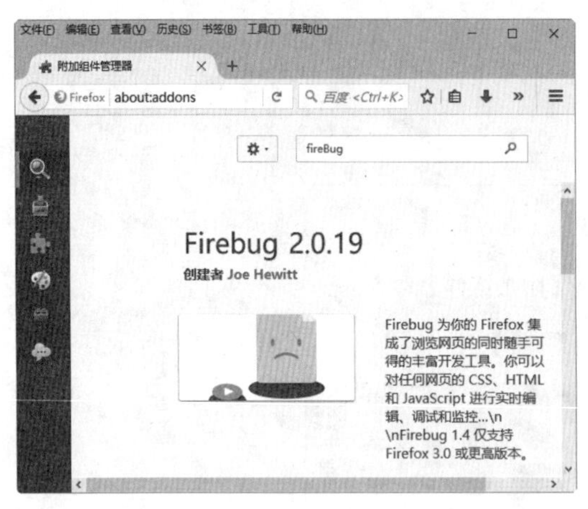

图 3-17　Firebug 的相关说明性信息

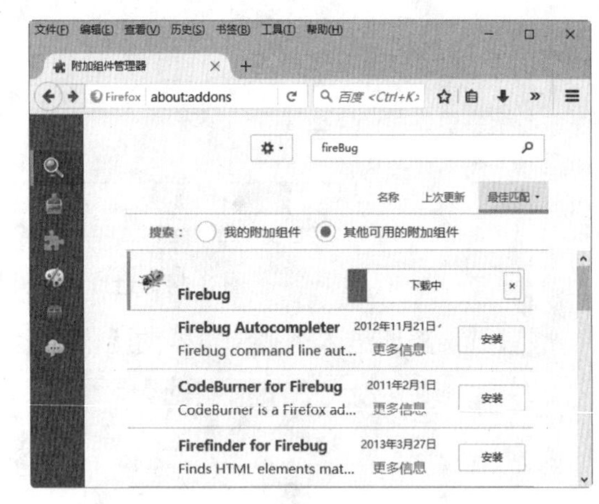

图 3-18　开始自动下载并安装 Firebug 插件

step 07 安装完毕后，重新启动火狐浏览器，选择【工具】→【Web 开发者】→【Firebug】→【打开 Firebug】命令，如图 3-19 所示。

step 08 这时，可以在火狐浏览器工作界面的下方显示 Firebug 的工作界面，包括 HTML、CSS、脚本(Script)、DOM、网络(Net)等标签，默认显示 Cookies 工作界面，如图 3-20 所示。

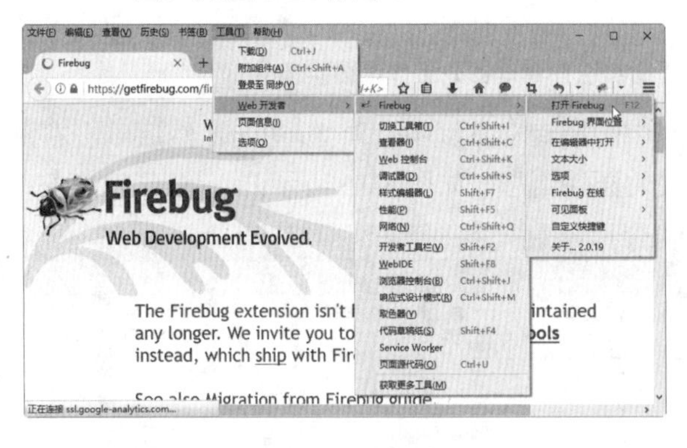

图 3-19　选择【打开 Firebug】命令

图 3-20　默认显示 Cookies 工作界面

step 09 选择【控制台】标签，进入控制台工作界面，在其中可以查看程序的错误和日志信息，如图 3-21 所示。

step 10 选择 HTML 标签，进入 HTML 工作界面，在其中可以看出程序的 HTML 相关代码信息，如图 3-22 所示。

step 11 选择 CSS 标签，进入 CSS 工作界面，在其中可以查看有关程序的 CSS 代码信息，如图 3-23 所示。

step 12 选择 DOM 标签，进入 DOM 工作界面，在其中可以查看有关程序的 DOM 代码信息，如图 3-24 所示。

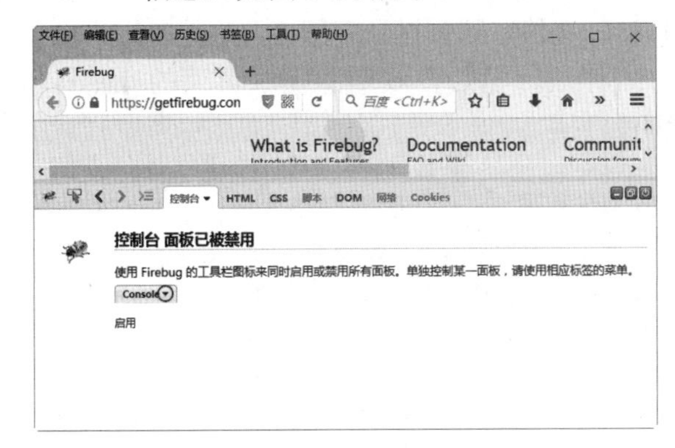

图 3-21　查看程序的错误和日志信息

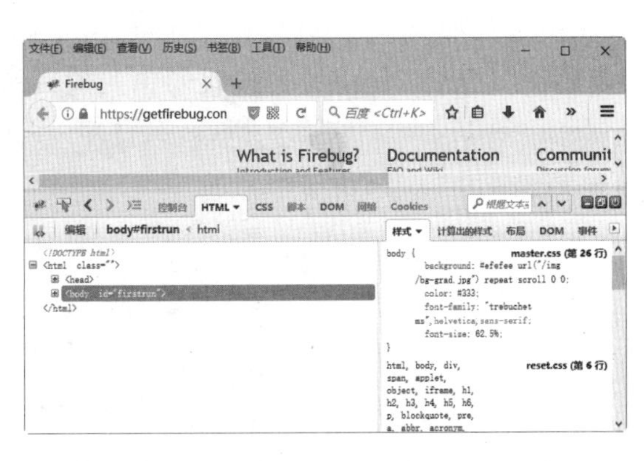

图 3-22　查看程序的 HTML 相关代码信息

图 3-23　查看有关程序的 CSS 代码信息

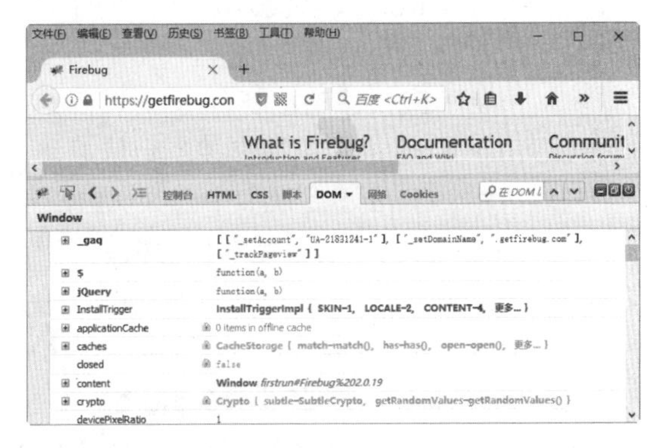

图 3-24　查看有关程序的 DOM 代码信息

Firebug 插件功能强大，而且它已经与 Firefox 浏览器无缝地结合在一起，使用简单直观。如果担心它会占用太多的系统资源，可以将其关闭，还可以对特定站点开启这个插件。

3.4.2　Blackbird

Blackbird 是一个开源的 JavaScript 库，提供了一种简单的记录日志的方式和一个控制台窗口。有了它之后，用户就可以抛弃 alert()了。如图 3-25 所示为 Blackbird 的工作界面。

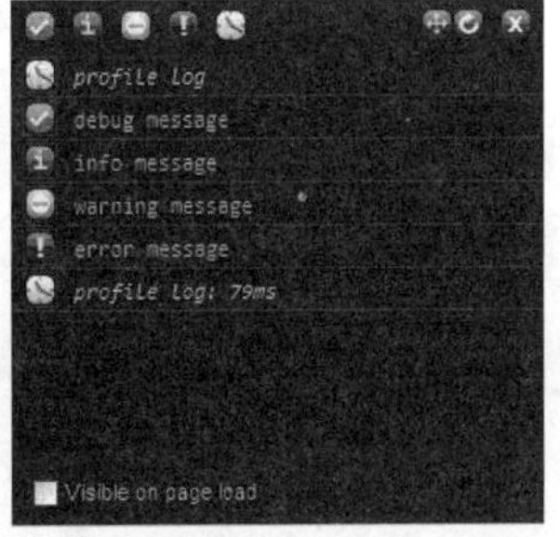

图 3-25　Blackbird 的工作界面

Blackbird 有 4 个文件，即 blackbird.css、blackbird.js、blackbird_icons.png 和 blackbird_panel.png。使用也非常简单，保持 CSS 文件和 PNG 文件在同一目录下即可。当然，用户也可以修改 CSS 文件，使之按我们想要的目录方式存放，然后在我们想调试页面的<head>和</head>标签之间加载该.js 和.css 文件即可，代码如下：

```
<html>
<head>
<script type="text/javascript" src="/PATH/TO/blackbird.js"></script>
<link type="text/css" rel="Stylesheet" href="/PATH/TO/blackbird.css" />
...
</head>
</html>
```

Blackbird 支持当前主流浏览器，如 IE 6+、Firefox 2+、Safari 2+、Opera 9.5 等，并支持快捷键操作。Blackbird 的快捷键详细说明如下。

- F2：显示和隐藏控制台。
- Shift + F2：移动控制台。
- Alt + Shift + F2：清空控制台信息。

同时，Blackbird 还提供多个公共 API，详细说明如下。

- log.toggle()：显示控制台面板。
- log.move()：移动控制台面板的位置。
- log.resize()：调整控制台面板的大小。
- log.clear()：清空控制台的内容。
- log.debug(message)：添加一个 Debug 信息。
- log.info(message)：添加一个 Info 信息。
- log.warn(message)：添加一个警告信息。
- log.error(message)：添加一个错误信息。
- log.profile(label)：计算两个 label 相同的两句语句之间的执行时间。

公共 API 的用法也很简单。例如想要在 JavaScript 代码中调用 Blackbird，代码如下：

```
log.debug('this is a debug message');
log.info('this is an info message');
log.warn('this is a warning message');
log.error('this is an error message');
```

下面是一个更为详细、具体的例子(计算消耗时间)：

```
log.profile('local anchors');
var anchors = document.getElementsByTagName('A');
for (var i=0; i<anchors.length; i++) {
    if (anchors[i].name) {
        log.debug(anchors[i].name);
    }
}
log.profile('local anchors');
```

3.4.3 jQueryPad

jQueryPad 是一个方便快捷的 JavaScript/HTML 编辑调试器。启动后，左边输入要操作的 HTML，右侧输入 jQuery 代码，按 F5 键，就可以看到结果。如图 3-26 所示为 jQueryPad 的工作界面。

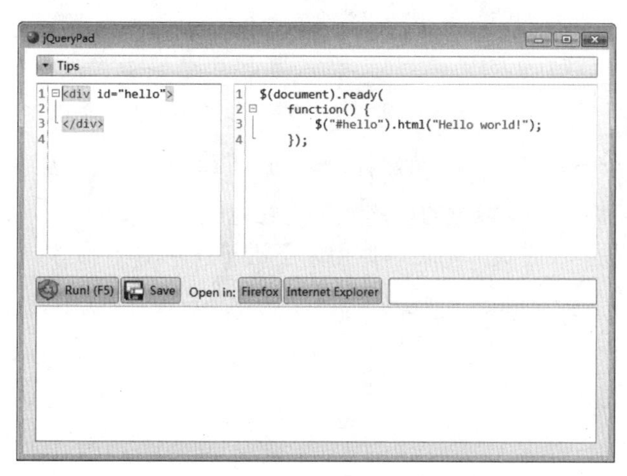

图 3-26　jQueryPad 的工作界面

这款软件的基本原理是：在调试时，将用户编写的 HTML 和 JavaScript 代码复制到一个文件中(当然，这个文件加载了 jQuery 框架，所有 jQuery 函数都可用)，然后显示。

对网页程序员来说，在代码编辑器和浏览器之间不停地使用 Alt+Tab 组合键来相互切换是家常便饭的事；而 jQueryPad 是一个整合 HTML/jQuery 代码编辑与测试的小软件，让程序员摆脱了来回切换的麻烦。但是 jQueryPad 也存在一些明显的问题，它没有任何帮助使用文档和基本的提示功能，也无法设断点 debug，有些可惜。

但是总体来讲，jQueryPad 算是一款方便实用的 jQuery 调试工具。

3.5 jQuery 与 CSS 3

对设计者来说，CSS 3 是一个非常灵活的工具，使用户不必再把复杂的样式定义编写在文档结构中，而将有关文档的样式内容全部脱离出来。这样做的最大优势就是在后期维护中只需修改代码即可。

3.5.1 CSS 3 构造规则

CSS 3 样式表是由若干条样式规则组成的，这些样式规则可以应用到不同的元素或文档，来定义它们显示的外观。每一条样式规则由 3 个部分构成：选择符(selector)、属性(property)和属性值(value)。基本语法格式如下：

```
selector{property: value}
```

(1) selector 选择符可以采用多种形式，可以是文档中的 HTML 标记，如<body>、<table>、<p>等，也可以是 XML 文档中的标记。

(2) property 属性则是选择符指定的标记所包含的属性。

(3) value 指定了属性的值。如果定义选择符的多个属性，则属性和属性值为一组，组与组之间用分号(;)隔开。基本语法格式如下：

```
selector{property1: value1; property2: value2; ...}
```

下面给出一条样式规则：

```
p{color: red}
```

该样式规则的选择符是 p，即为段落标记<p>提供样式；color 为指定文字颜色属性；red 为属性值。此样式表示标记<p>指定的段落文字为红色。

如果要为段落设置多种样式，则可以使用如下语句：

```
p{font-family:"隶书"; color:red; font-size:40px; font-weight:bold}
```

3.5.2 浏览器的兼容性

CSS 3 制定完成后，具有了很多新功能，即新样式，但这些新样式在浏览器中不能获得完全支持，其原因主要在于各个浏览器对 CSS 3 的细节处理上存在差异。例如，一种标记的某个属性，一种浏览器支持，而另一种浏览器不支持，或者两者浏览器都支持，但其显示效果不一样。

针对 CSS 3 与浏览器的兼容性，用户可以通过 http://www.css3.info 网站来测试自己所使用的浏览器版本对属性选择器的兼容性程度。

具体操作步骤如下。

step 01 打开 IE 浏览器，在地址栏中输入 http://www.css3.info，按 Enter 键，进入该网站的首页，选择 CSS SELECTORS TEST 选项卡，进入 CSS SELECTORS TEST 工作界面中，如图 3-27 所示。

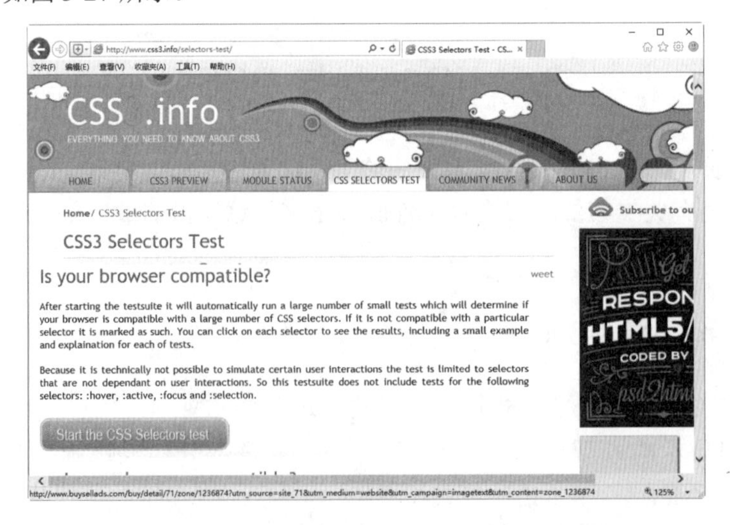

图 3-27 CSS SELECTORS TEST 工作界面

step 02 单击 Start the CSS Selectors test 按钮，即可开始测试本机浏览器版本(IE 11.0)与 CSS 属性选择的兼容性，其中红色部分说明兼容效果不好，绿色部分说明兼容效果好，如图 3-28 所示。

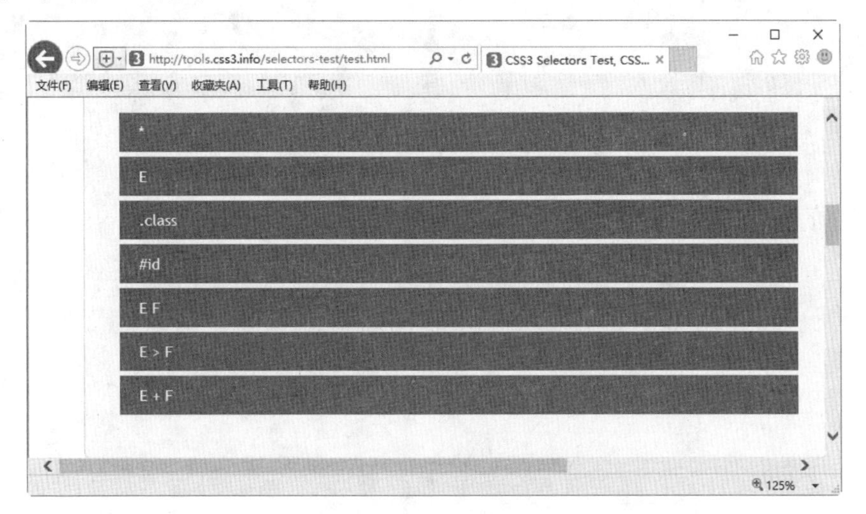

图 3-28 IE 11.0 的测试结果

3.5.3 jQuery 的引入

jQuery 的引入弥补了浏览器与 CSS 3 兼容性不好的缺陷，因为 jQuery 提供了几乎所有的 CSS 3 属性选择器，而且 jQuery 的兼容性很好，目前的主流浏览器几乎都可以完美实现。开发者只需按照以前的方法定义 CSS 类别，在引入 jQuery 后，通过 addClass()方法添加至指定元素中即可。

【例 3.3】(示例文件 ch03\3.3.html)

jQuery 的引入为 CSS 3 带来的便利：

```
<!DOCTYPE html>
<html>
<head>
<title>属性选择器</title>
<style type="text/css">
.NewClass{ /* 设定某个 CSS 类别 */
   background-color: #223344;
   color: #22ff37;
}
</style>
<script language="javascript" src="jquery.min.js"></script>
<script language="javascript">
$(function(){ /*先用 CSS 3 的选择器，然后添加样式风格*/
   $("a:nth-child4)").addClass("NewClass");
});
</script>
```

```
</head>
<body>
<a href="#">精选特卖</a>
<a href="#">51 特价</a>
<a href="#">满千降百</a>
<a href="#">精品荟萃</a>
<a href="#">特价包邮</a>
</body>
</html>
```

上述程序的运行结果如图 3-29 所示。

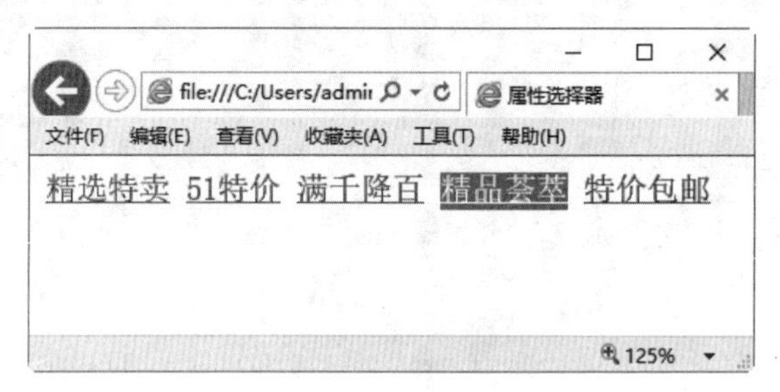

图 3-29　属性选择器

3.6　实战演练——我的第一个 jQuery 程序

开发 jQuery 程序其实很简单，只要先引入 jQuery 库，然后调用即可。下面制作一个简单的 jQuery 程序，来介绍如何引用 jQuery 库。

3.6.1　开发前的一些准备工作

由于 jQuery 是一个免费开源项目，任何人都可以在 jQuery 的官方网站 http://jquery.com 下载到最新版本的 jQuery 库文件。

jQuery 库文件有两种类型：完整版和压缩版。前者主要用于测试开发，后者主要用于项目应用。例如，jQuery 3.2.1 版本有 jquery-3.2.1.js 和 jquery-3.2.1.min.js 两个文件，它们分别对应完整版和压缩版。

下载完 jQuery 库之后，将其放置在具体的项目目录下，然后在 HTML 页面中引入该 jQuery 库文件，具体的代码如下：

```
<script language="javascript" src="../jquery.min.js"></script>
```

可以看出，在 HTML 页面中引入 jQuery 库文件和引入外部的 JavaScript 程序文件，形式上没有任何区别。同时，在 HTML 页面中直接插入 jQuery 代码或引入外部 jQuery 程序文件，需要符合的格式也跟 JavaScript 一样。

值得一提的是，外部 jQuery 程序文件是不同页面共享相同 jQuery 代码的一种高效方式。这样当修改 jQuery 代码时，只需要编辑一个外部文件，操作更为方便。此外，一旦载入某个外部 jQuery 文件，它就会存储在浏览器的缓存中，因此不同页面重复使用它时无须再次下载，从而加快了网页的访问速度。

3.6.2 具体的程序开发

环境配置好之后，下面就可以来开发程序了，这里以在记事本文件中开发程序为例。具体操作步骤如下。

step 01 打开记事本文件，在其中输入以下代码：

```
<html>
<head>
<title>第一个实例</title>
<script language="javascript" src="jquery-3.2.1.min.js"></script>
<script language="javascript">
$(document).ready(function(){
alert("Hello jQuery!");});
</script>
</head>
</body>
</html>
```

step 02 将记事本文件以.html 的格式进行保存，然后在 IE 11.0 中运行，结果如图 3-30 所示。

图 3-30 运行结果

3.7 疑 难 解 惑

疑问：jQuery 变量与普通 JavaScript 变量是否容易混淆？

答：jQuery 作为一个跨多个浏览器的 JavaScript 库，可以有助于写出高度兼容的代码。但其中有一点需要强调的是，jQuery 的函数调用返回的变量，与浏览器原生的 JavaScript 变量是有区别的，不可混用。例如以下代码是有问题的：

```
var a = $('#abtn');
a.click(function(){...});
```

可以这样理解，$('')选择器返回的变量属于“jQuery 变量”，通过复制给原生变量 a，将其转换为普通变量了，因而无法支持常见的 jQuery 操作。一个解决方法是将变量名加上$标记，使得其保持为“jQuery 变量”：

```
var $a = $('#abtn');
$a.click(function(){...});
```

除了上述例子之外，实际 jQuery 编程中也会有很多不经意间的转换，从而导致错误，需要读者根据这个原理仔细调试和修改。

第 4 章 jQuery 的选择器

在 JavaScript 中，要想获取网页的 DOM 元素，必须使用该元素的 ID 和 TagName。但是在 jQuery 库中却提供了许多功能强大的选择器，帮助开发人员获取页面上的 DOM 元素，而且获取到的每个对象都以 jQuery 包装集的形式返回。本章介绍如何应用 jQuery 的选择器选择匹配的元素。

4.1 jQuery 的$

$是 jQuery 中最常用的一个符号，用于声明 jQuery 对象。可以说，在 jQuery 中，无论使用哪种类型的选择器，都需要从一个$符号和一对“()”开始。在“()”中通常使用字符串参数，参数中可以包含任何 CSS 选择符表达式。

4.1.1 $符号的应用

$是 jQuery 选取元素的符号，用来选择某一类或者某一个元素。其通用语法格式如下：

```
$(selector)
```

$通常的用法有以下几种。

(1) 在参数中使用标记名，如$("div")，用于获取文档中全部的<div>。

(2) 在参数中使用 ID，如$("#usename")，用于获取文档中 ID 属性值为 usename 的一个元素。

(3) 在参数中使用 CSS 类名，如$(".btn_grey")，用于获取文档中使用 CSS 类名为 btn_grey 的所有元素。

【例 4.1】(示例文件 ch04\4.1.html)

选择文本段落中的奇数行：

```
<!DOCTYPE html>
<html>
<head>
<title>$符号的应用</title>
<script language="javascript" src="jquery-3.2.1.min.js"></script>
<script language="javascript">
window.onload = function(){
    var oElements = $("p:odd");      //选择匹配元素
    for(var i=0; i<oElements.length; i++)
        oElements[i].innerHTML = i.toString();
}
</script>
</head>
<body>
<div id="body">
<p>第一行</p>
<p>第二行</p>
<p>第三行</p>
<p>第四行</p>
<p>第五行</p>
</div>
</body>
</html>
```

上述程序的运行结果如图 4-1 所示。

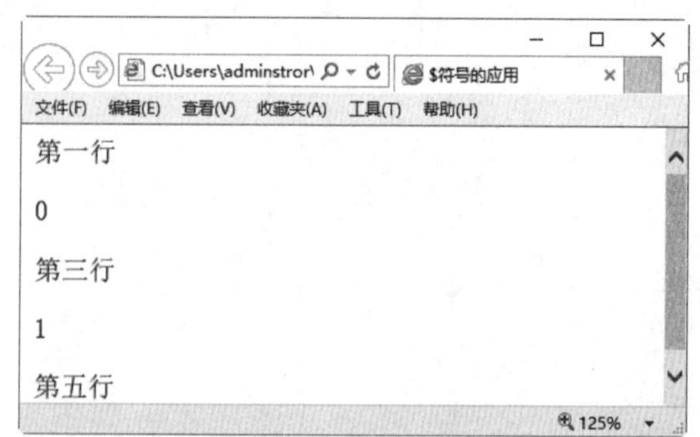

图 4-1 $符号的应用

4.1.2 功能函数的前缀

$是功能函数的前缀。例如，JavaScript 中没有提供清理文本框中空格的功能，但在引入 jQuery 后，开发者就可以直接调用 trim()函数来轻松地去掉文本框前后的空格，不过需要在函数前加上$符号。当然 jQuery 中这种函数还有很多，后面章节涉及的时候会继续介绍。

【例 4.2】(示例文件 ch04\4.2.html)

jQuery 的$.trim()函数的使用：

```
<!DOCTYPE html>
<html>
<head>
<title>$.trim()</title>
<script language="javascript" src="jquery-3.2.1.min.js"></script>
<script language="javascript">
var String = " Open in a new window ";
String = $.trim(String);
alert(String);
</script>
</head>
<body>
清除空格前“ Open in a new window ”
</body>
</html>
```

上述程序的运行结果如图 4-2 所示。可以看到，这段代码的功能是将字符串中首尾的空格全部去掉。

图 4-2 使用$.trim()函数

4.1.3 创建 DOM 元素

jQuery 可以使用$创建 DOM 元素。例如下面一段 JavaScript 就是用来创建 DOM 的代码：

```
var NewElement =
document.createElement("p");
var NewText = document.createTextNode
("Hello World!");
NewElement.appendChild(NewText);
```

其中，append()方法用于在节点之下加入新的文本。上面的一段代码在 jQuery 中可以直接简化为：

```
var NewElement = $("<p>Hello World!</p>");
```

【例 4.3】(示例文件 ch04\4.3.html)

创建 DOM 元素：

```
<!DOCTYPE html>
<html>
<head>
<title>创建 DOM 元素</title>
<script language="javascript" src="jquery-3.2.1.min.js"></script>
<script language="javascript">
$(document).ready(function(){
    var New = $("<a>(Open in a new window)</a>");        //创建 DOM 元素
    New.insertAfter("#target");     //insertAfter()方法
});
</script>
</head>
<body>
    <a id="target" href="https://www.google.com.hk/">Google</a>
    <a href="http://www.baidu.com">Baidu</a>
</body>
</html>
```

上述程序的运行结果如图 4-3 所示。

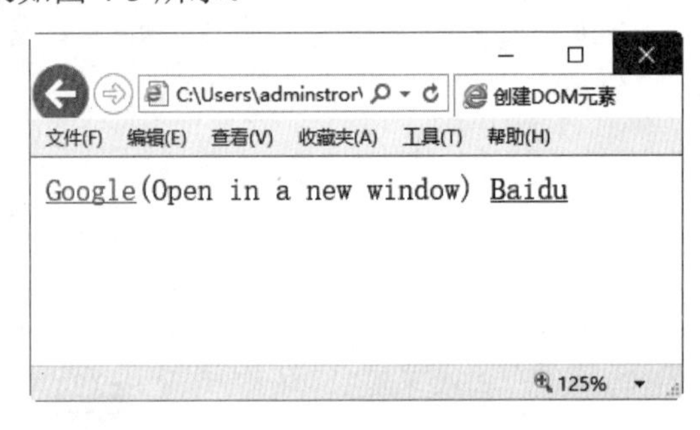

图 4-3 创建 DOM

4.2 基本选择器

jQuery 的基本选择器是应用最广泛的选择器，是其他类型选择器的基础，是 jQuery 选择器中最为重要的部分，这里建议读者重点掌握。jQuery 的基本选择器包括通配符选择器(*)、ID 选择器、类名选择器、元素选择器、复合选择器等。

4.2.1 通配符选择器(*)

通配符选择器(*)选择器选取文档中的每个单独的元素，包括 html、head 和 body。如果与其他元素(如嵌套选择器)一起使用，该选择器选取指定元素中的所有子元素。

*选择器的语法格式如下：

```
$(*)
```

【例 4.4】(示例文件 ch04\4.4.html)

选择<body>内的所有元素：

```
<!DOCTYPE html>
<html>
<head>
<script language="javascript" src="jquery-3.2.1.min.js"></script>
<script language="javascript">
$(document).ready(function(){
    $("body *").css("background-color","#B2E0FF");
});
</script>
</head>

<body>
<h1>欢迎光临我的网站主页</h1>
<p class="intro">网站管理员介绍</p>
<p>姓名：张三</p>
<p>性别：男</p>
<div id="choose">
兴趣爱好：
<ul>
<li>读书</li>
<li>听音乐</li>
<li>跑步</li>
</ul>
</div>
</body>
</html>
```

上述程序的运行结果如图 4-4 所示。可以看到，网页中用背景色显示出 body 中所有的元素内容。

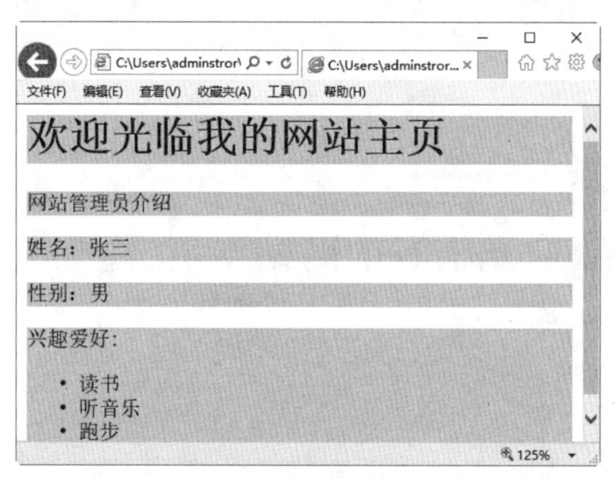

图 4-4　使用*选择器

4.2.2　ID 选择器(#id)

ID 选择器是利用 DOM 元素的 ID 属性值来筛选匹配的元素，并以 jQuery 包装集的形式返回给对象。ID 选择器的语法格式如下：

```
$("#id")
```

【例 4.5】(示例文件 ch04\4.5.html)

选择<body>中 id 为 choose 的所有元素：

```
<!DOCTYPE html>
<html>
<head>
<script language="javascript" src="jquery-3.2.1.min.js"></script>
<script language="javascript">
$(document).ready(function(){
    $("#choose").css("background-color","#B2E0FF");
});
</script>
</head>
<body>
<h1>欢迎光临我的网站主页</h1>
<p class="intro">网站管理员介绍</p>
<p>姓名：张三</p>
<p>性别：男</p>
<div id="choose">
兴趣爱好：
<ul>
<li>读书</li>
<li>听音乐</li>
<li>跑步</li>
</ul>
</div>
</body>
</html>
```

上述程序的运行结果如图 4-5 所示。可以看到，网页中只用背景色显示出 id 为 choose 的元素内容。

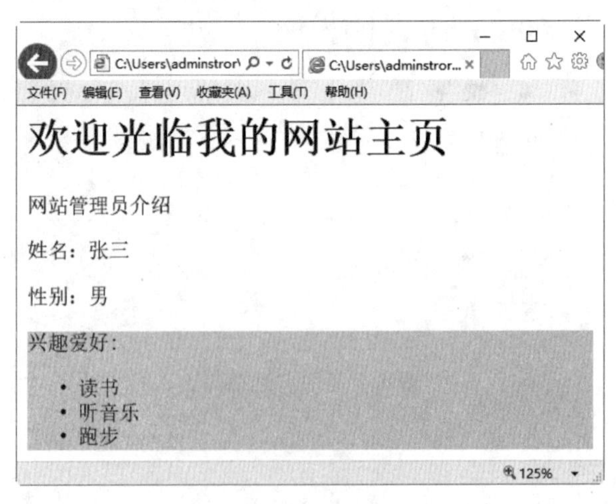

图 4-5　使用 ID 选择器

注意

不要使用数字开头的 ID 名称，因为在某些浏览器中可能出问题。

4.2.3 类名选择器(.class)

类名选择器是通过元素拥有的 CSS 类的名称查找匹配的 DOM 元素。与 ID 选择器不同，类名选择器常用于多个元素，这样就可以为带有相同类的任何 HTML 元素设置特定的样式了。

类名选择器的语法格式如下：

```
$(".class")
```

【例 4.6】(示例文件 ch04\4.6.html)

选择<body>中拥有指定 CSS 类名称的所有元素：

```
<!DOCTYPE html>
<html>
<head>
<script language="javascript" src="jquery-3.2.1.min.js"></script>
<script language="javascript">
$(document).ready(function(){
    $(".intro").css("background-color","#B2E0FF");
});
</script>
</head>
<body>
<h1>欢迎光临我的网站主页</h1>
<p class="intro">网站管理员介绍</p>
<p>姓名：张三</p>
<p>性别：男</p>
<div id="choose">
兴趣爱好:
<ul>
<li>读书</li>
<li>听音乐</li>
<li>跑步</li>
</ul>
</div>
</body>
</html>
```

上述程序的运行结果如图 4-6 所示。可以看到，网页中只突出显示拥有 CSS 类名称的匹配元素。

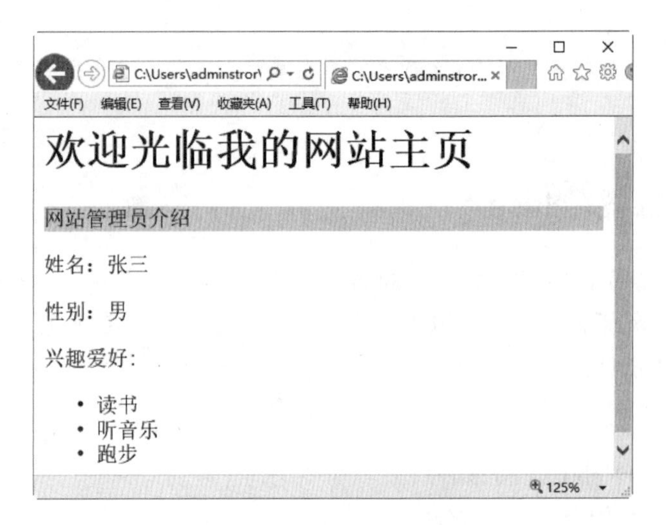

图 4-6　使用类名选择器

4.2.4　元素选择器(element)

元素选择器是根据元素名称匹配相应的元素。通俗地讲，元素选择器是根据选择的标记名来选择的，其中，标记名引用 HTML 标记的“<”与“>”之间的文本。多数情况下，元素选择器匹配的是一组元素。

元素选择器的语法格式如下：

```
$("element")
```

【例 4.7】(示例文件 ch04\4.7.html)

选择<body>中标记名为<p>的元素：

```
<!DOCTYPE html>
<html>
<head>
<script language="javascript" src="jquery-3.2.1.min.js"></script>
<script language="javascript">
$(document).ready(function(){
    $("p").css("background-color","#B2E0FF");
});
</script>
</head>
<body>
<h1>欢迎光临我的网站主页</h1>
<p class="intro">网站管理员介绍</p>
<p>姓名：张三</p>
<p>性别：男</p>
<div id="choose">
兴趣爱好：
<ul>
<li>读书</li>
<li>听音乐</li>
```

```
<li>跑步</li>
</ul>
</div>
</body>
</html>
```

上述程序的运行结果如图 4-7 所示。可以看到，网页中只突出显示标记名为<p>所对应的元素。

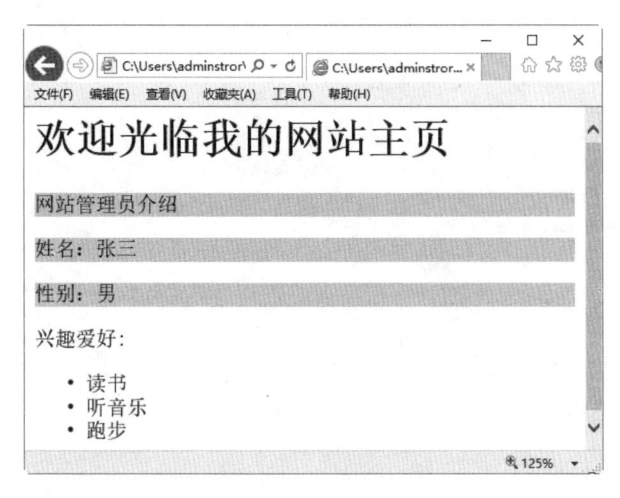

图 4-7 使用元素选择器

4.2.5 复合选择器

复合选择器是将多个选择器组合在一起，可以是 ID 选择器、类名选择器或元素选择器，它们之间用逗号分开，只要符合其中的任何一个筛选条件，就会匹配，并以集合的形式返回 jQuery 包装集。

复合选择器的语法格式如下：

```
$("selector1,selector2,selectorN")
```

参数的含义如下。

- selector1：一个有效的选择器，可以是 ID 选择器、元素选择器或者类名选择器等。
- selector2：另一个有效的选择器，可以是 ID 选择器、元素选择器或者类名选择器等。
- selectorN：任意多个选择器，可以是 ID 选择器、元素选择器或者类名选择器等。

【例 4.8】(示例文件 ch04\4.8.html)

获取<body>中 id 为 choose 和 CSS 类为 intro 的所有元素：

```
<!DOCTYPE html>
<html>
<head>
<script language="javascript" src="jquery-3.2.1.min.js"></script>
<script language="javascript">
$(document).ready(function(){
    $("#choose,.intro").css("background-color","#B2E0FF");
```

```
});
</script>
</head>
<body>
<h1>欢迎光临我的网站主页</h1>
<p class="intro">网站管理员介绍</p>
<p>姓名：张三</p>
<p>性别：男</p>
<div id="choose">
兴趣爱好：
<ul>
<li>读书</li>
<li>听音乐</li>
<li>跑步</li>
</ul>
</div>
</body>
</html>
```

上述程序的运行结果如图 4-8 所示。可以看到，网页中突出显示 id 为 choose 和 CSS 类为 intro 的元素内容。

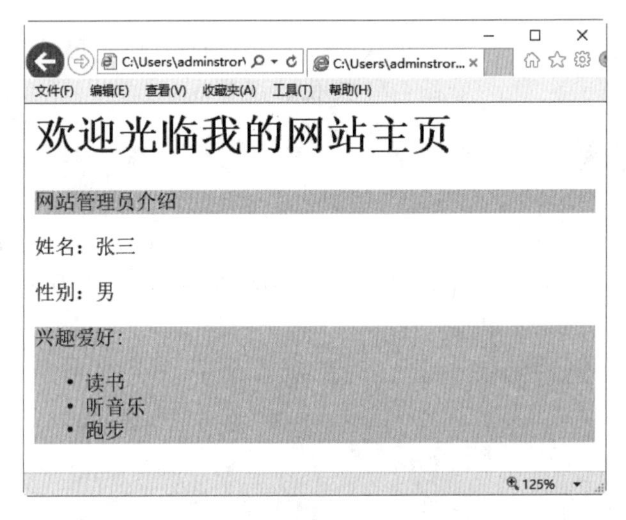

图 4-8　使用复合选择器

4.3　层级选择器

层级选择器是根据 DOM 元素之间的层次关系来获取特定的元素，如后代元素、子元素、相邻元素、兄弟元素等。

4.3.1　祖先后代选择器(ancestor descendant)

ancestor descendant 为祖先后代选择器，其中 ancestor 为祖先元素，descendant 为后代元

素，用于选取给定祖先元素下的所有匹配的后代元素。

ancestor descendant 的语法格式如下：

```
$("ancestor descendant")
```

参数的含义如下。

- ancestor：为任何有效的选择器。
- descendant：为用以匹配元素的选择器，并且是 ancestor 指定的元素的后代元素。

例如，想要获取 ul 元素下的全部 li 元素，就可以使用如下 jQuery 代码：

```
$("ul li")
```

【例 4.9】(示例文件 ch04\4.9.html)

使用 jQuery 为新闻列表设置样式：

```
<!DOCTYPE html>
<html>
<head>
<title>祖先后代选择器</title>
<style type="text/css">
body{
    margin: 0px;
}
#top{
    background-color: #B2E0FF;        /*设置背景颜色*/
    width: 450px;                     /*设置宽度*/
    height: 150px;                    /*设置高度*/
    clear: both;                      /*设置左右两侧无浮动内容*/
    padding-top: 10px;                /*设置顶边距*/
    font-size: 12pt;                  /*设置字体大小*/
}
.css{
    color: #FFFFFF;                   /*设置文字颜色*/
    line-height: 20px;                /*设置行高*/
}
</style>
<script type="text/javascript" src="jquery-3.2.1.min.js"></script>
<script type="text/javascript">
$(document).ready(function(){
    $("div ul").addClass("css");     //为 div 元素的子元素 ul 添加样式
});
</script>
</head>
<body>
<div id="top">
<ul>
    <li>贵阳北京现代瑞纳最高优惠 0.3 万 现车销售</li>
    <li>新宝来车型最高现金优惠 6000 元 现车供应</li>
    <li>2017 款荣威现车充足 优惠 1 万元送礼包</li>
```

```
    <li>世乒赛官方媒体指南力推国乒：定会蝉联 毫无弱点</li>
    <li>日女乒主帅：没福原爱四强都悬 乒超比日本联赛有钱</li>
    <li>俄美外长电话会谈 俄要求尽快叫停乌特别行动</li>

</ul>
</div>
<ul>
    <li>贵阳北京现代瑞纳最高优惠 0.3 万 现车销售</li>
    <li>新宝来车型最高现金优惠 6000 元 现车供应</li>
    <li>2017 款荣威现车充足 优惠 1 万元送礼包</li>
    <li>世乒赛官方媒体指南力推国乒：定会蝉联 毫无弱点</li>
    <li>日女乒主帅：没福原爱四强都悬 乒超比日本联赛有钱</li>
    <li>俄美外长电话会谈 俄要求尽快叫停乌特别行动</li>
</ul>
</body>
</html>
```

上述程序的运行结果如图 4-9 所示。其中，上面的新闻列表是通过 jQuery 添加的样式效果，下面的是默认的显示效果。

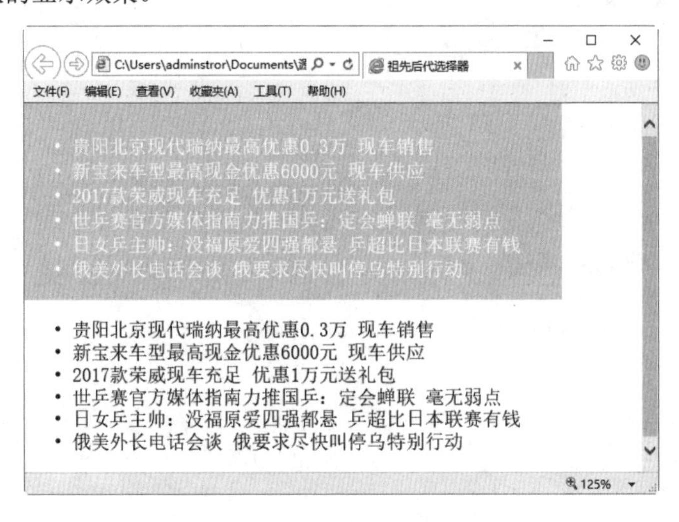

图 4-9 使用祖先后代选择器

代码中的 addClass()方法用于为元素添加 CSS 类。

4.3.2 父子选择器(parent>child)

父子选择器中的 parent 代表父元素，child 代表子元素，该选择器用于选择 parent 的直接子节点 child，child 必须包含在 parent 中，并且父类是 parent 元素。

parent>child 的语法格式如下：

```
$("Parent>child")
```

参数的含义如下。

- parent：指任何有效的选择器。
- child：用以匹配元素的选择器，是 parent 元素的子元素。

例如，想要获取表单中的所有元素的子元素 input，就可以使用如下 jQuery 代码：

```
$("form>input")
```

【例 4.10】(示例文件 ch04\4.10.html)

使用 jQuery 为表单元素添加背景色：

```
<!DOCTYPE html>
<html>
<head>
<title>父子选择器</title>
<style type="text/css">
input{
   margin: 5px;                             /*设置 input 元素的外边距为 5 像素*/
}
.input{
   font-size: 12pt;                         /*设置文字大小*/
   color: #333333;                          /*设置文字颜色*/
   background-color: #cef;                  /*设置背景颜色*/
   border: 1px solid #000000;               /*设置边框*/
}
</style>
<script type="text/javascript" src="jquery-3.2.1.min.js"></script>
<script type="text/javascript">
$(document).ready(function(){
   $("#change").ready(function(){
       //为表单元素的直接子元素 input 添加样式
       $("form>input").addClass("input");
   });
});
</script>
</head>
<body>
<h1>注册会员</h1>
<form id="form1" name="form1" method="post" action="">
 会员昵称：<input type="text" name="name" id="name" />
 <br />
 登录密码：<input type="password" name="password" id="password" />
 <br />
 确认密码：<input type="password" name="password" id="password" />
 <br />
 E-mail：<input type="text" name="email" id="email" />
 <br />
 <input type=submit value="同意协议并注册" class=button>
</form>
</body>
</html>
```

上述程序的运行结果如图 4-10 所示。可以看到，表单中直接子元素 input 都添加上了背景色。

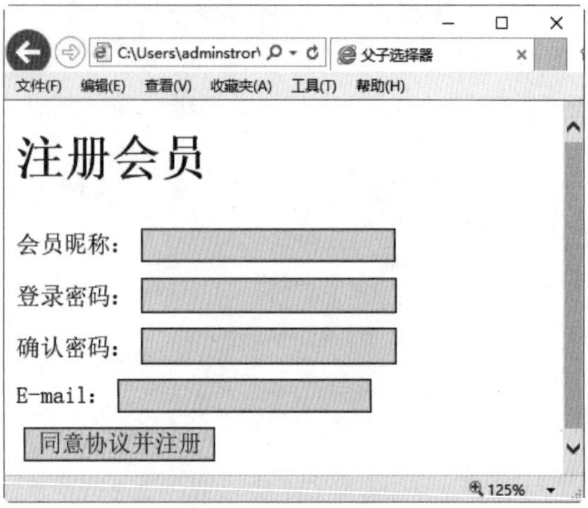

图 4-10　使用父子选择器

4.3.3　相邻元素选择器(prev+next)

相邻元素选择器用于获取所有紧跟在 prev 元素后的 next 元素，其中 prev 和 next 是两个同级别的元素。

prev+next 的语法格式如下：

```
$("prev+next")
```

参数的含义如下。

- prev：是指任何有效的选择器。
- next：是一个有效并紧接着 prev 的选择器。

例如，想要获取 div 标记后的<p>标记，就可以使用如下 jQuery 代码：

```
$("div+p")
```

【例 4.11】(示例文件 ch04\4.11.html)

使用 jQuery 制作隔行变色新闻列表：

```
<!DOCTYPE html>
<html>
<head>
<title>相邻元素选择器</title>
<style type="text/css">
    .background{background: #cef}
    body{font-size: 20px;}
</style>
<script type="text/javascript" src="jquery-3.2.1.min.js"></script>
<script type="text/javascript">
    $(document).ready(function() {
```

```
        $("label+p").addClass("background");
    });
</script>
</head>
<body>
<h2>新闻列表</h2>
    <label>贵阳北京现代瑞纳最高优惠 0.3 万现车销售</label>
    <p>新宝来车型最高现金优惠 6000 元 现车供应</p>
    <label>2016 款荣威现车充足 优惠 1 万元送礼包</label>
    <p>世乒赛官方媒体指南力推国乒：定会蝉联 毫无弱点</p>
    <label>日女乒主帅：没福原爱四强都悬 乒超比日本联赛有钱</label>
    <p>俄美外长电话会谈 俄要求尽快叫停乌特别行动</p>
</body>
</html>
```

上述程序的运行结果如图 4-11 所示。可以看到，页面中的新闻列表进行了隔行变色。

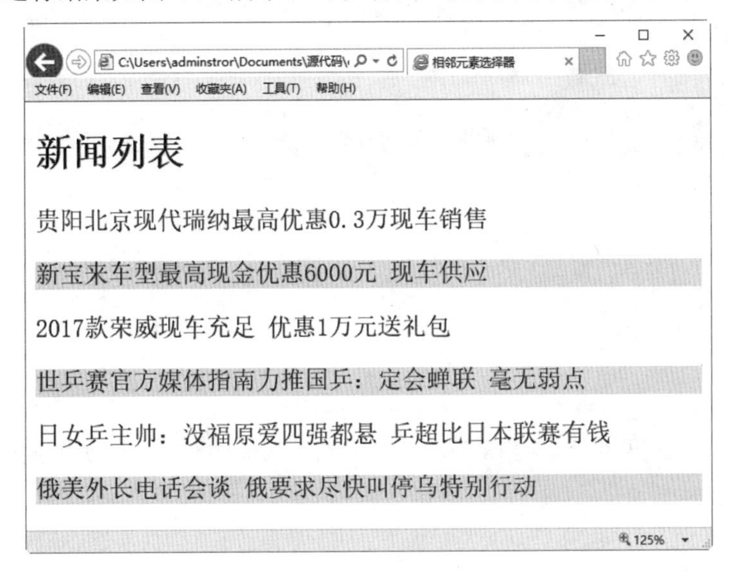

图 4-11 使用相邻元素选择器

4.3.4 兄弟选择器(prev~siblings)

兄弟选择器用于获取 prev 元素之后的所有 siblings。prev 和 siblings 是两个同辈的元素。prev~siblings 的语法格式如下：

```
$("prev~siblings");
```

参数的含义如下。

- prev：是指任何有效的选择器。
- siblings：是有效且并列跟随 prev 的选择器。

例如，想要获取与 div 标记同辈的 ul 元素，就可以使用如下 jQuery 代码：

```
$("div~ul")
```

【例 4.12】(示例文件 ch04\4.12.html)

使用 jQuery 筛选所需的新闻信息：

```
<!DOCTYPE html>
<html>
<head>
<title>兄弟元素选择器</title>
<style type="text/css">
    .background{background: #cef}
    body{font-size: 20px;}
</style>
<script type="text/javascript" src="jquery-3.2.1.min.js"></script>
<script type="text/javascript">
    $(document).ready(function() {
        $("div~p").addClass("background");
    });
</script>
</head>
<body>
<h2>新闻列表</h2>
<div>
    <p>贵阳北京现代瑞纳最高优惠 0.3 万现车销售</p>
    <p>新宝来车型最高现金优惠 6000 元 现车供应</p>
    <p>2017 款荣威现车充足 优惠 1 万元送礼包</p>
</div>
<p>世乒赛官方媒体指南力推国乒：定会蝉联 毫无弱点</p>
<p>日女乒主帅：没福原爱四强都悬 乒超比日本联赛有钱</p>
<p>俄美外长电话会谈 俄要求尽快叫停乌特别行动</p>
</body>
</html>
```

上述程序的运行结果如图 4-12 所示。可以看到，页面中与 div 同级别的<p>元素被筛选出来了。

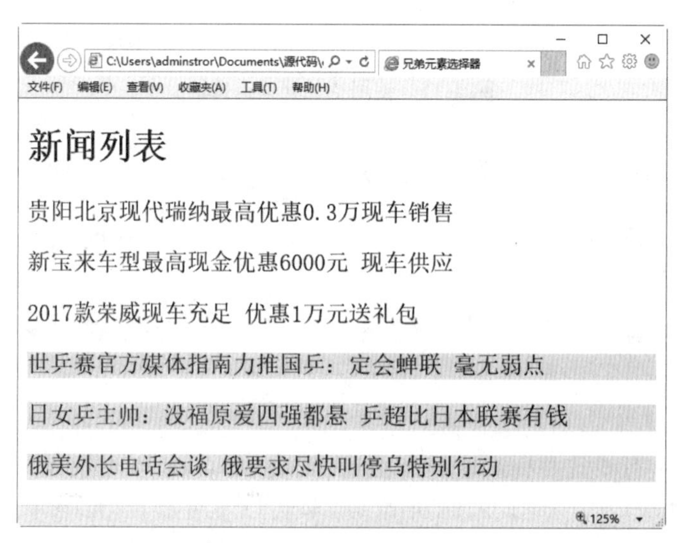

图 4-12　使用兄弟选择器

4.4 过滤选择器

jQuery 过滤选择器主要包括简单过滤器、内容过滤器、可见性过滤器、表单过滤器等。

4.4.1 简单过滤选择器

简单过滤选择器通常是以冒号开头，用于实现简单过滤效果的过滤器。常用的简单过滤选择器包括:first、:last、:even、:odd 等。

1. :first 选择器

:first 选择器用于选取第一个元素，最常见的用法就是与其他元素一起使用，选取指定组合中的第一个元素。

:first 选择器的语法格式如下：

```
$(":first")
```

例如，想要选取 body 中的第一个<p>元素，就可以使用如下 jQuery 代码：

```
$("p:first")
```

【例 4.13】(示例文件 ch04\4.13.html)

使用 jQuery 筛选新闻列表中的第一个信息：

```
<!DOCTYPE html>
<html>
<head>
<title>:first 选择器</title>
<style type="text/css">
    .background{background: #cef}
    body{font-size: 20px;}
</style>
<script type="text/javascript" src="jquery-3.2.1.min.js"></script>
<script type="text/javascript">
    $(document).ready(function() {
        $("p:first").addClass("background");
    });
</script>
</head>
<body>
<h2>新闻列表</h2>
    <p>贵阳北京现代瑞纳最高优惠 0.3 万现车销售</p>
    <p>新宝来车型最高现金优惠 6000 元 现车供应</p>
    <p>2017 款荣威现车充足 优惠 1 万元送礼包</p>
    <p>世乒赛官方媒体指南力推国乒：定会蝉联 毫无弱点</p>
    <p>日女乒主帅：没福原爱四强都悬 乒超比日本联赛有钱</p>
    <p>俄美外长电话会谈 俄要求尽快叫停乌特别行动</p>
</body>
</html>
```

上述程序的运行结果如图 4-13 所示。可以看到，页面中第一个<p>元素被筛选出来了。

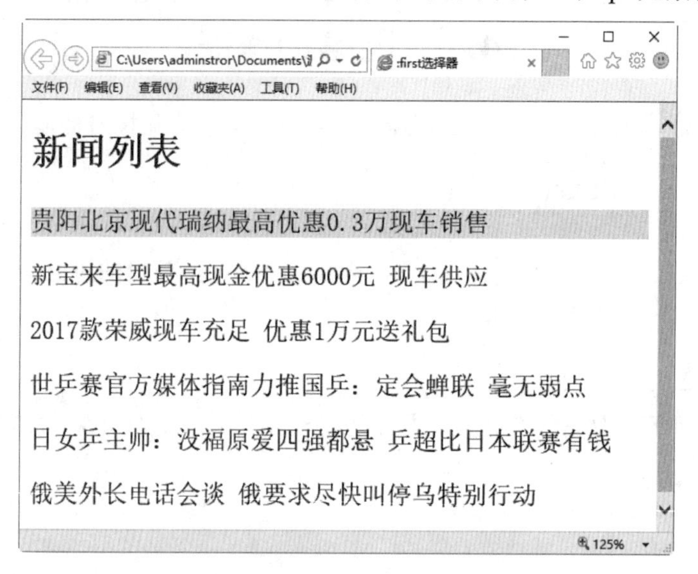

图 4-13　使用:first 选择器

2. :last 选择器

:last 选择器用于选取最后一个元素，最常见的用法就是与其他元素一起使用，选取指定组合中的最后一个元素。

:last 选择器的语法格式如下：

```
$(":last")
```

例如，想要选取 body 中的最后一个<p>元素，就可以使用如下 jQuery 代码：

```
$("p:last")
```

【例 4.14】(示例文件 ch04\4.14.html)

使用 jQuery 筛选新闻列表中的最后一个<p>元素信息：

```
<!DOCTYPE html>
<html>
<head>
<title>:last 选择器</title>
<style type="text/css">
    .background{background: #cef}
    body{font-size: 20px;}
</style>
<script type="text/javascript" src="jquery-3.2.1.min.js"></script>
<script type="text/javascript">
    $(document).ready(function() {
        $("p:last").addClass("background");
    });
</script>
</head>
```

```
<body>
<h2>新闻列表</h2>
    <p>贵阳北京现代瑞纳最高优惠 0.3 万现车销售</p>
    <p>新宝来车型最高现金优惠 6000 元 现车供应</p>
    <p>2017 款荣威现车充足 优惠 1 万元送礼包</p>
    <p>世乒赛官方媒体指南力推国乒：定会蝉联 毫无弱点</p>
    <p>日女乒主帅：没福原爱四强都悬 乒超比日本联赛有钱</p>
    <p>俄美外长电话会谈 俄要求尽快叫停乌特别行动</p>
</body>
</html>
```

上述程序的运行结果如图 4-14 所示。可以看到，页面中最后一个<p>元素被筛选出来了。

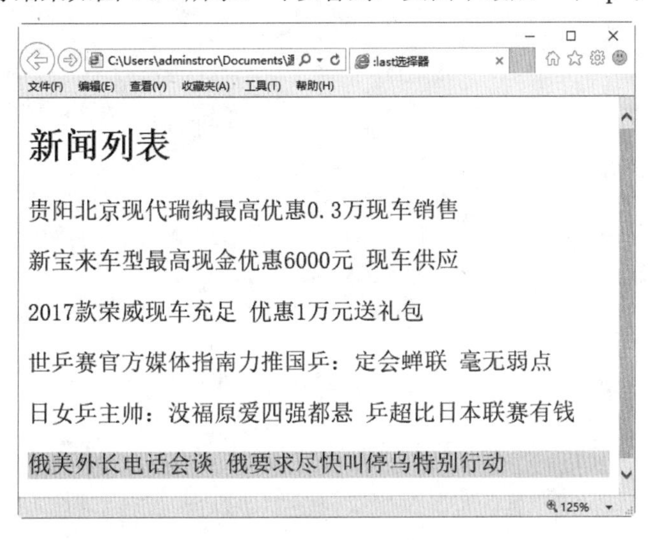

图 4-14 使用:last 选择器

3. :even 选择器

:even 选择器用于选取每个带有偶数索引值的元素(如 2、4、6)。索引值从 0 开始，所有第一个元素是偶数(0)。最常见的用法是与其他元素或选择器一起使用，来选择指定的组中偶数序号的元素。

:even 选择器的语法格式如下：

```
$(":even")
```

例如，想要选取表格中的所有偶数元素，就可以使用如下 jQuery 代码：

```
$("tr:even")
```

【例 4.15】(示例文件 ch04\4.15.html)

使用 jQuery 制作隔行(偶数行)变色的表格：

```
<!DOCTYPE html>
<html>
<head>
<script type="text/javascript" src="jquery-3.2.1.min.js "></script>
```

```
<script type="text/javascript">
$(document).ready(function(){
    $("tr:even").css("background-color", "#B2E0FF");
});
</script>
<style>
*{
  padding: 0px;
  margin: 0px;
}
body{
font-family: "黑体";
font-size: 20px;
}
table{
  text-align: center;
  width: 500px;
  border: 1px solid green;
}
td{
  border: 1px solid green;
  height: 30px;
}
h2{
  text-align: center;
}
</style>
</head>
<body>
<h2>学生成绩表</h2>
<table>
<tr>
<th>学号</th>
<th>姓名</th>
<th>语文</th>
<th>数学</th>
<th>英语</th>
</tr>

<tr>
<td>1</td>
<td>张三</td>
<td>87</td>
<td>68</td>
<td>89</td>
</tr>

<tr>
<td>2</td>
<td>李四</td>
```

```
<td>89</td>
<td>84</td>
<td>86 </td>
</tr>

<tr>
<td>3</td>
<td>王五</td>
<td>96</td>
<td>94</td>
<td>85</td>
</tr>

<tr>
<td>4</td>
<td>李六</td>
<td>98</td>
<td>87</td>
<td>67</td>
</tr>

</table>
</body>
</html>
```

上述程序的运行结果如图 4-15 所示。可以看到，表格中的偶数行被选取出来了。

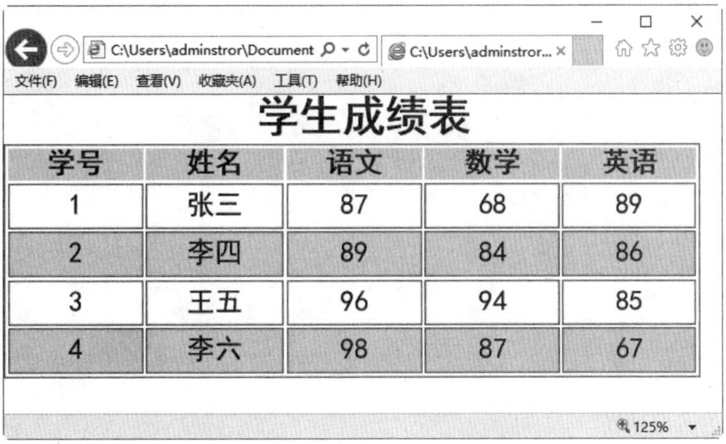

学生成绩表

学号	姓名	语文	数学	英语
1	张三	87	68	89
2	李四	89	84	86
3	王五	96	94	85
4	李六	98	87	67

图 4-15　使用:even 选择器

4. :odd 选择器

:odd 选择器用于选取每个带有奇数索引值的元素(如 1、3、5)。最常见的用法是与其他元素或选择器一起使用，来选择指定的组中奇数序号的元素。

:odd 选择器的语法格式如下：

```
$(":odd")
```

例如，想要选取表格中的所有奇数元素，就可以使用如下 jQuery 代码：

```
$("tr:odd")
```

【例 4.16】(示例文件 ch04\4.16.html)

使用 jQuery 制作隔行(奇数行)变色的表格：

```
<!DOCTYPE html>
<html>
<head>
<script type="text/javascript" src="jquery-3.2.1.min.js "></script>
<script type="text/javascript">
$(document).ready(function(){
    $("tr:odd").css("background-color","#B2E0FF");
});
</script>
<style>
*{
  padding: 0px;
  margin: 0px;
}
body{
  font-family: "黑体";
  font-size: 20px;
}
table{
  text-align: center;
  width: 500px;
  border: 1px solid green;
}
td{
  border: 1px solid green;
  height: 30px;
}
h2{
  text-align: center;
}
</style>
</head>
<body>
<h2>学生成绩表</h2>
<table>
<tr>
<th>学号</th>
<th>姓名</th>
<th>语文</th>
<th>数学</th>
<th>英语</th>
</tr>

<tr>
<td>1</td>
```

```
<td>张三</td>
<td>87</td>
<td>68</td>
<td>89</td>
</tr>

<tr>
<td>2</td>
<td>李四</td>
<td>89</td>
<td>84</td>
<td>86 </td>
</tr>

<tr>
<td>3</td>
<td>王五</td>
<td>96</td>
<td>94</td>
<td>85</td>
</tr>

<tr>
<td>4</td>
<td>李六</td>
<td>98</td>
<td>87</td>
<td>67</td>
</tr>

</table>
</body>
</html>
```

上述程序的运行结果如图 4-16 所示。可以看到，表格中的奇数行被选取出来了。

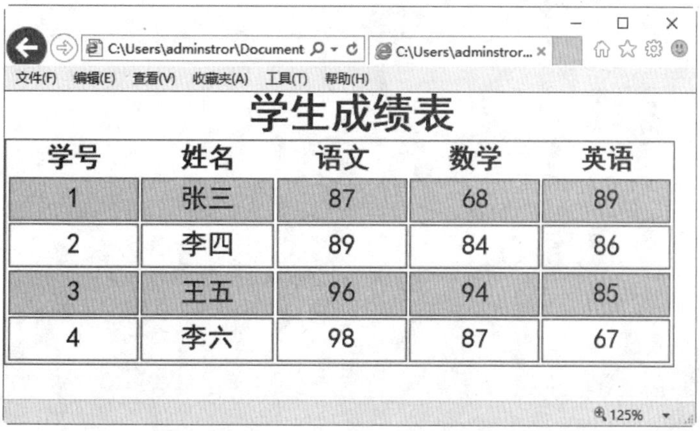

学生成绩表

学号	姓名	语文	数学	英语
1	张三	87	68	89
2	李四	89	84	86
3	王五	96	94	85
4	李六	98	87	67

图 4-16　使用:odd 选择器

4.4.2 内容过滤选择器

内容过滤选择器是通过 DOM 元素包含的文本内容以及是否含有匹配的元素来获取内容的。常见的内容过滤器有:contains(text)、:empty、:parent 等。

1. :contains(text)选择器

:contains 选择器选取包含指定字符串的元素，该字符串可以是直接包含在元素中的文本，也可以是被包含于子元素中的文本。该选择器经常与其他元素或选择器一起使用，来选择指定的组中包含指定文本的元素。

:contains(text)选择器的语法格式如下：

```
$(":contains(text)")
```

例如，想要选取所有包含 is 的<p>元素，就可以使用如下 jQuery 代码：

```
$("p:contains(is)")
```

【例 4.17】(示例文件 ch04\4.17.html)

选择学生成绩表中包含数字 9 的单元格：

```
<!DOCTYPE html>
<html>
<head>
<script type="text/javascript" src="jquery-3.2.1.min.js "></script>
<script type="text/javascript">
$(document).ready(function(){
    $("td:contains(9)").css("background-color","#B2E0FF");
});
</script>
<style>
*{
  padding: 0px;
  margin: 0px;
}
body{
  font-family: "黑体";
  font-size: 20px;
}
table{
  text-align: center;
  width: 500px;
  border: 1px solid green;
}
td{
  border: 1px solid green;
  height: 30px;
}
h2{
  text-align: center;
```

```
}
</style>
</head>
<body>
<h2>学生成绩表</h2>
<table>
<tr>
<th>学号</th>
<th>姓名</th>
<th>语文</th>
<th>数学</th>
<th>英语</th>
</tr>

<tr>
<td>1</td>
<td>张三</td>
<td>87</td>
<td>68</td>
<td>89</td>
</tr>

<tr>
<td>2</td>
<td>李四</td>
<td>89</td>
<td>84</td>
<td>86 </td>
</tr>

<tr>
<td>3</td>
<td>王五</td>
<td>96</td>
<td>94</td>
<td>85</td>
</tr>

<tr>
<td>4</td>
<td>李六</td>
<td>98</td>
<td>87</td>
<td>67</td>
</tr>

</table>
</body>
</html>
```

上述程序的运行结果如图 4-17 所示。可以看到，表格中包含数字 9 的单元格被选取出来了。

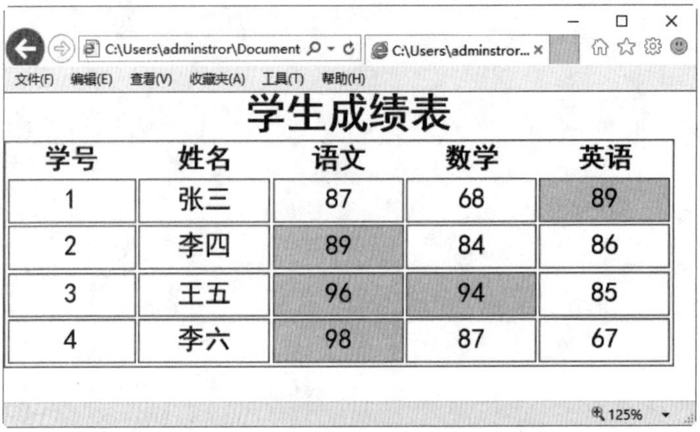
C:\Users\adminstror\Document
C:\Users\adminstror...
文件(F) 编辑(E) 查看(V) 收藏夹(A) 工具(T) 帮助(H)
学生成绩表
学号 姓名 语文 数学 英语
1 张三 87 68 89
2 李四 89 84 86
3 王五 96 94 85
4 李六 98 87 67
125%

图 4-17　使用:contains 选择器

2. :empty 选择器

:empty 选择器用于选取所有不包含子元素或者文本的空元素。:empty 选择器的语法格式如下：

```
$(":empty")
```

例如，想要选取表格中的所有空元素，就可以使用如下 jQuery 代码：

```
$("td:empty")
```

【例 4.18】(示例文件 ch04\4.18.html)

选择学生成绩表中无内容的单元格：

```
<!DOCTYPE html>
<html>
<head>
<script type="text/javascript" src="jquery-3.2.1.min.js "></script>
<script type="text/javascript">
$(document).ready(function(){
    $("td:empty").css("background-color","#B2E0FF");
});
</script>
<style>
*{
  padding: 0px;
  margin: 0px;
}
body{
  font-family: "黑体";
  font-size: 20px;
}
table{
  text-align: center;
  width: 500px;
```

```
  border: 1px solid green;
}
td{
  border: 1px solid green;
  height: 30px;
}
h2{
  text-align: center;
}
</style>
</head>
<body>
<h2>学生成绩表</h2>
<table>
<tr>
<th>学号</th>
<th>姓名</th>
<th>语文</th>
<th>数学</th>
<th>英语</th>
</tr>

<tr>
<td>1</td>
<td>张三</td>
<td>87</td>
<td></td>
<td></td>
</tr>

<tr>
<td>2</td>
<td>李四</td>
<td></td>
<td>84</td>
<td>86 </td>
</tr>

<tr>
<td>3</td>
<td>王五</td>
<td>96</td>
<td></td>
<td>85</td>
</tr>

<tr>
<td>4</td>
<td>李六</td>
<td>98</td>
```

```
<td>87</td>
<td></td>
</tr>

</table>
</body>
</html>
```

上述程序的运行结果如图 4-18 所示。可以看到，表格中无内容的单元格被选取出来了。

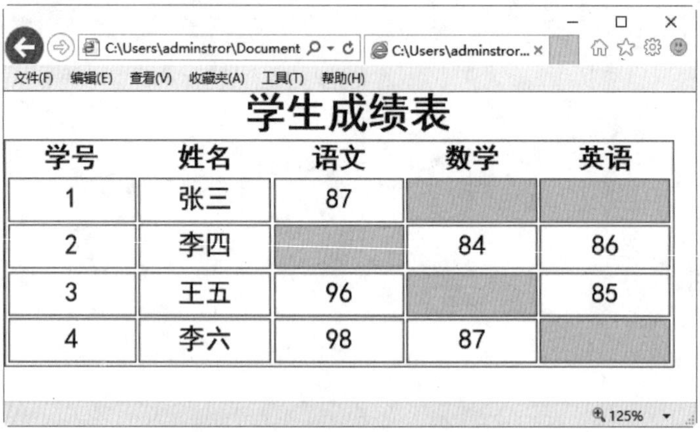
C:\Users\adminstror\Document
C:\Users\adminstror...
文件(F) 编辑(E) 查看(V) 收藏夹(A) 工具(T) 帮助(H)
学生成绩表
学号 姓名 语文 数学 英语
1 张三 87
2 李四 84 86
3 王五 96 85
4 李六 98 87
125%

图 4-18　使用:empty 选择器

3. :parent 选择器

:parent 用于选取包含子元素或文本的元素。:parent 选择器的语法格式如下：

```
$(":parent")
```

例如，想要选取表格中的所有包含子元素的内容，就可以使用如下 jQuery 代码：

```
$("td:parent")
```

【例 4.19】(示例文件 ch04\4.19.html)

选择学生成绩表中包含内容的单元格：

```
<!DOCTYPE html>
<html>
<head>
<script type="text/javascript" src="jquery-3.2.1.min.js"></script>
<script type="text/javascript">
$(document).ready(function(){
    $("td:parent").css("background-color","#B2E0FF");
});
</script>
<style>
*{
  padding: 0px;
  margin: 0px;
}
```

```
body{
  font-family: "黑体";
  font-size: 20px;
}
table{
  text-align: center;
  width: 500px;
  border: 1px solid green;
}
td{
  border: 1px solid green;
  height: 30px;
}
h2{
  text-align: center;
}
</style>
</head>
<body>
<h2>学生成绩表</h2>
<table>
<tr>
<th>学号</th>
<th>姓名</th>
<th>语文</th>
<th>数学</th>
<th>英语</th>
</tr>
<tr>
<td>1</td>
<td>张三</td>
<td>87</td>
<td></td>
<td></td>
</tr>
<tr>
<td>2</td>
<td>李四</td>
<td></td>
<td>84</td>
<td>86 </td>
</tr>
<tr>
<td>3</td>
<td>王五</td>
<td>96</td>
<td></td>
<td>85</td>
</tr>
<tr>
```

```
<td>4</td>
<td>李六</td>
<td>98</td>
<td>87</td>
<td></td>
</tr>
</table>
</body>
</html>
```

上述程序的运行结果如图 4-19 所示。可以看到，表格中包含内容的单元格被选取出来了。

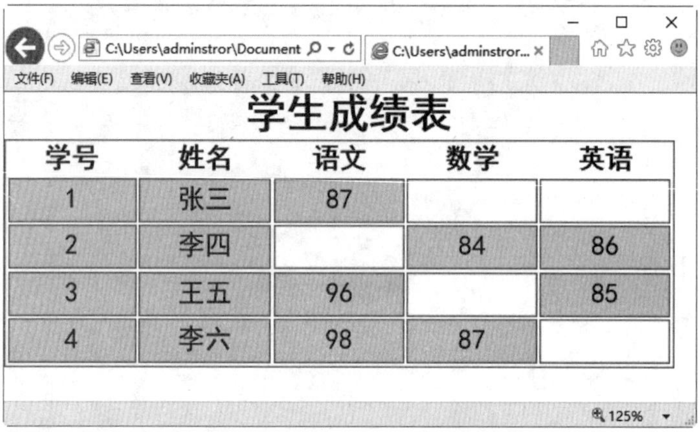

学生成绩表

学号	姓名	语文	数学	英语
1	张三	87		
2	李四		84	86
3	王五	96		85
4	李六	98	87	

图 4-19　使用:parent 选择器

4.4.3　可见性过滤器

元素的可见状态有隐藏和显示两种。可见性过滤器是利用元素的可见状态匹配元素的。因此，可见性过滤器也有两种，分别是用于显示元素的:visible 选择器和用于隐藏元素的:hidden 选择器。

1. :visible 选择器的语法格式如下：

```
$(":visible")
```

例如：想要获取页面中所有可见的表格元素，就可以使用如下 jQuery 代码：

```
$("table:visible")
```

2. :hidden 选择器的语法格式如下：

```
$(":hidden")
```

例如，想要获取页面中所有隐藏的<p>元素，就可以使用如下 jQuery 代码：

```
$("p:hidden")
```

【例 4.20】(示例文件 ch04\4.20.html)

选择学生成绩表中的所有表格元素：

```
<!DOCTYPE html>
<html>
<head>
<script type="text/javascript" src="jquery-3.2.1.min.js"></script>
<script type="text/javascript">
$(document).ready(function(){
    $("table:visible").css("background-color","#B2E0FF");
});
</script>
<style>
*{
  padding: 0px;
  margin: 0px;
}
body{
  font-family: "黑体";
  font-size: 20px;
}
table{
  text-align: center;
  width: 500px;
  border: 1px solid green;
}
td{
  border: 1px solid green;
  height: 30px;
}
h2{
  text-align: center;
}
</style>
</head>
<body>
<h2>学生成绩表</h2>
<table>
<tr>
<th>学号</th>
<th>姓名</th>
<th>语文</th>
<th>数学</th>
<th>英语</th>
</tr>

<tr>
<td>1</td>
<td>张三</td>
<td>87</td>
<td>68</td>
<td>89</td>
</tr>
```

```
<tr>
<td>2</td>
<td>李四</td>
<td>89</td>
<td>84</td>
<td>86 </td>
</tr>

<tr>
<td>3</td>
<td>王五</td>
<td>96</td>
<td>94</td>
<td>85</td>
</tr>

<tr>
<td>4</td>
<td>李六</td>
<td>98</td>
<td>87</td>
<td>67</td>
</tr>

</table>
</body>
</html>
```

上述程序的运行结果如图 4-20 所示。可以看到，表格中所有元素都被选取出来了。

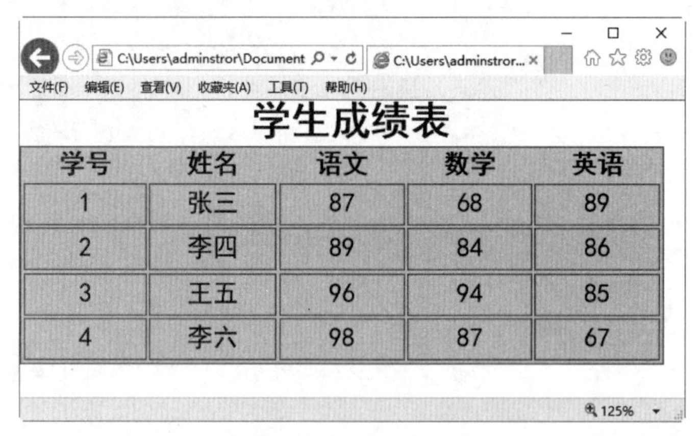

图 4-20　使用:visible 选择器

【例 4.21】(示例文件 ch04\4.21.html)

获取页面中所有隐藏的元素：

```
<!DOCTYPE html>
<html>
```

```
<head>
<title>显示隐藏元素</title>
<style>
div {
   width: 70px;
   height: 40px;
   background: #e7f;
   margin: 5px;
   float: left;
}
span {
   display: block;
   clear: left;
   color: black;
}
.starthidden {
   display: none;
}
</style>
<script type="text/javascript" src="jquery-3.2.1.min.js"></script>
</head>
<body>
<span></span>
<div></div>
<div style="display:none;">Hider!</div>
<div></div>
<div class="starthidden">Hider!</div>
<div></div>
<form>
  <input type="hidden">
  <input type="hidden">
  <input type="hidden">
</form>
<span></span>
<script>
var hiddenElements = $("body").find(":hidden").not("script");
$("span:first").text("发现" + hiddenElements.length + "个隐藏元素总量");
$("div:hidden").show(3000);
$("span:last").text("发现" + $("input:hidden").length + "个隐藏 input 元素");
</script>
</body>
</html>
```

上述程序的运行结果如图 4-21 所示。可以看到，网页中所有隐藏的元素都被显示出来了。

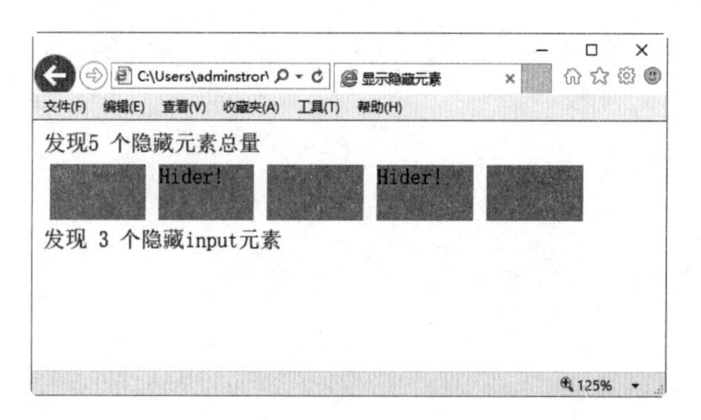

图 4-21 使用:hidden 选择器

4.4.4 表单过滤器

表单过滤器是通过表单元素的状态属性来选取元素的。表单元素的状态属性包括选中、不可用等。表单过滤器有 4 种，分别是:enabled、:disabled、:checked 和:selected。

1. :enabled

获取所有被选中的元素。:enabled 选择器的语法格式如下：

```
$(":enabled")
```

例如，想要获取所有 input 当中的可用元素，就可以使用如下 jQuery 代码：

```
$("input:enabled")
```

2. :disabled

获取所有不可用的元素。:disabled 选择器的语法格式如下：

```
$(":disabled")
```

例如，想要获取所有 input 当中的不可用元素，就可以使用如下 jQuery 代码：

```
$("input:disabled")
```

3. :checked

获取所有被选中元素(复选框、单选按钮等，不包括 select 中的 option)。:checked 选择器的语法格式如下：

```
$(":checked")
```

例如，想要查找所有选中的复选框元素，就可以使用如下 jQuery 代码：

```
$("input:checked")
```

4. :selected

获取所有选中的 option 元素。:selected 选择器的语法格式如下：

```
$(":selected")
```

例如，想要查找所有选中的选项元素，就可以使用如下 jQuery 代码：

```
$("select option:selected")
```

【例 4.22】(示例文件 ch04\4.22.html)

利用表单过滤器匹配表单中相应的元素：

```
<!DOCTYPE html>
<html>
<head>
<title>表单过滤器</title>
<script type="text/javascript" src="jquery-3.2.1.min.js"></script>
<script type="text/javascript">
$(document).ready(function() {
    $("input:checked").css("background-color","red");//设置选中的复选框的背景色
    $("input:disabled").val("不可用按钮");          //为灰色不可用按钮赋值
});
function selectVal(){                               //下拉列表框变化时执行的方法
    alert($("select option:selected").val());        //显示选中的值
}
</script>
</head>
<body>
<form>
  复选框 1：<input type="checkbox" checked="checked" value="复选框 1"/>
  复选框 2：<input type="checkbox" checked="checked" value="复选框 2"/>
  复选框 3：<input type="checkbox" value="复选框 3"/><br />
  不可用按钮：<input type="button" value="不可用按钮" disabled><br />
  下拉列表框：
  <select onchange="selectVal()">
    <option value="列表项 1">列表项 1</option>
    <option value="列表项 2">列表项 2</option>
    <option value="列表项 3">列表项 3</option>
  </select>
</form>
</body>
</html>
```

上述程序的运行结果如图 4-22 所示。当在下拉列表框中选择【列表 3】选项时，弹出提示信息框。

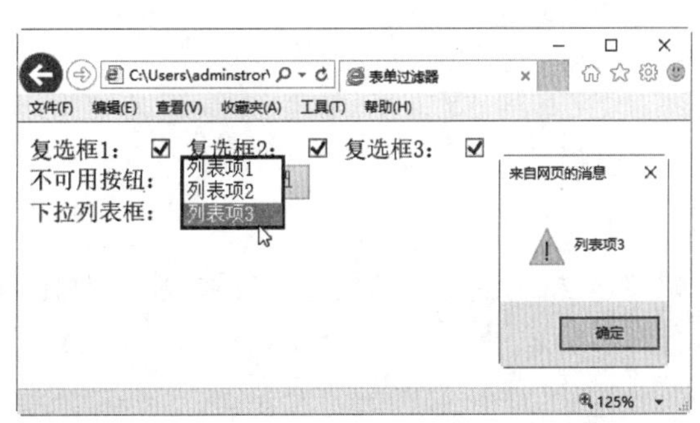

图 4-22　利用表单过滤器匹配表单中相应的元素

4.5 表单选择器

表单选择器用于选取经常在表单内出现的元素。不过，选取的元素并不一定在表单之中。jQuery 提供的表单选择器主要有以下几种。

4.5.1 :input 选择器

:input 选择器用于选取表单元素。该选择器的语法格式如下：

```
$(":input")
```

例如，想要选取页面中的所有<input>元素，就可以使用如下 jQuery 代码：

```
$(":input")
```

【例 4.23】(示例文件 ch04\4.23.html)

为页面中所有的表单元素添加背景色：

```
<!DOCTYPE html>
<html>
<head>
<script type="text/javascript" src="jquery-3.2.1.min.js"></script>
<script type="text/javascript">
$(document).ready(function(){
    $(":input").css("background-color","#B2E0FF");
});
</script>
</head>
<body>
<form action="">
姓名: <input type="text" name="姓名" />
<br />
密码: <input type="password" name="密码" />
<br />
<button type="button">按钮 1</button>
<input type="button" value="按钮 2" />
<br />
<input type="reset" value="重置" />
<input type="submit" value="提交" />
<br />
</form>
</body>
</html>
```

上述程序的运行结果如图 4-23 所示。可以看到，网页中表单元素都被添加上了背景色，而且从代码中可以看出该选择器也适用于<button>元素。

4.5.2 :text 选择器

:text 选择器选取类型为 text 的所有<input>元素。该选择器的语法格式如下：

```
$(":text")
```

例如，想要选取页面中类型为 text 的所有的<input>元素，就可以使用如下 jQuery 代码：

```
$(":text")
```

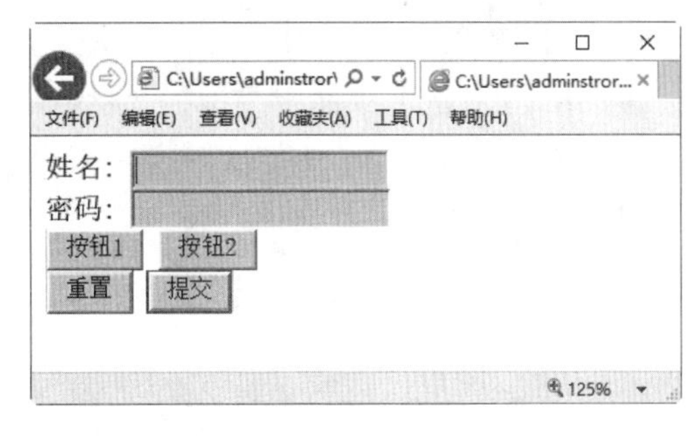

图 4-23 使用:input 选择器

【例 4.24】(示例文件 ch04\4.24.html)

为页面中类型为 text 的所有<input>元素添加背景色：

```
<!DOCTYPE html>
<html>
<head>
<script type="text/javascript" src="jquery-3.2.1.min.js"></script>
<script type="text/javascript">
$(document).ready(function(){
    $(":text").css("background-color","#B2E0FF");
});
</script>
</head>
<body>
<form action="">
姓名: <input type="text" name="姓名" />
<br />
密码: <input type="password" name="密码" />
<br />
<button type="button">按钮 1</button>
<input type="button" value="按钮 2" />
<br />
<input type="reset" value="重置" />
<input type="submit" value="提交" />
<br />
</form>
</body>
</html>
```

上述程序的运行结果如图 4-24 所示。可以看到，网页中表单类型为 text 的元素被添加上了背景色。

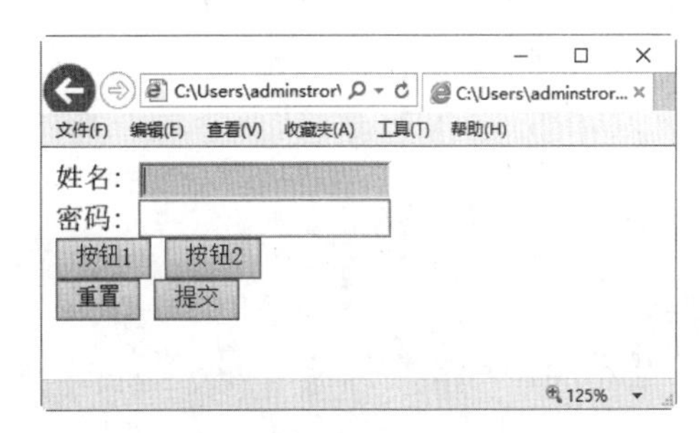

图 4-24 使用:text 选择器

4.5.3 :password 选择器

:password 选择器选取类型为 password 的所有<input>元素。该选择器的语法格式如下：

```
$(":password")
```

例如，想要选取页面中类型为 password 的所有<input>元素，可以使用如下 jQuery 代码：

```
$(":password")
```

【例 4.25】(示例文件 ch04\4.25.html)

为页面中类型为 password 的所有<input>元素添加背景色：

```
<!DOCTYPE html>
<html>
<head>
<script type="text/javascript" src="jquery-3.2.1.min.js"></script>
<script type="text/javascript">
$(document).ready(function(){
    $(":password").css("background-color","#B2E0FF");
});
</script>
</head>
<body>
<form action="">
姓名: <input type="text" name="姓名" />
<br />
密码: <input type="password" name="密码" />
<br />
<button type="button">按钮 1</button>
<input type="button" value="按钮 2" />
<br />
<input type="reset" value="重置" />
<input type="submit" value="提交" />
<br />
</form>
</body>
</html>
```

上述代码的运行结果如图 4-25 所示。可以看到，网页中表单类型为 password 的元素已经被添加上了背景色。

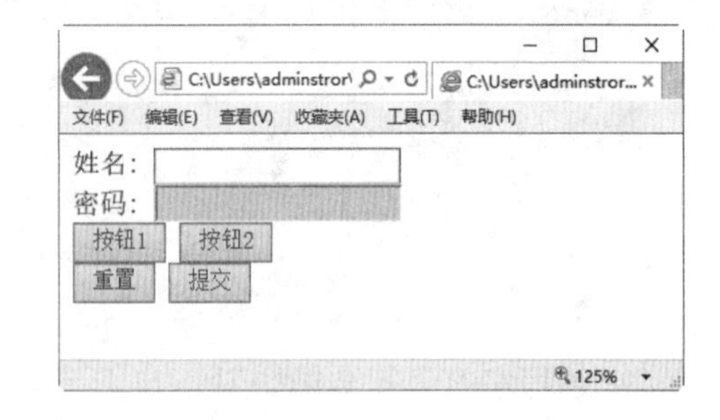

图 4-25　使用:password 选择器

4.5.4　:radio 选择器

:radio 选择器选取类型为 radio 的<input>元素。该选择器的语法格式如下：

```
$(":radio")
```

例如，想要隐藏页面中的单选按钮，就可以使用如下 jQuery 代码：

```
$(":radio").hide()
```

【例 4.26】(示例文件 ch04\4.26.html)

隐藏页面中的单选按钮：

```
<!DOCTYPE html>
<html>
<head>
<title>选择感兴趣的图书</title>
<script type="text/javascript" src="jquery-3.2.1.min.js"></script>
<script type="text/javascript">
$(document).ready(function(){
    $(".btn1").click(function(){
        $(":radio").hide();
    });
});
</script>
</head>
<body>
<form >
请选择您感兴趣的图书类型：
<br>
<input type="radio" name="book" value = "Book1">网站编程<br>
<input type="radio" name="book" value = "Book2">办公软件<br>
<input type="radio" name="book" value = "Book3">设计软件<br>
<input type="radio" name="book" value = "Book4">网络管理<br>
<input type="radio" name="book" value = "Book5">黑客攻防<br>
</form>
<button class="btn1">隐藏单选按钮</button>
</body>
</html>
```

上述程序的运行结果如图 4-26 所示，可以看到网页中的单选按钮。然后单击【隐藏单选按钮】按钮，就可以隐藏页面中的单选按钮，如图 4-27 所示。

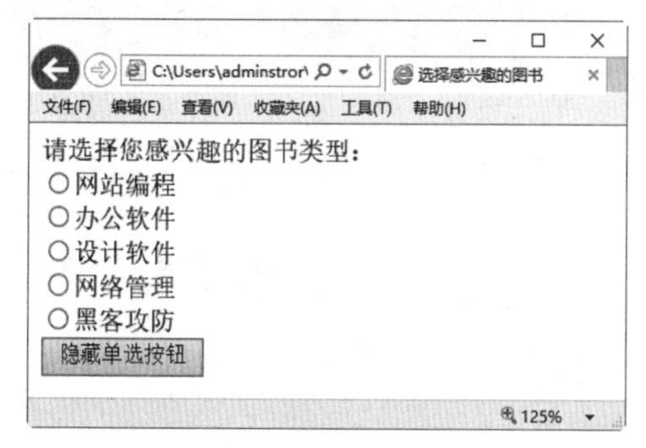

图 4-26　初始运行结果

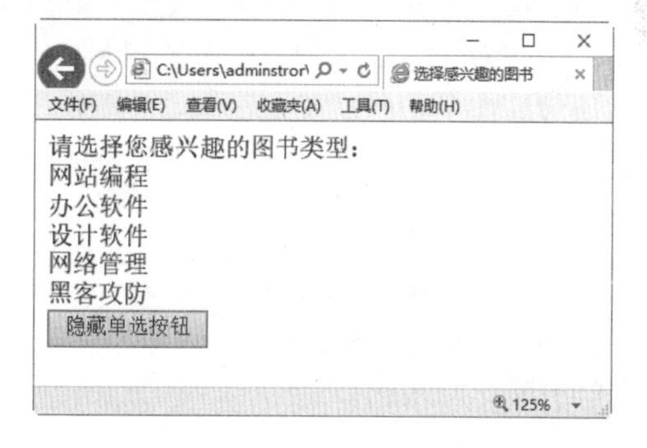

图 4-27　通过:radio 选择器隐藏单选按钮

4.5.5　:checkbox 选择器

:checkbox 选择器选取类型为 checkbox 的<input>元素。该选择器的语法格式如下：

```
$(":checkbox")
```

例如，想要隐藏页面中的复选框，就可以使用如下 jQuery 代码：

```
$(":checkbox").hide()
```

【例 4.27】(示例文件 ch04\4.27.html)

隐藏页面中的复选框：

```
<!DOCTYPE html>
<html>
<head>
<title>选择感兴趣的图书</title>
<script type="text/javascript" src="jquery-3.2.1.min.js"></script>
<script type="text/javascript">
$(document).ready(function(){
    $(".btn1").click(function(){
        $(":checkbox").hide();
    });
});
</script>
</head>
<body>
<form>
请选择您感兴趣的图书类型：
<br>
<input type="checkbox" name="book" value = "Book1">网站编程<br>
<input type="checkbox" name="book" value = "Book2">办公软件<br>
<input type="checkbox" name="book" value = "Book3">设计软件<br>
<input type="checkbox" name="book" value = "Book4">网络管理<br>
<input type="checkbox" name="book" value = "Book5">黑客攻防<br>
</form>
<button class="btn1">隐藏复选框</button>
</body>
</html>
```

上述程序的运行结果如图 4-28 所示，可以看到网页中的复选框。然后单击【隐藏单选按钮】，就可以隐藏页面中的复选框，如图 4-29 所示。

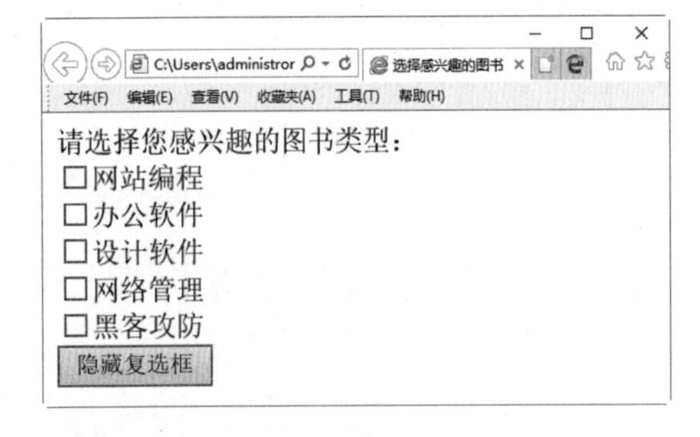

图 4-28　初始运行效果

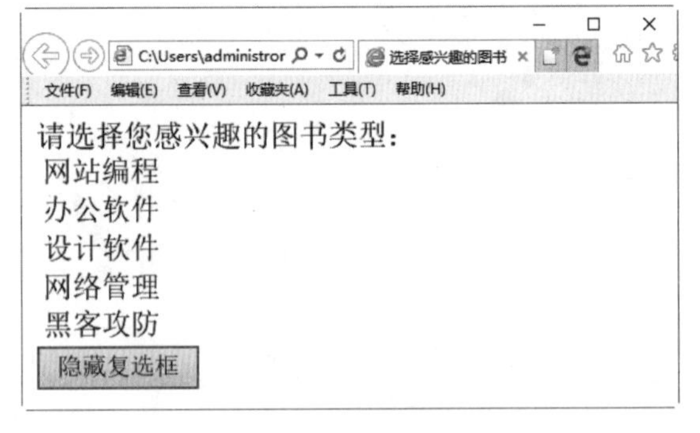

图 4-29　通过:checkbox 选择器隐藏复选框

4.5.6　:submit 选择器

:submit 选择器选取类型为 submit 的<button>和<input>元素。如果<button>元素没有定义类型，大多数浏览器会把该元素当作类型为 submit 的按钮。该选择器的语法格式如下：

```
$(":submit")
```

例如，想要选取页面中类型为 submit 的所有<input>和<button>元素，就可以使用如下 jQuery 代码：

```
$(":submit")
```

【例 4.28】(示例文件 ch04\4.28.html)

为页面中类型为 submit 的所有<input>和<button>元素添加背景色：

```
<!DOCTYPE html>
<html>
<head>
<script type="text/javascript" src="jquery-3.2.1.min.js"></script>
<script type="text/javascript">
$(document).ready(function(){
    $(":submit").css("background-color","#B2E0FF");
});
</script>
</head>
<body>
<form action="">
姓名: <input type="text" name="姓名" />
<br />
密码: <input type="password" name="密码" />
<br />
<button type="button">按钮 1</button>
<input type="button" value="按钮 2" />
<br />
<input type="reset" value="重置" />
<input type="submit" value="提交" />
<br />
</form>
</body>
</html>
```

上述程序的运行结果如图 4-30 所示。可以看到，网页中表单类型为 submit 的元素被添加上了背景色。

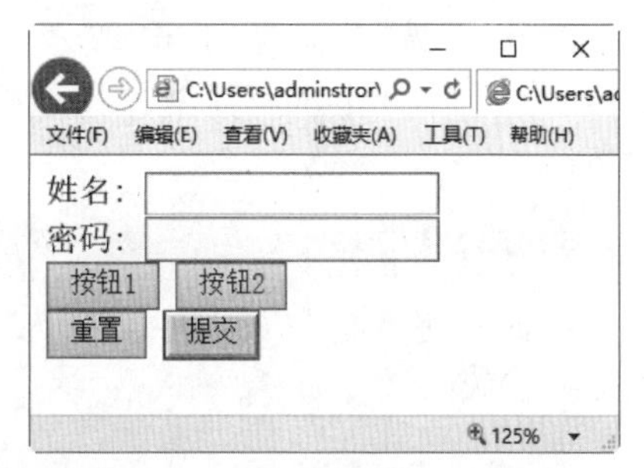

图 4-30　使用:submit 选择器

4.5.7　:reset 选择器

:reset 选择器选取类型为 reset 的<button>和<input>元素。该选择器的语法格式为：

```
$(":reset")
```

例如，想要选取页面中类型为 reset 的所有<input>和<button>元素，就可以使用如下 jQuery 代码：

```
$(":reset")
```

【例 4.29】(示例文件 ch04\4.29.html)

为页面中类型为 reset 的所有<input>和<button>元素添加背景色：

```
<!DOCTYPE html>
<html>
<head>
<script type="text/javascript" src="jquery-3.2.1.min.js"></script>
<script type="text/javascript">
$(document).ready(function(){
    $(":reset").css("background-color","#B2E0FF");
});
</script>
</head>
<body>
<form action="">
姓名: <input type="text" name="姓名" />
<br />
密码: <input type="password" name="密码" />
<br />
<button type="button">按钮 1</button>
<input type="button" value="按钮 2" />
<br />
<input type="reset" value="重置" />
<input type="submit" value="提交" />
<br />
</form>
</body>
</html>
```

上述程序的运行结果如图 4-31 所示。可以看到，网页中表单类型为 reset 的元素被添加上了背景色。

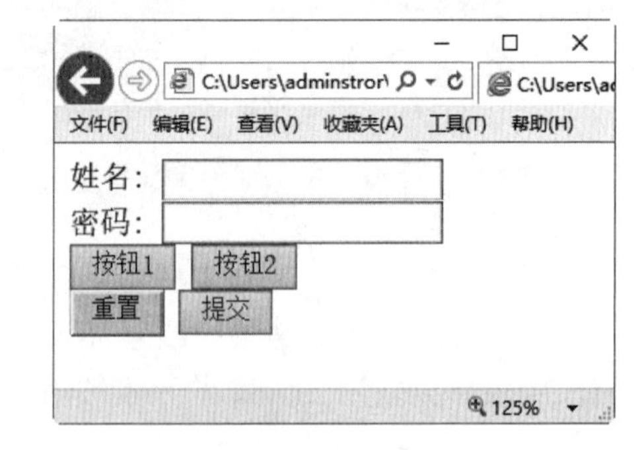

图 4-31 使用:reset 选择器

4.5.8 :button 选择器

:button 选择器用于选取类型为 button 的<button>元素和<input>元素。该选择器的语法格式如下：

```
$(":button")
```

例如，想要选取页面中类型为 button 的所有<input>和<button>元素，就可以使用如下 jQuery 代码：

```
$(":button")
```

【例 4.30】(示例文件 ch04\4.30.html)

为页面中类型为 button 的所有<input>和<button>元素添加背景色：

```
<!DOCTYPE html>
<html>
<head>
<script type="text/javascript" src="jquery-3.2.1.min.js"></script>
<script type="text/javascript">
$(document).ready(function(){
   $(":button").css("background-color","#B2E0FF");
});
</script>
</head>
<body>
<form action="">
姓名: <input type="text" name="姓名" />
<br />
密码: <input type="password" name="密码" />
<br />
<button type="button">按钮 1</button>
<input type="button" value="按钮 2" />
<br />
<input type="reset" value="重置" />
<input type="submit" value="提交" />
<br />
</form>
</body>
</html>
```

上述程序的运行结果如图 4-32 所示。可以看到，表单类型为 button 的元素被添加上了背景色。

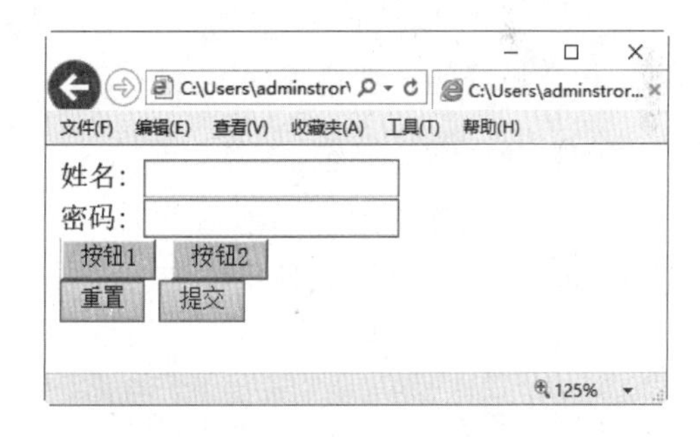

图 4-32 使用:button 选择器

4.5.9 :image 选择器

:image 选择器选取类型为 image 的<input>元素。该选择器的语法格式如下：

```
$(":image")
```

例如，想要选取页面中类型为 image 的所有<input>元素，就可以使用如下 jQuery 代码：

```
$(":image")
```

【例 4.31】(示例文件 ch04\4.31.html)

使用 jQuery 为图像域添加图片：

```
<!DOCTYPE html>
<html>
<head>
<script type="text/javascript" src="jquery-3.2.1.min.js"></script>
```

```
<script type="text/javascript">
$(document).ready(function(){
    $(":image").attr("src","1.jpg");
});
</script>
</head>
<body>
<form action="">
姓名: <input type="text" name="姓名" />
<br />
密码: <input type="password" name="密码" />
<br />
<button type="button">按钮 1</button>
<input type="button" value="按钮 2" />
<br />
<input type="reset" value="重置" />
<input type="submit" value="提交" />
<br />
<input type="image" />
</form>
</body>
</html>
```

上述程序的运行结果如图 4-33 所示。可以看到，网页中的图像域中添加了图片。

图 4-33　使用:image 选择器

4.5.10　:file 选择器

:file 选择器选取类型为 file 的<input>元素。该选择器的语法格式如下：

```
$(":file")
```

例如，想要选取页面中类型为 image 的所有<input>元素，就可以使用如下 jQuery 代码：

```
$(":file")
```

【例 4.32】(示例文件 ch04\4.32.html)

为页面中类型为 file 的所有<input>元素添加背景色：

```
<!DOCTYPE html>
<html>
<head>
<script type="text/javascript" src="jquery-3.2.1.min.js"></script>
<script type="text/javascript">
$(document).ready(function(){
    $(":file").css("background-color","#B2E0FF");
});
</script>
</head>
```

```
<body>
<form action="">
姓名：<input type="text" name="姓名" />
<br />
密码：<input type="password" name="密码" />
<br />
<button type="button">按钮 1</button>
<input type="button" value="按钮 2" />
<br />
<input type="reset" value="重置" />
<input type="submit" value="提交" />
<br />
文件域：<input type="file">
</form>
</body>
</html>
```

上述程序的运行结果如图 4-34 所示。可以看到，网页中表单类型为 file 的元素被添加上了背景色。

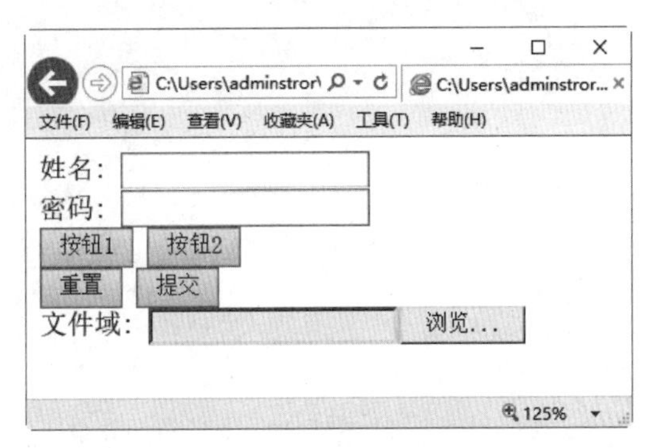

图 4-34 使用:file 选择器

4.6 属性选择器

属性选择器是将元素的属性作为过滤条件来筛选对象的选择器。常见的属性选择器主要有以下几种。

4.6.1 [attribute]选择器

[attribute]用于选择每个带有指定属性的元素，可以选取带有任何属性的元素，而且对于指定的属性没有限制。[attribute]选择器的语法格式如下：

```
$("[attribute]")
```

例如，想要选择页面中带有 id 属性的所有元素，就可以使用如下 jQuery 代码：

```
$("[id]")
```

【例 4.33】(示例文件 ch04\4.33.html)

选择页面中带有 id 属性的所有元素，并为其添加背景色：

```
<!DOCTYPE html>
<html>
<head>
<script language="javascript" src="jquery-3.2.1.min.js"></script>
<script language="javascript">
$(document).ready(function(){
    $("[id]").css("background-color","#B2E0FF");
});
</script>
</head>
<body>
<h1>欢迎光临我的网站主页</h1>
<p class="intro">网站管理员介绍</p>
<p>姓名：张三</p>
<p>性别：男</p>
<div id="choose">
兴趣爱好：
<ul>
<li>读书</li>
<li>听音乐</li>
<li>跑步</li>
</ul>
</div>
</body>
</html>
```

上述程序的运行结果如图 4-35 所示。可以看到，网页中带有 id 属性的所有元素都被添加上了背景色。

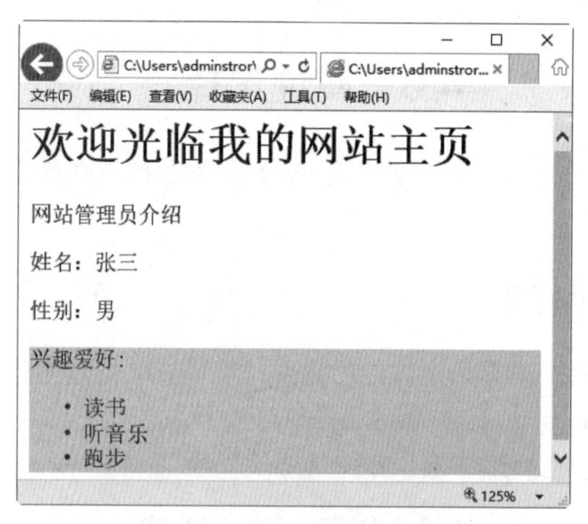

图 4-35　使用[attribute]选择器

4.6.2 [attribute=value]选择器

[attribute=value]选择器选取每个带有指定属性和值的元素。[attribute=value]选择器的语法格式如下：

```
$("[attribute=value]")
```

参数含义说明如下。

- attribute：必需，规定要查找的属性。
- value：必需，规定要查找的值。

例如，想要选择页面中每个 id=choose 的元素，就可以使用如下 jQuery 代码：

```
$("[id=choose]")
```

【例 4.34】(示例文件 ch04\4.34.html)

选择页面中带有 id=choose 属性的所有元素，并为其添加背景色：

```
<!DOCTYPE html>
<html>
<head>
<script language="javascript" src="jquery-3.2.1.min.js">
</script>
<script language="javascript">
$(document).ready(function(){
   $("[id=choose]").css("background-color","#B2E0FF");
});
</script>
</head>
<body>
<h1>欢迎光临我的网站主页</h1>
<p class="intro">网站管理员介绍</p>
<p>姓名：张三</p>
<p>性别：男</p>
<div id="choose">
兴趣爱好：
<ul>
<li>读书</li>
<li>听音乐</li>
<li>跑步</li>
</ul>
</div>
</body>
</html>
```

上述程序的运行结果如图 4-36 所示。可以看到，网页中带有 id=choose 属性的所有元素都被添加上了背景色。

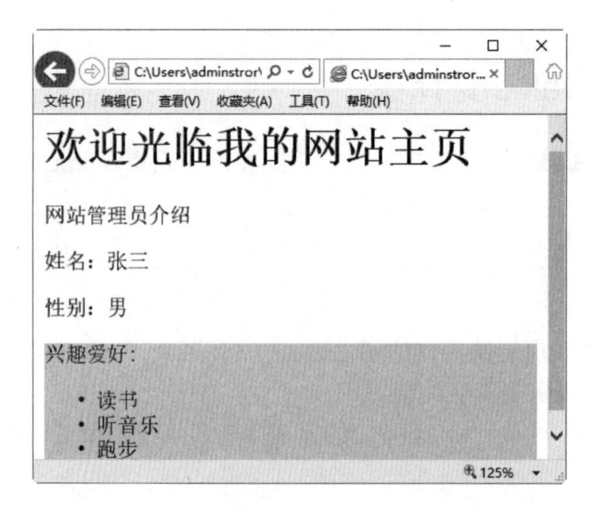

图 4-36 使用[attribute=value]选择器

4.6.3 [attribute!=value]选择器

[attribute!=value]选择器选取每个不带有指定属性及值的元素。不过，带有指定的属性，但不带有指定的值的元素，也会被选择。

[attribute!=value]选择器的语法格式如下：

```
$("[attribute!=value]")
```

参数含义说明如下。

- attribute：必需，规定要查找的属性。
- value：必需，规定要查找的值。

例如，想要选择 body 标签中不包含 id=choose 的元素，就可以使用如下 jQuery 代码：

```
$("body[id!=choose]")
```

【例 4.35】(示例文件 ch04\4.35.html)

选择页面中不包含 id=header 属性的所有元素，并为其添加背景色：

```
<!DOCTYPE html>
<html>
<head>
<script language="javascript" src="jquery-3.2.1.min.js"></script>
<script language="javascript">
$(document).ready(function(){
    $("body [id!=header]").css("background-color","#B2E0FF");
});
</script>
</head>
<body>
<h1 id="header">欢迎光临我的网站主页</h1>
<p class="intro">网站管理员介绍</p>
```

```
<p>姓名：张三</p>
<p>性别：男</p>
<div id="choose">
兴趣爱好：
<ul>
<li>读书</li>
<li>听音乐</li>
<li>跑步</li>
</ul>
</div>
</body>
</html>
```

上述程序的运行结果如图 4-37 所示。可以看到，网页中不包含 id=header 属性的所有元素都被添加上了背景色。

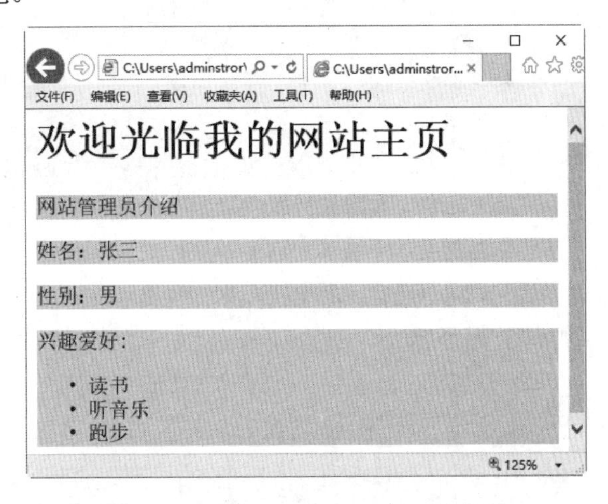

图 4-37　使用[attribute!=value]选择器

4.6.4　[attribute$=value]选择器

[attribute$=value]选择器选取每个带有指定属性且以指定字符串结尾的元素。

[attribute$=value]选择器的语法格式如下：

```
$("[attribute$=value]")
```

参数含义说明如下。

- attribute：必需，规定要查找的属性。
- value：必需，规定要查找的值。

例如，选择所有带 id 属性且属性值以 header 结尾的元素，可使用如下 jQuery 代码：

```
$("[id$=header]")
```

【例 4.36】(示例文件 ch04\4.36.html)

选择所有带有 id 属性且属性值以 header 结尾的元素，并为其添加背景色：

```
<!DOCTYPE html>
<html>
<head>
<script language="javascript" src="jquery-3.2.1.min.js"></script>
<script language="javascript">
$(document).ready(function(){
    $("[id$=header]").css("background-color","#B2E0FF");
});
</script>
</head>
<body>
<h1 id="header">欢迎光临我的网站主页</h1>
<p class="intro">网站管理员介绍</p>
<p>姓名：张三</p>
<p>性别：男</p>
<div id="choose">
兴趣爱好：
<ul>
<li>读书</li>
<li>听音乐</li>
<li>跑步</li>
</ul>
</div>
</body>
</html>
```

上述程序的运行结果如图 4-38 所示。可以看到，网页中所有带有 id 属性且属性值以 header 结尾的元素都被添加上了背景色。

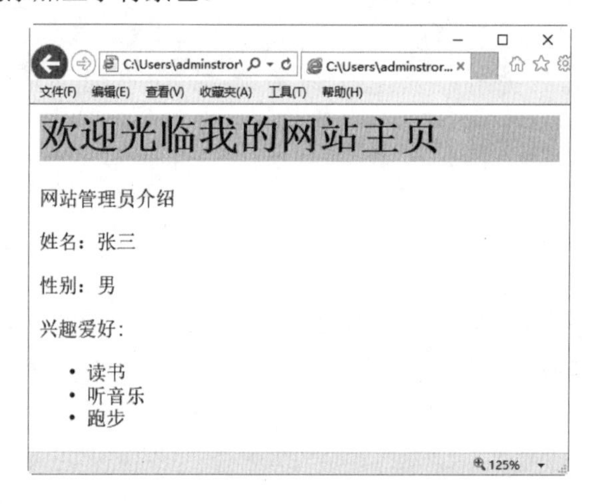

图 4-38　使用[attribute$=value]选择器

4.7　实战演练——匹配表单中的元素并实现不同的操作

本实例主要是通过匹配表单中的不同元素，从而实现不同的操作。具体代码如下：

```
<!DOCTYPE html>
<html>
```

```
<head>
<title>表单选择器的综合使用</title>
<script language="javascript" src="jquery-3.2.1.min.js"></script>
<script language="javascript">
    $(document).ready(function() {
       $(":checkbox").attr("checked","checked");                 //勾选复选框
       $(":radio").attr("checked","true");                       //选中单选按钮
       $(":image").attr("src","images/fish1.jpg");               //设置图片路径
       $(":file").hide();                                        //隐藏文件域
       $(":password").val("123");                                //设置密码域的值
       $(":text").val("文本框");                                 //设置文本框的值
       $(":button").attr("disabled","disabled");                 //设置按钮不可用
       $(":reset").val("重置按钮");                              //设置重置按钮的值
       $(":submit").val("提交按钮");                             //设置提交按钮的值
       $("#testDiv").append($("input:hidden:eq(1)").val()); //显示隐藏域的值
     });
</script>
</head>
<body>
<form>
   复选框：<input type="checkbox"/>
   单选按钮：<input type="radio"/>
   图像域：<input type="image"/><br>
   文件域：<input type="file"/><br>
   密码域：<input type="password" width="150px"/><br>
   文本框：<input type="text" width="150px"/><br>
   按  钮：<input type="button" value="按钮"/><br>
   重  置：<input type="reset" value=""/><br>
   提  交：<input type="submit" value=""><br>
   隐藏域： <input type="hidden" value="这是隐藏的元素">
  <div id="testDiv"><font color="blue">隐藏域的值：</font></div>
</form>
</body>
</html>
```

上述程序的运行结果如图4-39所示。

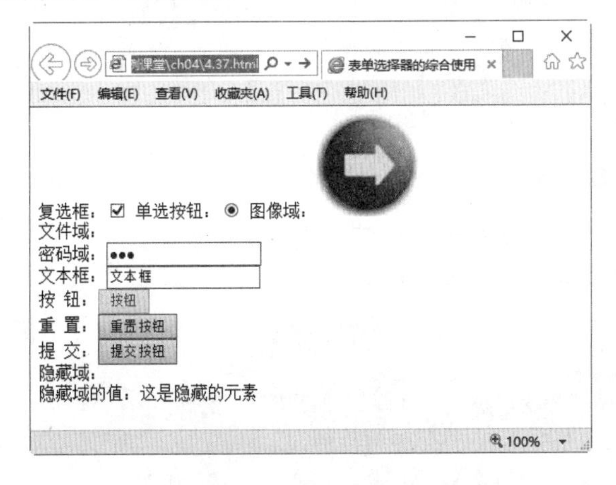

图4-39 表单选择器的综合应用

4.8 疑难解惑

疑问 1：如何实现鼠标指向后变色的表格？

答：一些清单通常以表格的形式展示，在数据比较多的情况下，很容易看串行。此时，如果能让鼠标指针指向的行变色，则可以很容易解决上述问题。

用户可以先为表格定义样式，例如以下代码：

```
<style type="text/css">
table{ border:0;border-collapse:collapse;}                    /*设置表格整体样式*/
td{font:normal 12px/17px Arial;padding:2px;width:100px;}    /*设置单元格的样式*/
th{ /*设置表头的样式*/
    font:bold 12px/17px Arial;
    text-align:left;
    padding:4px;
    border-bottom:1px solid #333;
}
.odd{background:#cef;}            /*设置奇数行样式*/
.even{background:#ffc;}           /*设置偶数行样式*/
.light{background:#00A1DA;}       /*设置鼠标移动到行的样式*/
</style>
```

定义完样式后，即可定义 jQuery 代码，实现表格的各行换色并且让鼠标指针移动到的行变色的效果。代码如下：

```
<script type="text/javascript">
$(document).ready(function(){
  $("tbody tr:even").addClass("odd");   //为偶数行添加样式
  $("tbody tr:odd").addClass("even");   //为奇数行添加样式
  $("tbody tr").hover(                  //为表格主体每行绑定 hover 方法
      function() {$(this).addClass("light");},
      function() {$(this).removeClass("light");}
  );
});
</script>
```

疑问 2：如何通过选择器实现一个带表头的双色表格？

答：通过过滤选择器，可以实现一个带表头的双色表格。

首先可以定义样式风格，例如以下代码：

```
<style type="text/css">
    td{
        font-size:12px;       /*设置单元格的样式*/
        padding:3px;          /*设置内边距*/
    }
    .th{
        background-color:#B6DF48;    /*设置背景颜色*/
```

```
        font-weight:bold;           /*设置文字加粗显示*/
        text-align:center;          /*文字居中对齐*/
    }
    .even{
        background-color:#E8F3D1;   /*设置偶数行的背景颜色*/
    }
    .odd{
        background-color:#F9FCEF;   /*设置奇数行的背景颜色*/
    }
</style>
```

定义完样式后，即可定义 jQuery 代码，实现带表头的双色表格效果。代码如下：

```
<script type="text/javascript">
    $(document).ready(function() {
        $("tr:even").addClass("even");       //设置奇数行所用的 CSS 类
        $("tr:odd").addClass("odd");         //设置偶数行所用的 CSS 类
        $("tr:first").removeClass("even");   //移除 even 类
        $("tr:first").addClass("th");        //添加 th 类
    });
</script>
```

第 2 篇

核 心 技 术

- 第 5 章　用 jQuery 控制页面
- 第 6 章　jQuery 的动画特效
- 第 7 章　jQuery 的事件处理
- 第 8 章　jQuery 的功能函数
- 第 9 章　jQuery 与 Ajax 技术的应用
- 第 10 章　jQuery 插件的开发与使用

第 5 章 用 jQuery 控制页面

在网页制作的过程中，jQuery 具有强大的功能。从本章开始，将陆续讲解 jQuery 的实用功能。本章主要介绍 jQuery 如何控制页面，对标记的属性进行操作、对表单元素进行操作和对元素的 CSS 样式进行操作等。

5.1 对页面的内容进行操作

jQuery 提供了对元素内容进行操作的方法。元素内容是指定义元素的起始标记和结束标记中间的内容，又可以分为文本内容和 HTML 内容。

5.1.1 对文本内容进行操作

jQuery 提供了 text()和 text(val)两种方法，用于对文本内容进行操作，主要作用是设置或返回所选元素的文本内容。其中，text()用来获取全部匹配元素的文本内容，text(val)方法用来设置全部匹配元素的文本内容。

1. 获取文本内容

下面通过例子来理解如何获取文本的内容。

【例 5.1】(示例文件 ch05\5.1.html)

获取文本内容：

```
<!DOCTYPE html>
<html>
<head>
<meta http-equiv="Content-Type" content="text/html; charset=gb2312" />
<script src="jquery.min.js">
</script>

<script>
$(document).ready(function(){
$("#btn1").click(function(){
alert("文本内容为: " + $("#test").text());
});
});
</script>

</head>
<body>
<p id="test">床前明月光，疑是地上霜。</p>
<button id="btn1">获取文本内容</button>
</body>
</html>
```

在 IE 11.0 中浏览页面，单击【获取文本内容】按钮，效果如图 5-1 所示。

2. 修改文本内容

下面通过例子来理解如何修改文本的内容。

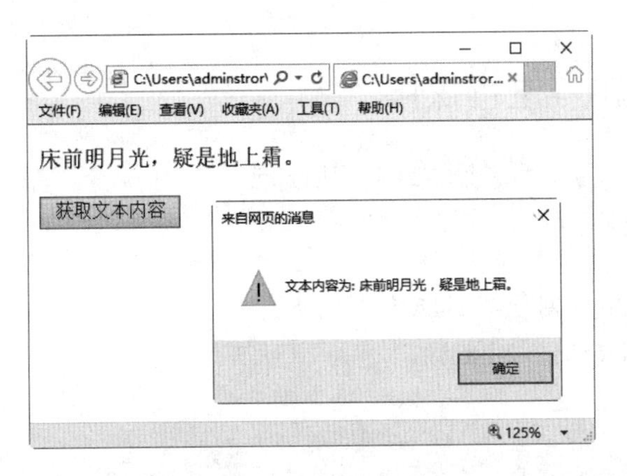

图 5-1 获取文本内容

【例 5.2】(示例文件 ch05\5.2.html)

修改文本内容：

```
<!DOCTYPE html>
<html>
<head>
<script src="jquery.min.js"></script>
<script>
$(document).ready(function(){
$("#btn1").click(function(){
$("#test1").text("清极不知寒");
});
});
</script>

</head>
<body>
<p id="test1">香中别有韵</p>
<button id="btn1">修改文本内容</button>
</body>
</html>
```

在 IE 11.0 中浏览页面，效果如图 5-2 所示。单击【修改文本内容】按钮，最终效果如图 5-3 所示。

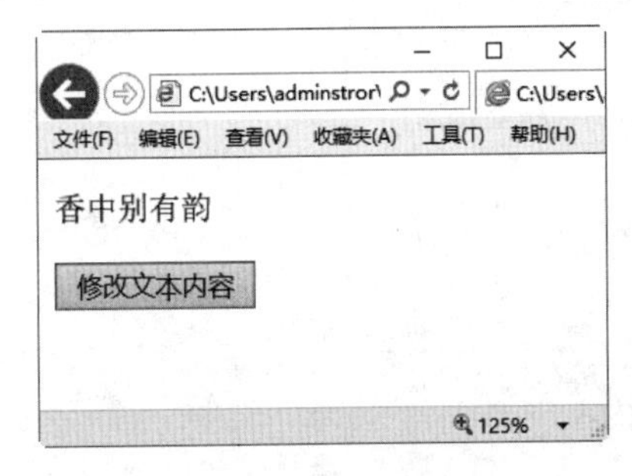

图 5-2 程序初始结果

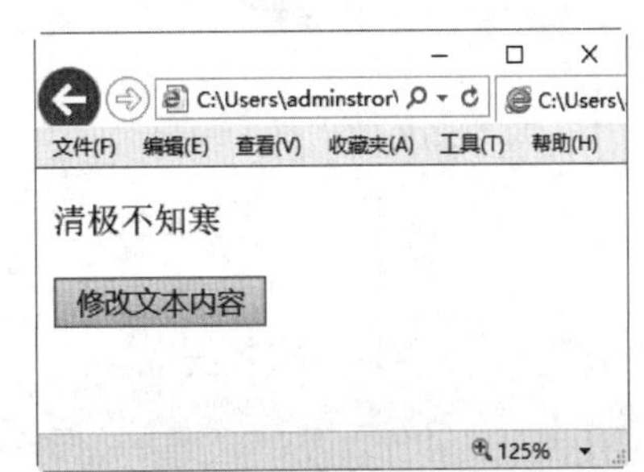

图 5-3 单击按钮后修改的结果

5.1.2 对 HTML 内容进行操作

jQuery 提供的 html()方法用于设置或返回所选元素的内容，这里包括 HTML 标记。

1. 获取 HTML 内容

下面通过例子来理解如何获取 HTML 的内容。

【例 5.3】(示例文件 ch05\5.3.html)

获取 HTML 内容：

```
<!DOCTYPE html>
<html>
<head>
<meta http-equiv="Content-Type" content="text/html; charset=gb2312" />
<script src="jquery.min.js"></script>

<script>
$(document).ready(function(){
$("#btn1").click(function(){
alert("HTML 内容为: " + $("#test").html());
});
});
</script>

</head>
<body>
<p id="test">床前明月光，<b>疑是地上霜</b> </p>
<button id="btn1">获取 HTML 内容</button>
</body>
</html>
```

在 IE 11.0 中浏览页面，单击【获取 HTML 内容】按钮，效果如图 5-4 所示。

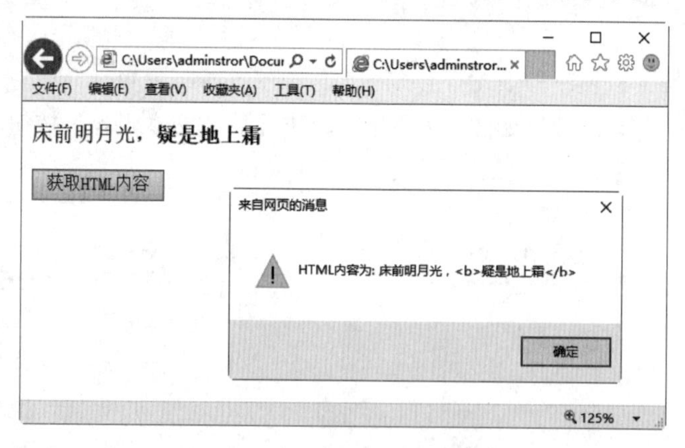

图 5-4 获取 HTML 内容

2. 修改 HTML 内容

下面通过例子来理解如何修改 HTML 的内容。

【例 5.4】(示例文件 ch05\5.4.html)

修改 HTML 内容：

```
<!DOCTYPE html>
<html>
<head>
<meta http-equiv="Content-Type" content="text/html; charset=gb2312" />
<script src="jquery.min.js"></script>
<script>
$(document).ready(function(){
$("#btn1").click(function(){
$("#test1").html("<b>清极不知寒</b> ");
});
});
</script>
</head>
<body>
<p id="test1">香中别有韵</p>
<button id="btn1">修改 HTML 内容</button>
</body>
</html>
```

在 IE 11.0 中浏览页面，效果如图 5-5 所示。单击【修改 HTML 内容】按钮，效果如图 5-6 所示，可见不仅内容发生了变化，而且字体也修改为粗体了。

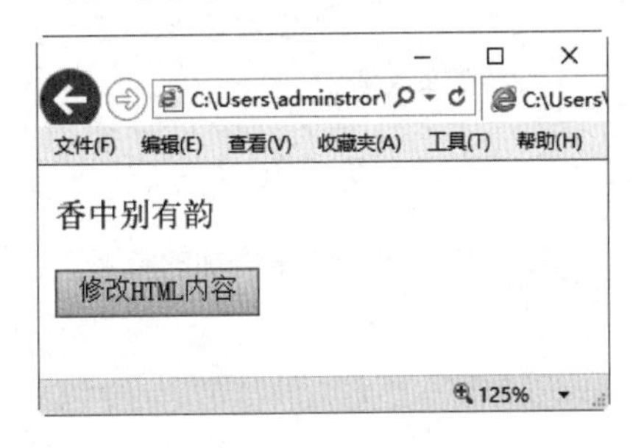

图 5-5 程序初始结果

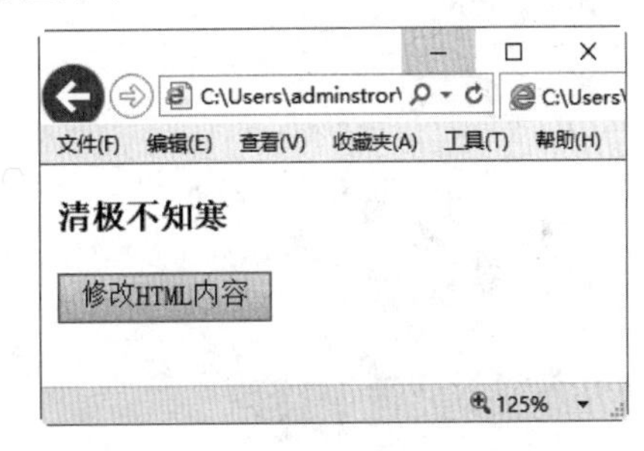

图 5-6 单击按钮后修改的结果

5.1.3 移动和复制页面内容

jQuery 提供的 append()方法和 appendTo()方法主要用于向匹配的元素内部追加内容。append()和 appendTo()方法执行的任务相同。不同之处在于内容的位置和选择器。

下面通过使用 append()方法的例子来理解相关内容。

【例 5.5】(示例文件 ch05\5.5.html)

使用 append()方法：

```
<!DOCTYPE html>
<html>
<head>
<meta http-equiv="Content-Type" content="text/html; charset=gb2312" />
```

```
<script src="jquery.min.js"></script>
<script>
$(document).ready(function(){
$("button").click(function(){
$("<b> 春风花草香。</b>").append("p");
});
});
</script>
</head>
<body>
<p>迟日江山丽，</p>
<p>泥融飞燕子，</p>
<button>每个 p 元素都添加</button>
</body>
</html>
```

在 IE 11.0 中浏览页面，效果如图 5-7 所示。单击【每个 p 元素都添加】按钮，效果如图 5-8 所示。

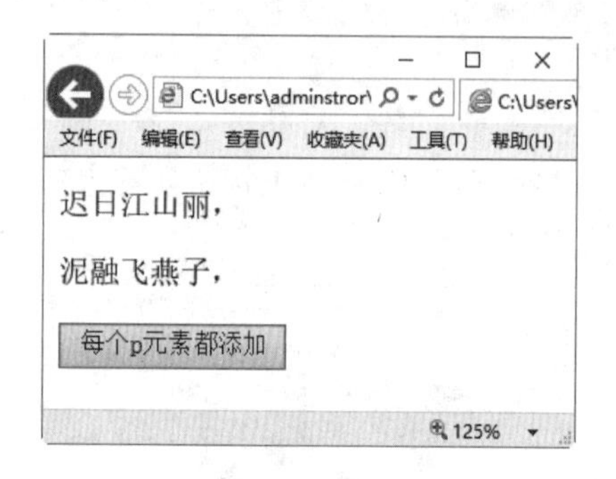

图 5-7　程序初始结果

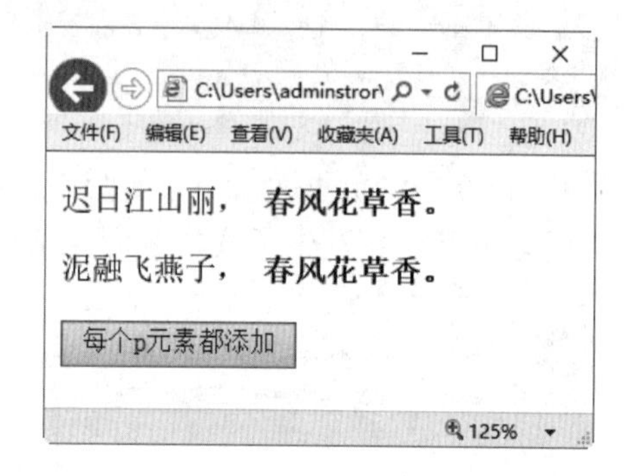

图 5-8　单击按钮后添加的结果

5.1.4　删除页面内容

jQuery 提供的 remove()方法用于移除被选元素，包括所有文本和子节点。该方法不会把匹配的元素从 jQuery 对象中删除，因而可以在将来再使用这些匹配的元素。但除了这个元素本身得以保留之外，remove()不会保留元素的 jQuery 数据。其他比如绑定的事件、附加的数据等都会被移除。

【例 5.6】(示例文件 ch05\5.6.html)

使用 remove()方法：

```
<!DOCTYPE html>
<html>
<head>
<meta http-equiv="Content-Type" content="text/html; charset=gb2312" />
<script src="jquery.min.js"></script>
<script>
$(document).ready(function(){
$("button").click(function(){
```

```
$("p").remove();
});
});
</script>
</head>
<body>
<p>迟日江山丽，春风花草香。泥融飞燕子，沙暖睡鸳鸯。</p>
<button>删除页面 p 元素的内容</button>
</body>
</html>
```

在 IE 11.0 中浏览页面，效果如图 5-9 所示。单击【删除页面 p 元素的内容】按钮，最终效果如图 5-10 所示。

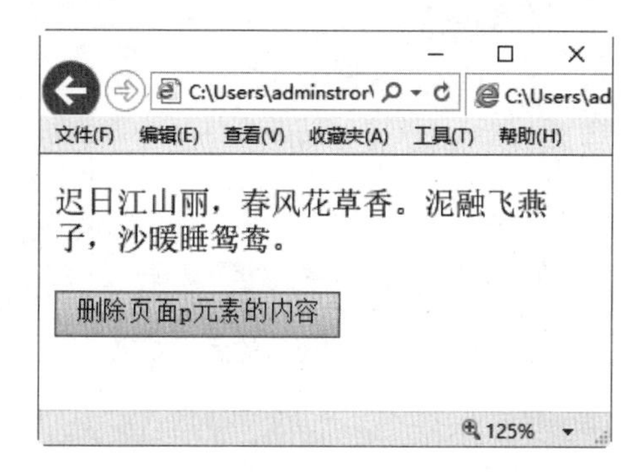

图 5-9　程序初始结果

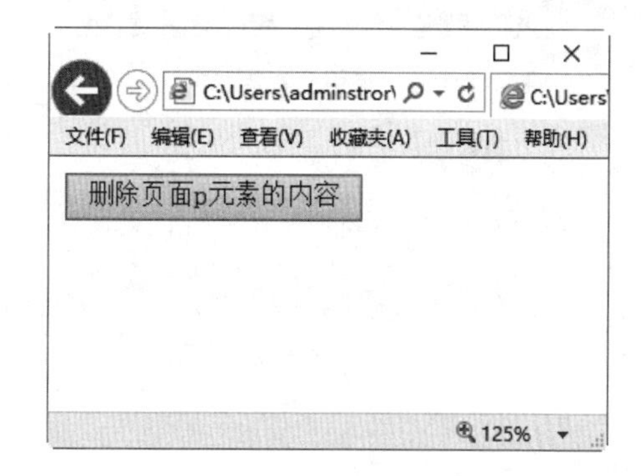

图 5-10　单击按钮后的结果

5.1.5　克隆页面内容

jQuery 提供的 clone()方法主要用于生成被选元素的副本，包含子节点、文本和属性。

【例 5.7】(示例文件 ch05\5.7.html)

使用 clone()方法：

```
<!DOCTYPE html>
<html>
<head>
<meta http-equiv="Content-Type" content="text/html; charset=gb2312" />
<script src="jquery.min.js"></script>
<script>
$(document).ready(function(){
   $("button").click(function(){
$("body").append($("p:first").clone(true));
   });
   $("p").click(function(){
$(this).animate({fontSize:"+=1px"});
   });
});
</script>
</head>
```

```
<body>
<p>谁言寸草心，报得三春晖。</p>
<button>克隆内容</button>
</body>
</html>
```

在 IE 11.0 中浏览页面，效果如图 5-11 所示。反复单击 3 次【克隆内容】按钮，最终效果如图 5-12 所示。当然，这个例子中还为 p 标记做了单击动画效果。

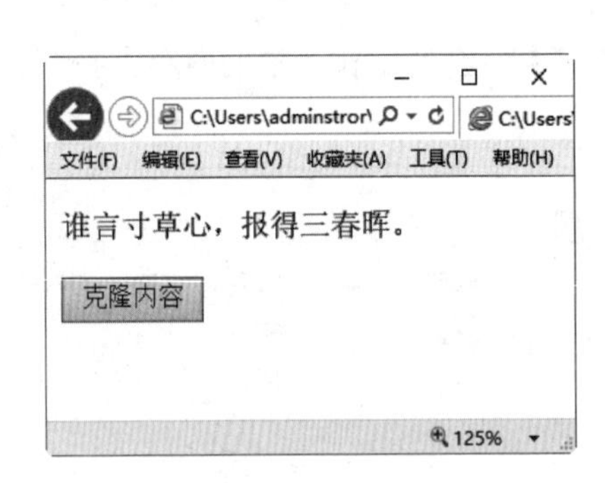

图 5-11 程序初始结果

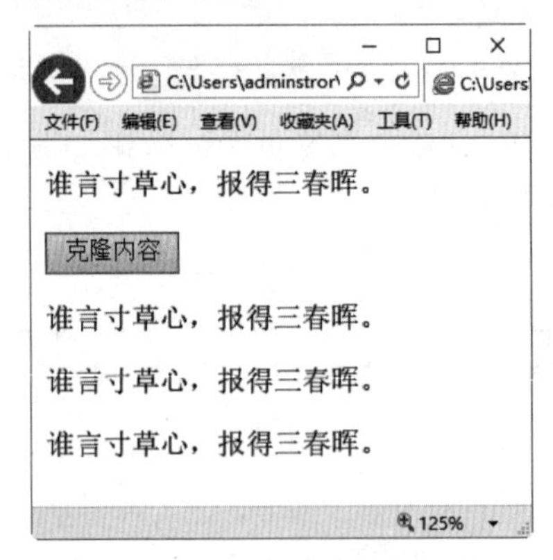

图 5-12 单击按钮后的运行结果

5.2 对标记的属性进行操作

jQuery 提供了对标记的属性进行操作的方法。

5.2.1 获取属性的值

jQuery 提供的 prop()方法主要用于设置或返回被选元素的属性值。

【**例 5.8**】(示例文件 ch05\5.8.html)

获取属性的值：

```
<!DOCTYPE html>
<html>
<head>
<script src="jquery.min.js"></script>
<script>
$(document).ready(function(){
$("button").click(function(){
alert("图像宽度为: " + $("img").prop("width"));
});
});
</script>
</head>
<body>
<img src="123.jpg" />
```

```
<br />
<button>查看图像的宽度</button>
</body>
</html>
```

在 IE 11.0 中浏览页面，单击【查看图像的宽度】按钮，效果如图 5-13 所示。

图 5-13 获取属性的值

5.2.2 设置属性的值

prop()方法除了可以获取元素属性的值之外，还可以通过它设置属性的值。其语法格式如下：

```
prop(name,value);
```

该方法将元素的 name 属性的值设置为 value。

attr(name,value)方法也可以设置元素的属性值。读者可以自行测试效果。

【例 5.9】(示例文件 ch05\5.9.html)

设置属性的值：

```
<!DOCTYPE html>
<html>
<head>
<meta http-equiv="Content-Type" content="text/html; charset=gb2312" />
<script src="jquery.min.js"></script>
<script>
$(document).ready(function(){
$("button").click(function(){
$("img").prop("width","300");
});
});
</script>
```

```
</head>
<body>
<img src="123.jpg" />
<br />
<button>修改图像的宽度</button>
</body>
</html>
```

在 IE 11.0 中浏览页面，效果如图 5-14 所示。单击【修改图像的宽度】按钮，最终结果如图 5-15 所示。

图 5-14　程序初始结果

图 5-15　单击按钮后的结果

5.2.3　删除属性的值

jQuery 提供的 removeAttr(name)方法可以用来删除属性的值。

【例 5.10】(示例文件 ch05\5.10.html)

删除属性的值：

```
<!DOCTYPE html>
<html>
<head>
<meta http-equiv="Content-Type" content="text/html; charset=gb2312" />
<script src="jquery.min.js"></script>
<script type="text/javascript">
$(document).ready(function(){
$("button").click(function(){
$("p").removeAttr("style");
});
});
</script>
</head>
<body>
<h1>观沧海</h1>
<p style="font-size:120%;color:red">东临碣石，以观沧海。</p>
<p>水何澹澹，山岛竦峙。</p>
```

```
<button>删除所有 p 元素的 style 属性</button>
</body>
</html>
```

在 IE 11.0 中浏览页面，效果如图 5-16 所示。单击【删除所有 p 元素的 style 属性】按钮，最终结果如图 5-17 所示。

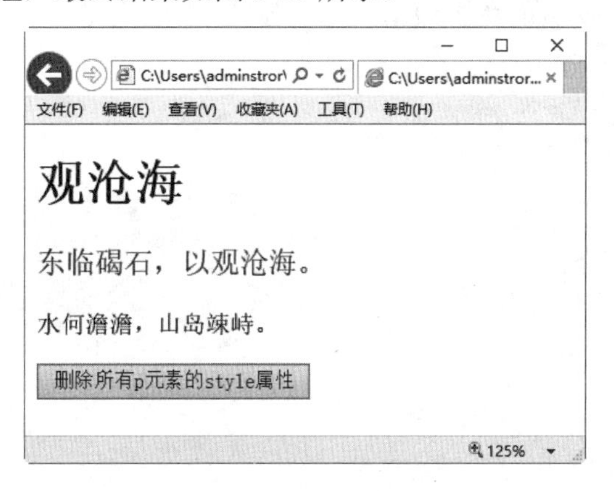

图 5-16　程序初始结果

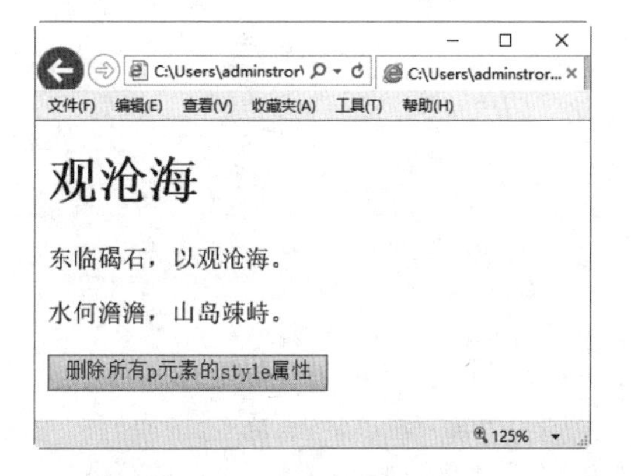

图 5-17　单击按钮后的结果

5.3　对表单元素进行操作

jQuery 提供了对表单元素进行操作的方法。

5.3.1　获取表单元素的值

val()方法返回或设置被选元素的值。元素的值是通过 value 属性设置的。该方法大多用于表单元素。如果该方法未设置参数，则返回被选元素的当前值。

【例 5.11】(示例文件 ch05\5.11.html)

获取表单元素的值：

```
<!DOCTYPE html>
<html>
<head>
<meta http-equiv="Content-Type" content="text/html; charset=gb2312" />
<script src="jquery.min.js"></script>
<script type="text/javascript">
$(document).ready(function(){
$("button").click(function(){
alert($("input:text").val());
});
});
</script>
</head>
<body>
```

```
名称: <input type="text" name="fname" value="冰箱" /><br />
类别: <input type="text" name="lname" value="电器" /><br /><br />
<button>获得第一个文本域的值</button>
</body>
</html>
```

在 IE 11.0 中浏览页面。单击【获得第一个文本域的值】按钮，结果如图 5-18 所示。

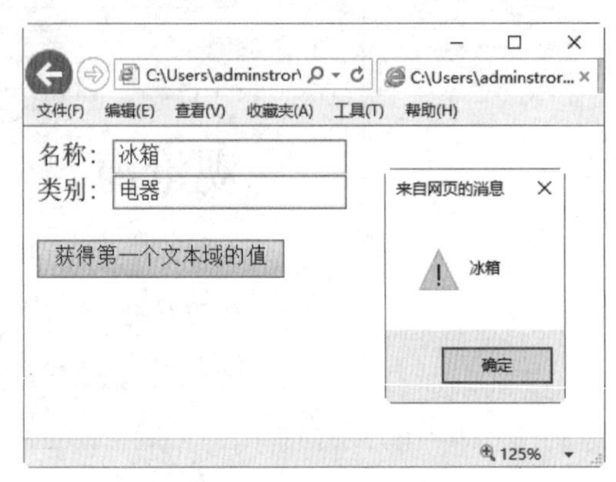

图 5-18　获取表单元素的值

5.3.2　设置表单元素的值

val()方法也可以设置表单元素的值。其语法格式如下：

```
$("selector").val(value);
```

【例 5.12】(示例文件 ch05\5.12.html)

设置表单元素的值：

```
<!DOCTYPE html>
<html>
<head>
<meta http-equiv="Content-Type" content="text/html; charset=gb2312" />
<script src="jquery.min.js"></script>
<script type="text/javascript">
$(document).ready(function(){
$("button").click(function(){
$(":text").val("冰箱");
});
});
</script>
</head>
<body>
<p>电器名称: <input type="text" name="user" value="洗衣机" /></p>
<button>改变文本域的值</button>
</body>
</html>
```

在 IE 11.0 中浏览页面，效果如图 5-19 所示。单击【改变文本域的值】按钮，最终结果如图 5-20 所示。

图 5-19　程序初始结果

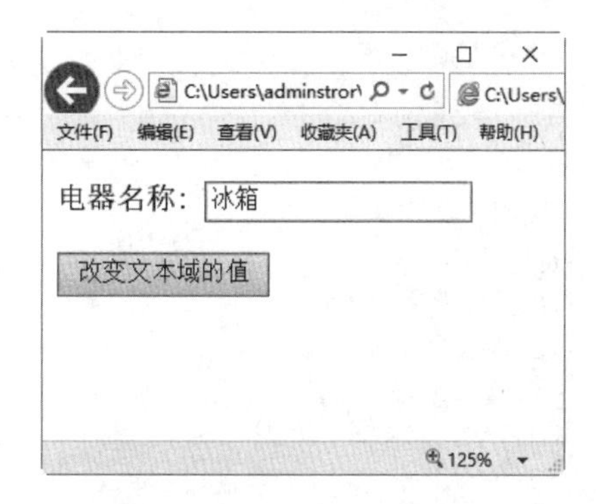

图 5-20　单击按钮后的结果

5.4　对元素的 CSS 样式进行操作

通过 jQuery，用户可以很容易地对 CSS 样式进行操作。

5.4.1　添加 CSS 类

addClass()方法主要是向被选元素添加一个或多个类。

下面的例子展示如何向不同的元素添加 class 属性。当然，在添加类时，也可以选取多个元素。

【例 5.13】(示例文件 ch05\5.13.html)

向不同的元素添加 class 属性：

```
<!DOCTYPE html>
<html>
<head>
<meta http-equiv="Content-Type" content="text/html; charset=gb2312" />
<script src="jquery.min.js"></script>
<script>
$(document).ready(function(){
$("button").click(function(){
$("h1,h2,p").addClass("blue");
$("div").addClass("important");
});
});
</script>
<style type="text/css">
.important
{
font-weight: bold;
font-size: xx-large;
```

```
}
.blue
{
color: blue;
}
</style>
</head>
<body>
<h1>梅雪</h1>
<h2>梅雪争春未肯降</h2>
<p>骚人阁笔费评章</p>
<p>梅须逊雪三分白</p>
<div>雪却输梅一段香</div>
<br>
<button>向元素添加 CSS 类</button>
</body>
</html>
```

在 IE 11.0 中浏览页面，效果如图 5-21 所示。单击【向元素添加 CSS 类】按钮，最终结果如图 5-22 所示。

图 5-21　程序初始结果

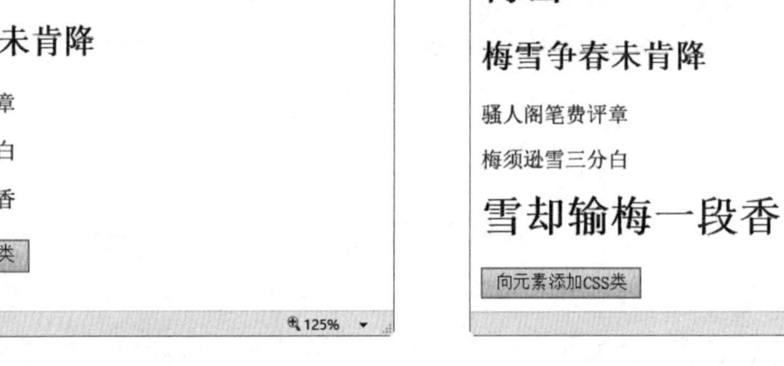

图 5-22　单击按钮后的结果

addClass()方法也可以同时添加多个 CSS 类。

【例 5.14】(示例文件 ch05\5.14.html)

同时添加多个 CSS 类：

```
<!DOCTYPE html>
<html>
<head>
<meta http-equiv="Content-Type" content="text/html; charset=gb2312" />
<script src="jquery.min.js"></script>
<script>
$(document).ready(function(){
$("button").click(function(){
```

```
$("#div1").addClass("important blue");
});
});
</script>
<style type="text/css">
.important
{
font-weight: bold;
font-size: xx-large;
}
.blue
{
color: blue;
}
</style>
</head>
<body>
<div id="div1">梅须逊雪三分白</div>
<div id="div2">雪却输梅一段香</div>
<br>
<button>向第一个 div 元素添加多个 CSS 类</button>
</body>
</html>
```

在 IE 11.0 中浏览页面，效果如图 5-23 所示。单击【向第一个 div 元素添加多个 CSS 类】按钮，最终结果如图 5-24 所示。

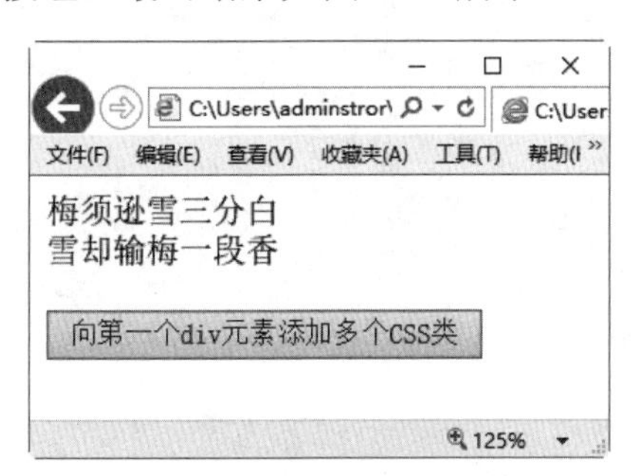

图 5-23　程序初始结果

图 5-24　单击按钮后的结果

5.4.2　删除 CSS 类

removeClass()方法主要是从被选元素删除一个或多个类。

【例 5.15】(示例文件 ch05\5.15.html)

删除 CSS 类：

```
<!DOCTYPE html>
<html>
<head>
<meta http-equiv="Content-Type" content="text/html; charset=gb2312" />
<script src="jquery.min.js"></script>
```

```
<script>
$(document).ready(function(){
$("button").click(function(){
$("h1,h2,p").removeClass("important blue");
});
});
</script>

<style type="text/css">
.important
{
font-weight: bold;
font-size: xx-large;
}
.blue
{
color: blue;
}
</style>

</head>

<body>
 <h1 class="blue">梅雪</h1>
 <h2 class="blue">梅雪争春未肯降</h2>
 <p class="blue">骚人阁笔费评章</p>
 <p>雪却输梅一段香</p>
 <br>
 <button>从元素上删除 CSS 类</button>
</body>
</html>
```

在 IE 11.0 中浏览页面，效果如图 5-25 所示。单击【从元素上删除 CSS 类】按钮，最终结果如图 5-26 所示。

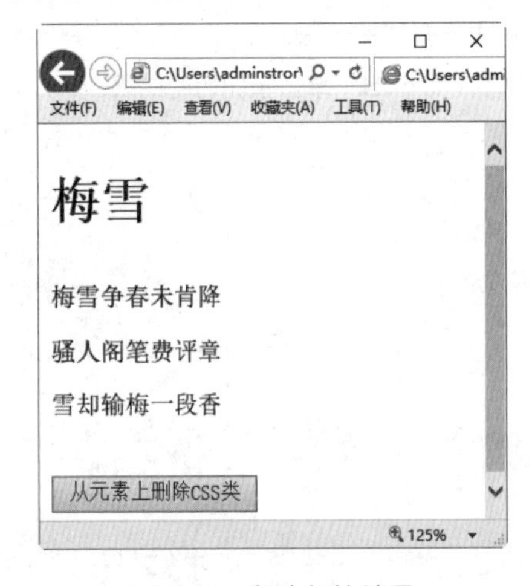

图 5-25　程序初始结果

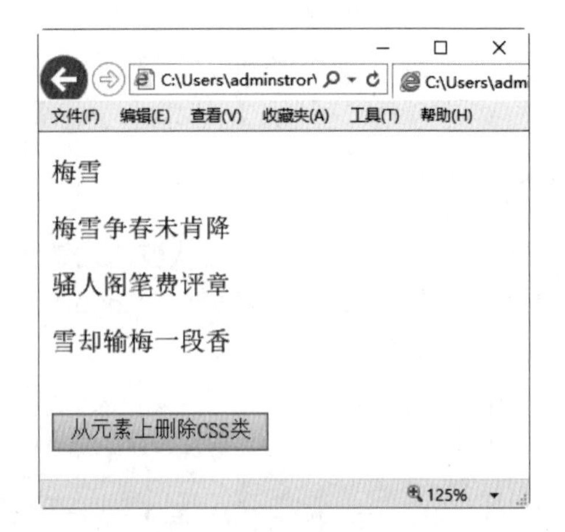

图 5-26　单击按钮后的结果

5.4.3 动态切换 CSS 类

jQuery 提供的 toggleClass()方法主要作用是对设置或移除被选元素的一个或多个 CSS 类进行切换。该方法检查每个元素中指定的类。如果不存在则添加类；如果已设置则删除之。这就是所谓的切换效果。不过，通过使用 switch 参数，我们能够规定只删除或只添加类。使用的语法格式如下：

```
$(selector).toggleClass(class,switch)
```

其中 class 是必需的。规定添加或移除 class 的指定元素。如需规定多个 class，使用空格来分隔类名。switch 是可选的布尔值，确定是否添加或移除 class。

【例 5.16】 (示例文件 ch05\5.16.html)

动态切换 CSS 类：

```
<!DOCTYPE html>
<html>
<head>
<meta http-equiv="Content-Type" content="text/html; charset=gb2312" />
<script src="jquery.min.js"></script>
<script>
$(document).ready(function(){
$("button").click(function(){
$("p").toggleClass("main");
});
});
</script>
<style type="text/css">
.main
{
font-size: 120%;
color: red;
}
</style>
</head>
<body>
<h1 id="h1">望岳</h1>
<p>会当凌绝顶</p>
<p>一览众山小</p>
<button class="btn1">切换段落的"main" 类</button>
</body>
</html>
```

在 IE 11.0 中浏览页面，效果如图 5-27 所示。单击【切换段落的“main”类】按钮，最终结果如图 5-28 所示。再次单击上面的按钮，则会在两个不同的效果之间切换。

图 5-27　程序初始结果

图 5-28　单击按钮后的结果

5.4.4　获取和设置 CSS 样式

jQuery 提供的 css()方法，可以用来获取或设置匹配的元素的一个或多个样式属性。下面通过 css(name)来获得某种样式的值。

【例 5.17】(示例文件 ch05\5.17.html)

获取 CSS 样式：

```
<!DOCTYPE html>
<html>
<head>
<meta http-equiv="Content-Type" content="text/html; charset=gb2312" />
<script src="jquery.min.js"></script>
<script>
$(document).ready(function(){
$("button").click(function(){
alert($("p").css("color"));
});
});
</script>
</head>
<body>
<p style="color:red">相见时难别亦难，东风无力百花残</p>
<button type="button">返回段落的颜色</button>
</body>
</html>
```

在 IE 11.0 中浏览页面。单击【返回段落的颜色】按钮，结果如图 5-29 所示。

图 5-29　获取 CSS 样式

下面通过 css(name,value)来设置元素的样式。

【例 5.18】(示例文件 ch05\5.18.html)

设置 CSS 样式：

```
<!DOCTYPE html>
<html>
<head>
<meta http-equiv="Content-Type" content="text/html; charset=gb2312" />
<script src="jquery.min.js"></script>
<script>
$(document).ready(function(){
$("button").click(function(){
$("p").css("color","red");
});
});
</script>
</head>
<body>
<p>相见时难别亦难，东风无力百花残</p>
<p>春蚕到死丝方尽，蜡炬成灰泪始干</p>
<button type="button">改变段落的颜色</button>
</body>
</html>
```

在 IE 11.0 中浏览页面，效果如图 5-30 所示。单击【改变段落的颜色】按钮，最终结果如图 5-31 所示。

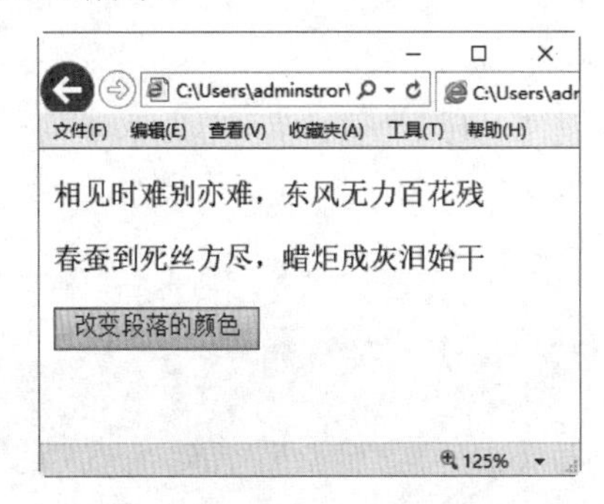

图 5-30　程序初始结果

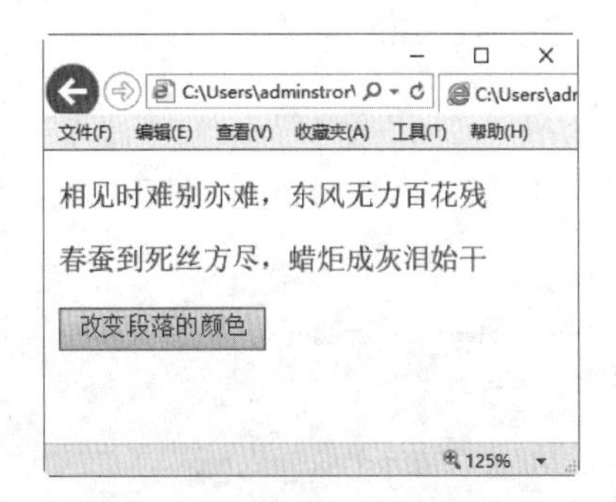

图 5-31　单击按钮后的结果

5.5 实战演练——制作奇偶变色的表格

在网站制作中，经常需要制作奇偶变色的表格。通过 jQuery 可以轻松地实现该效果。

step 01 制作含有表格的网页。代码如下：

```
<html>
<head>
<meta http-equiv="Content-Type" content="text/html; charset=gb2312" />
<title>jquery 奇偶变色</title>
<script src="jquery.min.js"></script>
<script>
$(document).ready(function() {
$('tr').addClass('odd');
$('tr:even').addClass('even'); //奇偶变色，添加样式
});
</script>
</head>
<body>
<table width="182" height="164" border="3" id="hacker">
<tr>
<td>商品名称</td>
<td>销量</td>
</tr>
<tr>
<td>冰箱</td>
<td>185620</td>
</tr>
<tr>
<td>洗衣机</td>
<td>562030</td>
</tr>
<tr>
<td>冰箱</td>
<td>568210</td>
</tr>
<tr>
<td>空调</td>
<td>380010</td>
</tr>
<tr>
<td>电视机</td>
<td>965420</td>
</tr>
<tr>
<td>电脑</td>
<td>56000</td>
</tr>
```

```
</table>
</body>
</html>
```

step 02 运行上述代码，效果如图 5-32 所示。

step 03 添加 CSS 样式。代码如下：

```
<style>
#hacker tr:hover{
background-color: red; //使用 CSS 伪类实现鼠标移入行变色的效果
}
.odd {
background-color: #ffc; /* pale yellow for odd rows */
}
.even {
background-color: #cef; /* pale blue for even rows */
}
</style>
```

添加代码后，运行程序，效果如图 5-33 所示。

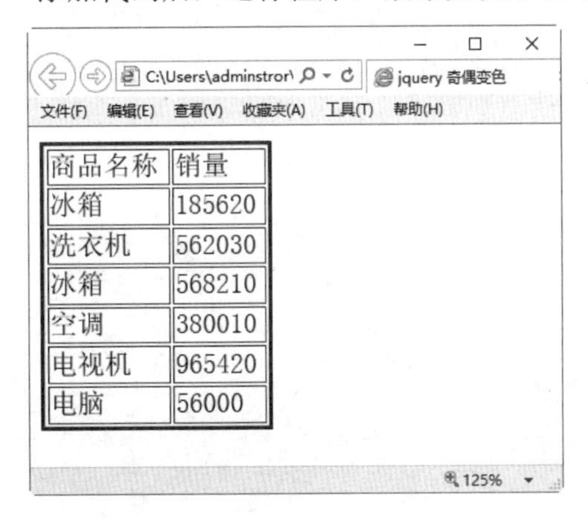

图 5-32　程序初始结果

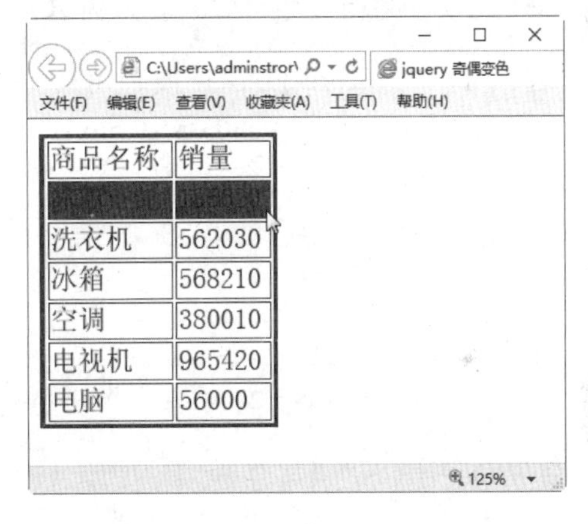

图 5-33　添加 CSS 后的结果

step 04 添加 jQuery 代码，实现奇偶变色的效果。代码如下：

```
<script src="jquery.min.js"></script>
<script>
$(document).ready(function() {
$('tr').addClass('odd');
$('tr:even').addClass('even'); //奇偶变色，添加样式
});
</script>
```

step 05 添加 jQuery 代码后，运行结果如图 5-34 所示。

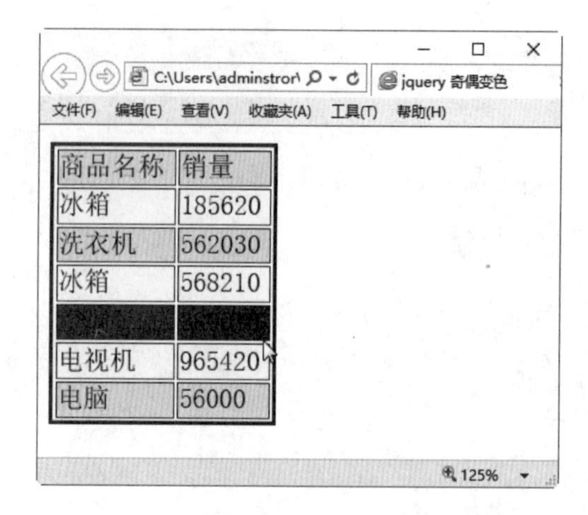

图 5-34 添加 jQuery 代码后的结果

5.6 疑难解惑

疑问 1：如何向指定内容前插入内容？

答：可以用 before()方法在被选元素前插入指定的内容。

【例 5.19】(示例文件 ch05\5.19.html)

向指定内容前插入内容：

```
<!DOCTYPE html>
<html>
<head>
<meta http-equiv="Content-Type" content="text/html; charset=gb2312" />
<script src="jquery.min.js"></script>
<script>
$(document).ready(function(){
$(".btn1").click(function(){
$("p").before("<p>孤舟蓑笠翁，</p>");
});
});
</script>
</head>
<body>
<p>独钓寒江雪</p>
<button class="btn1">在段落前面插入新的内容</button>
</body>
</html>
```

在 IE 11.0 中浏览页面，效果如图 5-35 所示。单击【在段落前面插入新的内容】按钮，最终结果如图 5-36 所示。

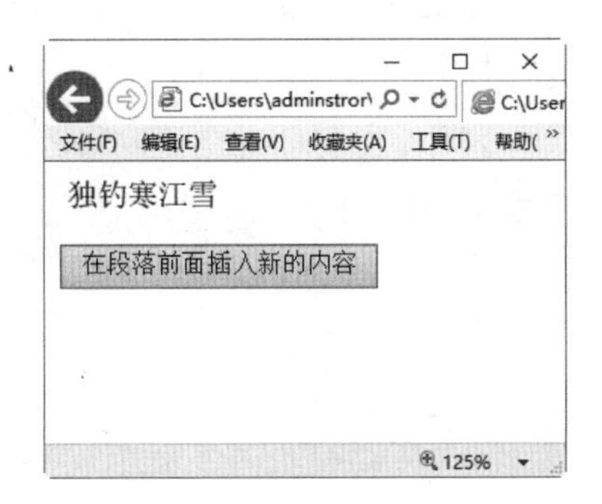

图 5-35 程序初始结果

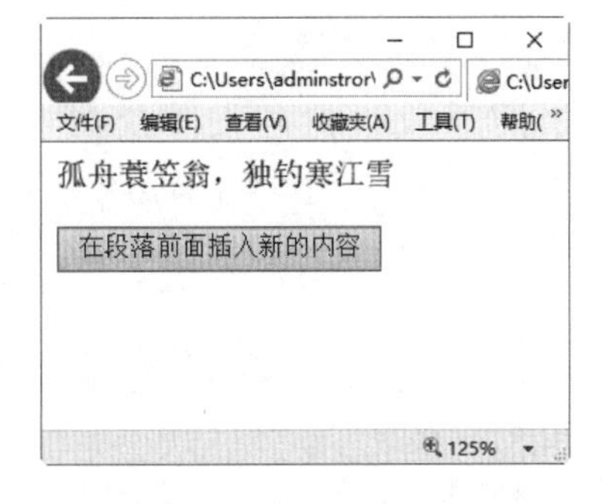

图 5-36 单击按钮后的结果

疑问 2：如何检查段落中是否添加了指定的 CSS 类？

答：可以用 hasClass()方法来检查被选元素是否包含指定的 CSS 类。

【例 5.20】(示例文件 ch05\5.20.html)

检查被选元素是否包含指定的 CSS 类：

```
<!DOCTYPE html>
<html>
<head>
<script src="jquery.min.js"></script>
<script type="text/javascript">
$(document).ready(function(){
$("button").click(function(){
alert($("p:first").hasClass("class1"));
});
});
</script>
<style type="text/css">
.class1
{
font-size: 120%;
color: red;
}
</style>
</head>
<body>
<p class="class1">青青河边草</p>
<p>绵绵到海角</p>
<button>检查第一个段落是否拥有类 "class1"</button>
</body>
</html>
```

在 IE 11.0 中浏览页面，单击【检查第一个段落是否拥有类"class1"】按钮，结果如图 5-37 所示。

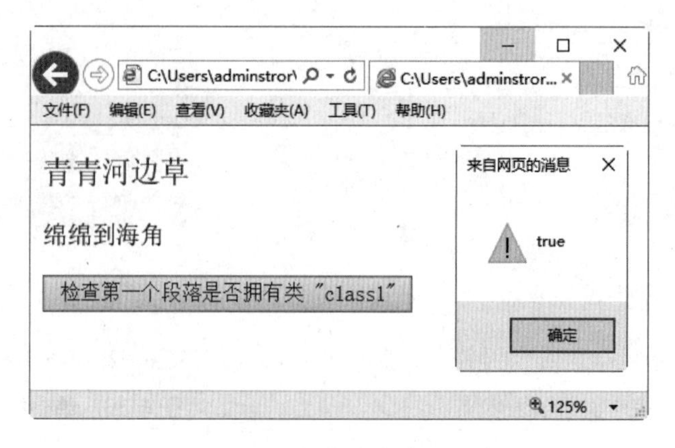

图 5-37　单击按钮后的结果

第 6 章 jQuery 的动画特效

jQuery 能在页面上实现绚丽的动画效果。jQuery 本身对页面动态效果提供了一些有限的支持，如动态显示和隐藏页面的元素、淡入淡出动画效果、滑动动画效果等。本章就来介绍如何使用 jQuery 制作动画特效。

6.1 jQuery 的基本动画效果

显示与隐藏是 jQuery 实现的基本动画效果。在 jQuery 中，提供了两种显示与隐藏元素的方法：一是分别显示和隐藏网页元素；二是切换显示与隐藏元素。

6.1.1 隐藏元素

在 jQuery 中，使用 hide()方法来隐藏匹配元素。hide()方法相当于将元素的 CSS 样式属性 display 的值设置为 none。

1. 简单隐藏

在使用 hide()方法隐藏匹配元素的过程中，当 hide()方法不带有任何参数时，就实现了元素的简单隐藏。其语法格式如下：

```
hide()
```

例如，想要隐藏页面当中的所有文本元素，就可以使用如下 jQuery 代码：

```
$("p").hide()
```

【例 6.1】(示例文件 ch06\6.1.html)

网页元素的简单隐藏：

```
<!DOCTYPE html>
<html>
<head>
<script src="jquery.min.js">
</script>
<script>
$(document).ready(function(){
$("p").click(function(){
$(this).hide();
});
});
</script>
</head>
<body>
<p>如果点击我，我会隐藏。</p>
<p>如果点击我，我也会隐藏。</p>
<p>如果点击我，我也会隐藏哦。</p>
</body>
</html>
```

程序运行结果如图 6-1 所示，单击页面中的文本段，该文本段就会隐藏，这就实现了元素的简单隐藏动画效果。

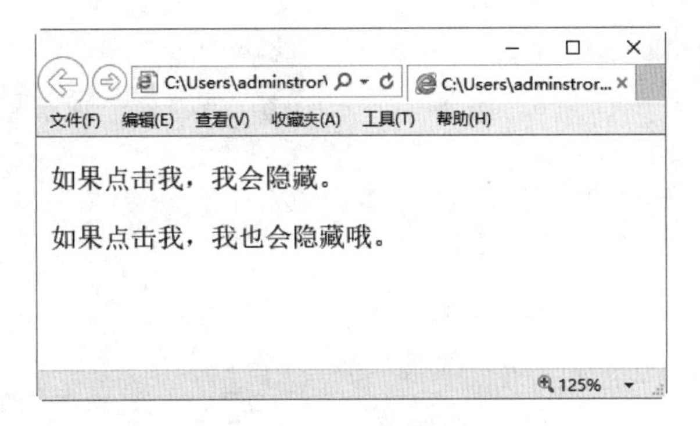

图 6-1 网页元素的简单隐藏

2. 部分隐藏

使用 hide()方法，除了可以对网页当中的内容一次性全部进行隐藏外，还可以对网页内容进行部分隐藏。

【例 6.2】(示例文件 ch06\6.2.html)

网页元素的部分隐藏：

```
<!DOCTYPE html>
<html>
<head>
<script src="jquery.min.js"></script>
<script type="text/javascr2ipt">
$(document).ready(function(){
$(".ex .hide").click(function(){
$(this).parents(".ex").hide();
});
});
</script>
<style type="text/css">
div .ex
{
background-color: #e5eecc;
padding: 7px;
border: solid 1px #c3c3c3;
}
</style>
</head>
<body>
<h3>总经理</h3>
<div class="ex">
<button class="hide" type="button">隐藏</button>
<p>姓名：张三<br />
电话：13512345678<br />
公司地址：北京西路 20 号</p>
</div>

<h3>办公室主任</h3>
```

```
<div class="ex">
<button class="hide" type="button">隐藏</button>
<p>姓名：李四<br />
电话：13012345678<br />
公司地址：北京西路 20 号</p>
</div>
</body>
</html>
```

程序运行结果如图 6-2 所示，单击页面中的【隐藏】按钮，即可将下方的联系人信息隐藏。

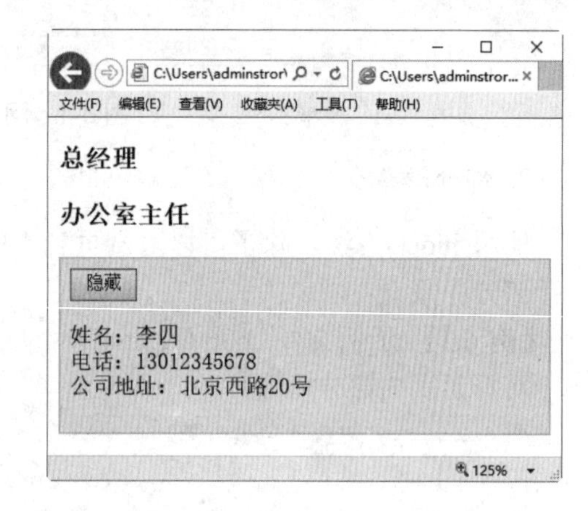

图 6-2　网页元素的部分隐藏

3. 设置隐藏参数

带有参数的 hide()隐藏方式，可以实现不同方式的隐藏效果，其语法格式如下：

```
$(selector).hide(speed,callback);
```

参数含义说明如下。

- speed：可选的参数，规定隐藏的速度，可以取 slow、fast 或毫秒等参数。
- callback：可选的参数，规定隐藏完成后所执行的函数名称。

【例 6.3】(示例文件 ch06\6.3.html)

设置网页元素的隐藏参数：

```
<!DOCTYPE html>
<html>
<head>
<script src="jquery.min.js"></script>
<script type="text/javascript">
$(document).ready(function(){
$(".ex .hide").click(function(){
$(this).parents(".ex").hide("3000");
});
});
</script>
<style type="text/css">
div .ex
{
background-color: #e5eecc;
padding: 7px;
border: solid 1px #c3c3c3;
}
</style>
</head>
<body>
<h3>总经理</h3>
```

```
<div class="ex">
<button class="hide" type="button">隐藏</button>
<p>姓名：张三<br />
电话：13512345678<br />
公司地址：北京西路 20 号</p>
</div>

<h3>办公室主任</h3>
<div class="ex">
<button class="hide" type="button">隐藏</button>
<p>姓名：李四<br />
电话：13012345678<br />
公司地址：北京西路 20 号</p>
</div>
</body>
</html>
```

程序运行结果如图 6-3 所示，单击页面中的【隐藏】按钮，即可将下方的联系人信息慢慢地隐藏起来。

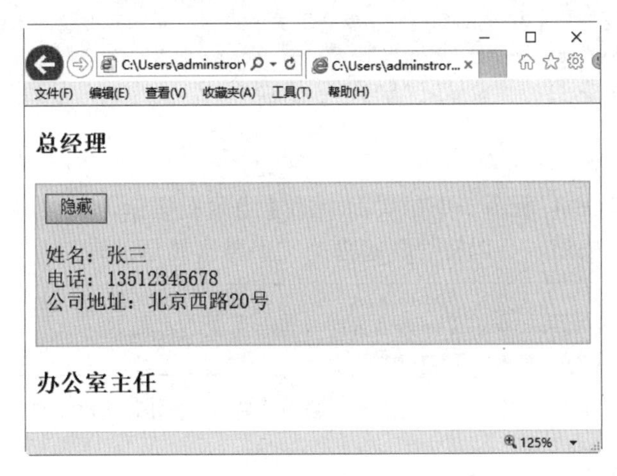

图 6-3　设置网页元素的隐藏参数

6.1.2　显示元素

使用 show()方法可以显示匹配的网页元素。show()方法有两种语法格式：一是不带有参数的形式；二是带有参数的形式。

1. 不带有参数的格式

不带有参数的格式，用以实现不带有任何效果的显示匹配元素。其语法格式如下：

```
show()
```

例如，想要显示页面中的所有文本元素，就可以使用如下 jQuery 代码：

```
$("p").show()
```

【例 6.4】(示例文件 ch06\6.4.html)

显示或隐藏网页中的元素：

```
<!DOCTYPE html>
<html>
<head>
<script src="jquery.min.js"></script>
<script type="text/javascript">
$(document).ready(function(){
$("#hide").click(function(){
$("p").hide();
});
$("#show").click(function(){
$("p").show();
});
});
</script>
</head>
<body>
<p id="p1">点击【隐藏】按钮，本段文字就会消失；点击【显示】按钮，本段文字就会显示。</p>
<button id="hide" type="button">隐藏</button>
<button id="show" type="button">显示</button>
</body>
</html>
```

程序运行结果如图 6-4 所示，单击页面中的【隐藏】按钮，就会将网页中的文字隐藏起来，然后单击【显示】按钮，可以将隐藏起来的文字再次显示出来。

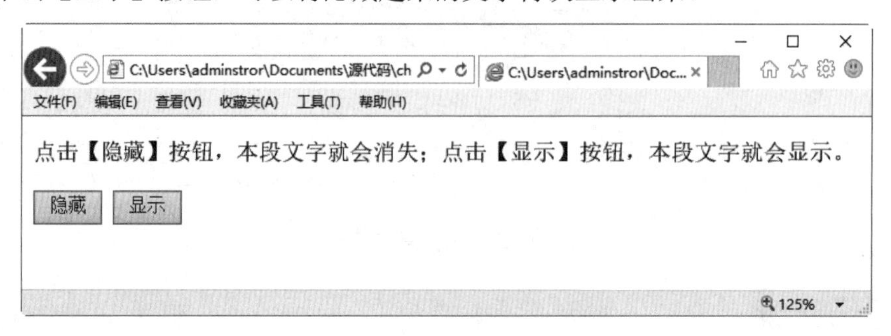

图 6-4　显示或隐藏网页中的元素

2. 带有参数的格式

带有参数的格式用来实现以优雅的动画方式显示网页中的元素，并在隐藏完成后可选择地触发一个回调函数。其语法格式如下：

```
$(selector).show(speed,callback);
```

参数说明如下。

- speed：可选的参数，规定显示的速度，可以取 slow、fast 或毫秒等参数。
- callback：可选的参数，规定显示完成后所执行的函数名称。

例如，想要在 300 毫秒内显示网页中的 p 元素，就可以使用如下 jQuery 代码：

```
$("p").show(300);
```

【例 6.5】(示例文件 ch06\6.5.html)

在 3000 毫秒内显示或隐藏网页中的元素：

```
<!DOCTYPE html>
<html>
<head>
<script src="jquery.min.js"></script>
<script type="text/javascript">
$(document).ready(function(){
$("#hide").click(function(){
$("p").hide("3000");
});
$("#show").click(function(){
$("p").show("3000");
});
});
</script>
</head>
<body>
<p id="p1">点击【隐藏】按钮，本段文字就会消失；点击【显示】按钮，本段文字就会显示。</p>
<button id="hide" type="button">隐藏</button>
<button id="show" type="button">显示</button>
</body>
</html>
```

程序运行结果如图 6-5 所示，单击页面中【隐藏】按钮，就会将网页中的文字在 3000 毫秒内慢慢隐藏起来，然后单击【显示】按钮，又可以将隐藏起来的文字在 3000 毫秒内慢慢地显示出来。

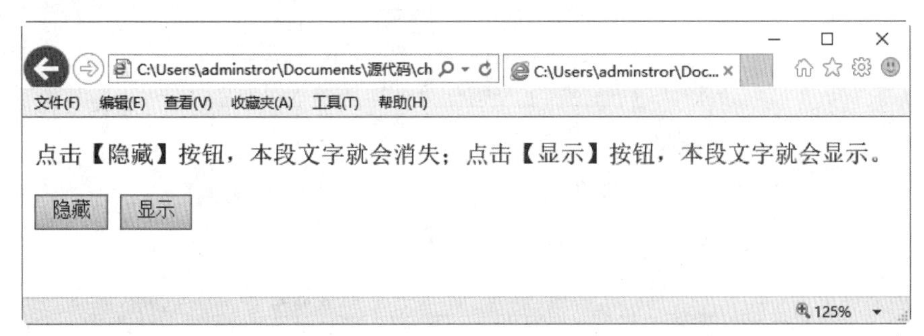

图 6-5 在 3000 毫秒内显示或隐藏网页中的元素

6.1.3 状态切换

使用 toggle()方法可以切换元素的可见(显示与隐藏)状态。简单地说，就是当元素为显示状态时，使用 toggle()方法可以将其隐藏起来；反之，可以将其显示出来。

toggle()方法的语法格式如下：

```
$(selector).toggle(speed,callback);
```

参数说明如下。

- speed：可选的参数，规定隐藏/显示的速度，可以取 slow、fast 或毫秒等参数。
- callback：可选的参数，是 toggle()方法完成后所执行的函数名称。

【例 6.6】(示例文件 ch06\6.6.html)

切换(隐藏/显示)网页中的元素：

```
<!DOCTYPE html>
<html>
<head>
<script src="jquery.min.js"></script>
<script type="text/javascript">
$(document).ready(function(){
$("button").click(function(){
$("p").toggle();
});
});
</script>
</head>
<body>
<button type="button">切换</button>
<p>清明时节雨纷纷，</p>
<p>路上行人欲断魂。</p>
</body>
</html>
```

程序运行结果如图 6-6 所示，单击页面中的【切换】按钮，可以实现网页文字段落的显示与隐藏的切换效果。

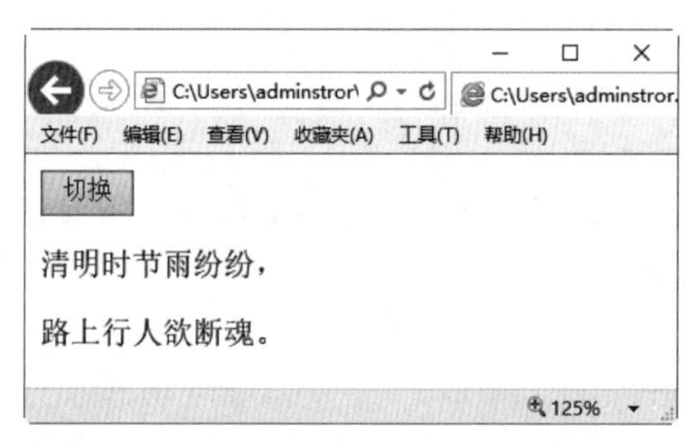

图 6-6　切换(隐藏/显示)网页中的元素

6.2　淡入淡出的动画效果

通过 jQuery 可以实现元素的淡入淡出动画效果。实现淡入淡出效果的方法主要有 fadeIn()、fadeOut()、fadeToggle()、fadeTo()。

6.2.1 淡入隐藏元素

fadeIn()是通过增大不透明度来实现匹配元素淡入效果的方法。该方法的语法格式如下：

```
$(selector).fadeIn(speed,callback);
```

参数说明如下。

- speed：可选的参数，规定淡入效果的时长，可以取 slow、fast 或毫秒等参数。
- callback：可选的参数，是 fadeIn()方法完成后所执行的函数名称。

【例 6.7】(示例文件 ch06\6.7.html)

以不同效果淡入网页中的矩形：

```
<!DOCTYPE html>
<html>
<head>
<script src="jquery.min.js"></script>
<script>
$(document).ready(function(){
$("button").click(function(){
$("#div1").fadeIn();
$("#div2").fadeIn("slow");
$("#div3").fadeIn(3000);
});
});
</script>
</head>
<body>
<p>以不同参数方式淡入网页元素</p>
<button>单击按钮，使矩形以不同的方式淡入</button><br><br>
<div id="div1"
  style="width:80px;height:80px;display:none;background-color:red;">
</div><br>
<div id="div2"
  style="width:80px;height:80px;display:none;background-color:green;">
</div><br>
<div id="div3"
  style="width:80px;height:80px;display:none;background-color:blue;">
</div>
</body>
</html>
```

程序运行结果如图 6-7 所示，单击页面中的按钮，网页中的矩形会以不同的方式淡入显示。

图 6-7　以不同效果淡入网页中的矩形

6.2.2　淡出可见元素

fadeOut()是通过减小不透明度来实现匹配元素淡出效果的方法。fadeOut()方法的语法格式如下：

```
$(selector).fadeOut(speed,callback);
```

参数说明如下。

- speed：可选的参数，规定淡出效果的时长，可以取 slow、fast 或毫秒等参数。
- callback：可选的参数，是 fadeOut()方法完成后所执行的函数名称。

【例 6.8】(示例文件 ch06\6.8.html)

以不同效果淡出网页中的矩形：

```
<!DOCTYPE html>
<html>
<head>
<script src="jquery.min.js"></script>
<script type="text/javascript">
$(document).ready(function(){
$("button").click(function(){
$("#div1").fadeOut();
$("#div2").fadeOut("slow");
$("#div3").fadeOut(3000);
});
});
</script>
</head>
<body>
<p>以不同参数方式淡出网页元素</p>
```

```
<button>单击按钮，使矩形以不同的方式淡出</button><br><br>
<div id="div1" style="width:80px;height:80px;background-color:red;"></div>
<br>
<div id="div2" style="width:80px;height:80px;background-color:green;">
</div><br>
<div id="div3" style="width:80px;height:80px;background-color:blue;"></div>
</body></html>
```

程序运行结果如图 6-8 所示，单击页面中的按钮，网页中的矩形就会以不同的方式淡出。

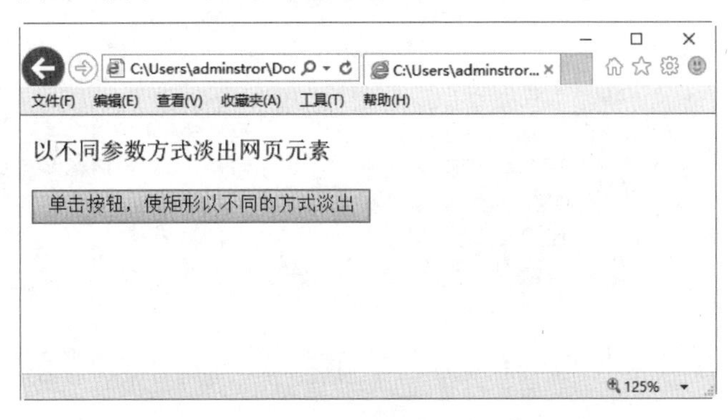

图 6-8 以不同效果淡出网页中的矩形

6.2.3 切换淡入淡出元素

fadeToggle()方法可以在 fadeIn()与 fadeOut()方法之间进行切换。也就是说，如果元素已淡出，则 fadeToggle()会向元素添加淡入效果；如果元素已淡入，则 fadeToggle()会向元素添加淡出效果。

fadeToggle()方法的语法格式如下：

```
$(selector).fadeToggle(speed,callback);
```

参数说明如下。

- speed：可选的参数，规定淡入淡出效果的时长，可以取 slow、fast 或毫秒等参数。
- callback：可选的参数，是 fadeToggle()方法完成后所执行的函数名称。

【例 6.9】(示例文件 ch06\6.9.html)

实现网页元素的淡入淡出效果：

```
<!DOCTYPE html>
<html>
<head>
<script src="jquery.min.js"></script>
<script>
$(document).ready(function(){
$("button").click(function(){
$("#div1").fadeToggle();
$("#div2").fadeToggle("slow");
$("#div3").fadeToggle(3000);
});
});
```

```
</script>
</head>

<body>
<p>以不同参数方式淡入淡出网页元素</p>
<button>单击按钮，使矩形以不同的方式淡入淡出</button>
<br><br>
<div id="div1" style="width:80px;height:80px;background-color:red;">
</div>
<br>
<div id="div2" style="width:80px;height:80px;background-color:green;">
</div>
<br>
<div id="div3" style="width:80px;height:80px;background-color:blue;">
</div>
</body>
</body>
</html>
```

程序运行结果如图 6-9 所示，单击页面中的按钮，网页中的矩形就会以不同的方式淡入淡出。

图 6-9　切换淡入淡出效果

6.2.4　淡入淡出元素至指定数值

使用 fadeTo()方法可以将网页元素淡入/淡出至指定不透明度，不透明度的值在 0~1 之间。fadeTo()方法的语法格式如下：

```
$(selector).fadeTo(speed,opacity,callback);
```

参数说明如下。

- speed：可选的参数，规定淡入淡出效果的时长，可以取 slow、fast 或毫秒等参数。
- opacity：必需的参数，参数将淡入淡出效果设置为给定的不透明度(0~1 之间)。
- callback：可选的参数，是该函数完成后所执行的函数名称。

【例 6.10】(示例文件 ch06\6.10.html)

实现网页元素的淡出至指定数值：

```
<!DOCTYPE html>
<html>
<head>
<script src="jquery.min.js"></script>
<script>
$(document).ready(function(){
$("button").click(function(){
$("#div1").fadeTo("slow",0.6);
$("#div2").fadeTo("slow",0.4);
$("#div3").fadeTo("slow",0.7);
});
});
</script>
</head>
<body>
<p>以不同参数方式淡出网页元素</p>
<button>单击按钮，使矩形以不同的方式淡出至指定参数</button>
<br><br>
<div id="div1" style="width:80px;height:80px;background-color:red;"></div>
<br>
<div id="div2" style="width:80px;height:80px;background-
color:green;"></div>
<br>
<div id="div3" style="width:80px;height:80px;background-color:blue;"></div>
</body>
</html>
```

程序运行结果如图 6-10 所示，单击页面中的按钮，网页中的矩形就会以不同的方式淡出至指定参数值。

图 6-10　淡出至指定数值

6.3 滑动效果

通过 jQuery，可以在元素上创建滑动效果。jQuery 中用于创建滑动效果的方法有 slideDown()、slideUp()、slideToggle()。

6.3.1 滑动显示匹配的元素

使用 slideDown()方法可以向下增加元素高度，动态显示匹配的元素。slideDown()方法会逐渐向下增加匹配的隐藏元素的高度，直到元素完全显示为止。

slideDown()方法的语法格式如下：

```
$(selector).slideDown(speed,callback);
```

参数说明如下。

- speed：可选的参数，规定效果的时长，可以取 slow、fast 或毫秒等参数。
- callback：可选的参数，是滑动完成后所执行的函数名称。

【例 6.11】(示例文件 ch06\6.11.html)

滑动显示网页元素：

```
<!DOCTYPE html>
<html>
<head>
<script src="jquery.min.js"></script>
<script type="text/javascript">
$(document).ready(function(){
$(".flip").click(function(){
$(".panel").slideDown("slow");
});
});
</script>

<style type="text/css">
div.panel,p.flip
{
margin: 0px;
padding: 5px;
text-align: center;
background: #e5eecc;
border: solid 1px #c3c3c3;
}
div.panel
{
height: 120px;
display: none;
}
</style>
```

```
</head>
<body>
<div class="panel">
<p>小荷才露尖尖角，</p>
<p>早有蜻蜓立上头。</p>
</div>
<p class="flip">请点击这里</p>
</body>
</html>
```

程序运行结果如图 6-11 所示，单击页面中的【请点击这里】文字，网页中隐藏的元素就会以滑动的方式显示出来。

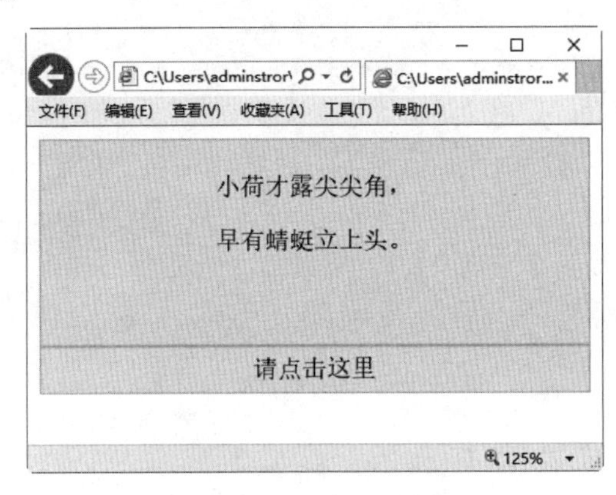

图 6-11 滑动显示网页元素

6.3.2 滑动隐藏匹配的元素

使用 slideUp()方法可以向上减少元素高度，动态隐藏匹配的元素。slideUp()方法会逐渐向上减少匹配的显示元素的高度，直到元素完全隐藏为止。slideUp()方法的语法格式如下：

```
$(selector).slideUp(speed,callback);
```

参数说明如下。

- speed：可选的参数，规定效果的时长，可以取 slow、fast 或毫秒等参数。
- callback：可选的参数，是滑动完成后所执行的函数名称。

【例 6.12】(示例文件 ch06\6.12.html)

滑动隐藏网页元素：

```
<!DOCTYPE html>
<html>
<head>
<script src="jquery.min.js"></script>
<script type="text/javascript">
$(document).ready(function(){
$(".flip").click(function(){
$(".panel").slideUp("slow");
});
```

```
});
</script>
<style type="text/css">
div.panel,p.flip
{
margin: 0px;
padding: 5px;
text-align: center;
background: #e5eecc;
border: solid 1px #c3c3c3;
}
div.panel
{
height: 120px;
}
</style>
</head>
<body>
<div class="panel">
<p>小荷才露尖尖角，</p>
<p>早有蜻蜓立上头。</p>
</div>
<p class="flip">请点击这里</p>
</body>
</html>
```

程序运行结果如图 6-12 所示，单击页面中的【请点击这里】文字，网页中显示的元素就会以滑动的方式隐藏起来。

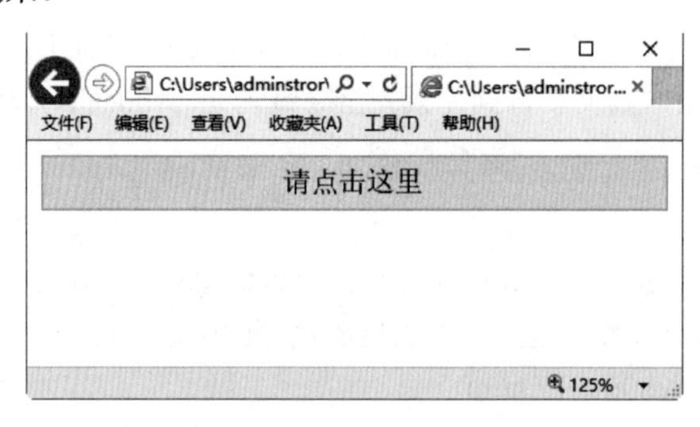

图 6-12　滑动隐藏网页元素

6.3.3　通过高度的变化动态切换元素的可见性

通过 slideToggle()方法可以实现通过高度的变化动态切换元素的可见性。也就是说，如果元素是可见的，就通过减少高度使元素全部隐藏；如果元素是隐藏的，就可以通过增加高度使元素最终全部可见。

slideToggle()方法的语法格式如下：

```
$(selector).slideToggle(speed,callback);
```

参数说明如下。

- speed：可选的参数，规定效果的时长，可以取 slow、fast 或毫秒等参数。
- callback：可选的参数，是滑动完成后所执行的函数名称。

【例 6.13】(示例文件 ch06\6.13.html)

通过高度的变化动态切换网页元素的可见性：

```
<!DOCTYPE html>
<html>
<head>
<script src="jquery.min.js"></script>
<script type="text/javascript">
$(document).ready(function(){
$(".flip").click(function(){
$(".panel"). slideToggle("slow");
});
});
</script>
<style type="text/css">
div.panel,p.flip
{
margin: 0px;
padding: 5px;
text-align: center;
background: #e5eecc;
border: solid 1px #c3c3c3;
}
div.panel
{
height: 120px;
display: none;
}
</style>
</head>
<body>
<div class="panel">
<p>小荷才露尖尖角，</p>
<p>早有蜻蜓立上头。</p>
</div>
<p class="flip">请点击这里</p>
</body>
</html>
```

程序运行结果如图 6-13 所示，单击页面中的【请点击这里】文字，网页中显示的元素就可以在显示与隐藏之间进行切换。

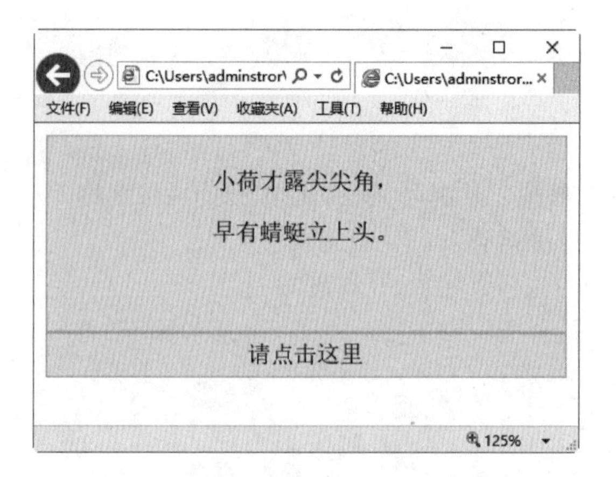

图 6-13 通过高度的变化动态切换网页元素的可见性

6.4 自定义的动画效果

有时程序预设的动画效果并不能满足用户的需求，这时就需要采取高级的自定义动画来解决这个问题。在 jQuery 中，要实现自定义动画效果，主要使用 animate()方法创建自定义动画，使用 stop()方法停止动画。

6.4.1 创建自定义动画

使用 animate()方法创建自定义动画的方法更加自由，可以随意控制元素的参数，实现更为绚丽的动画效果。animate()方法的基本语法格式如下：

```
$(selector).animate({params},speed,callback);
```

参数说明如下。

- params：必需的参数，定义形成动画的 CSS 属性。
- speed：可选的参数，规定效果的时长，可以取 slow、fast 或毫秒等参数。
- callback：可选的参数，是动画完成后所执行的函数名称。

在默认情况下，所有 HTML 元素都有一个静态位置，且无法移动。如果需要对位置进行操作，要记得首先把元素的 CSS position 属性设置为 relative、fixed 或 absolute。

【例 6.14】(示例文件 ch06\6.14.html)

创建自定义动画效果：

```
<!DOCTYPE html>
<html>
<head>
<script src="jquery.min.js"></script>
<script>
$(document).ready(function(){
```

```
$("button").click(function(){
var div = $("div");
div.animate({left:'100px'},"slow");
div.animate({fontSize:'3em'},"slow");
});
});
</script>
</head>
<body>
<button>开始动画</button>
<div
  style="background:#98bf21;height:100px;width:200px;position:absolute;">
  HELLO</div>
</body>
</html>
```

程序运行结果如图 6-14 所示，单击页面中的【开始动画】按钮，网页中显示的元素就会以设定的动画效果运行。

图 6-14　创建自定义动画效果

6.4.2　停止动画

stop()方法用于停止动画或效果。stop()方法适用于所有 jQuery 效果函数，包括滑动、淡入淡出和自定义动画。默认地，stop()会清除在被选元素上指定的当前动画。

stop()方法的语法格式如下：

```
$(selector).stop(stopAll,goToEnd);
```

- stopAll：可选的参数，规定是否应该清除动画队列。默认是 false，即仅停止活动的动画，允许任何排入队列的动画向后执行。
- goToEnd：可选的参数，规定是否立即完成当前动画。默认是 false。

【例 6.15】(示例文件 ch06\6.15.html)

停止动画效果：

```
<!DOCTYPE html>
<html>
```

```
<head>
<script src="jquery.min.js"></script>
<script>
$(document).ready(function(){
$("#flip").click(function(){
$("#panel").slideDown(5000);
});
$("#stop").click(function(){
$("#panel").stop();
});
});
</script>
<style type="text/css">
#panel,#flip
{
padding: 5px;
text-align: center;
background-color: #e5eecc;
border: solid 1px #c3c3c3;
}
#panel
{
padding: 50px;
display: none;
}
</style>
</head>
<body>
<button id="stop">停止滑动</button>
<div id="flip">点击这里，向下滑动面板</div>
<div id="panel">Hello jQuery!</div>
</body>
</html>
```

程序运行结果如图 6-15 所示，单击页面中的【点击这里，向下滑动面板】文字，下面的网页元素开始慢慢滑动以显示隐藏的元素。在滑动的过程中，如果想要停止滑动，可以单击【停止滑动】按钮，从而停止滑动。

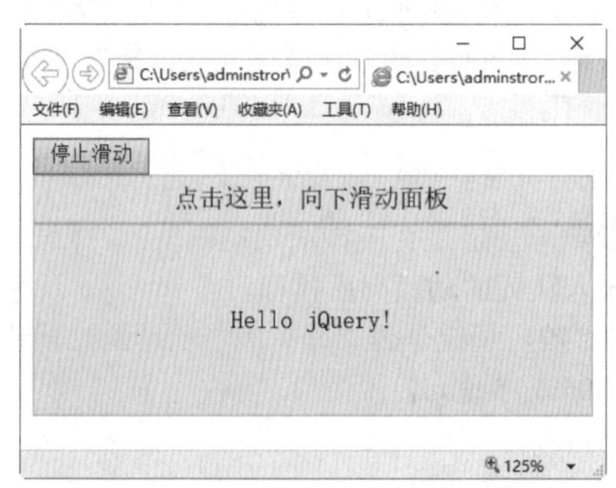

图 6-15　停止动画效果

6.5 疑难解惑

疑问 1：淡入淡出的工作原理是什么？

答：让元素在页面中不可见，常用的办法就是通过设置样式的 display:none。除此之外，还有一些类似的办法可以达到这个目的。设置元素透明度为 0，可以让元素不可见。透明度的参数是 0~1 之间的值，通过改变这个值可以让元素有一个透明度的效果。本章中讲述的淡入淡出动画 fadeIn()和 fadeOut()方法正是这样的原理。

疑问 2：通过 CSS 如何实现隐藏元素的效果？

答：hide()方法是隐藏元素最简单的方法。如果没有参数，匹配的元素将被立即隐藏，没有动画。这大致相当于调用.css('display', 'none')。其中 display 属性值保存在 jQuery 的数据缓存中，所以 display 可以方便以后恢复到其初始值。如果一个元素的 display 属性值为 inline，那么隐藏再显示时，这个元素将再次显示 inline。

第 7 章

jQuery 的事件处理

脚本语言有了事件就有了“灵魂”，可见事件对于脚本语言是多么重要。这是因为事件使页面具有了动态性和响应性。如果没有事件，将很难完成页面与用户之间的交互。本章主要介绍 jQuery 的事件处理。

7.1 jQuery 的事件机制概述

jQuery 有效地简化了 JavaScript 的编程。jQuery 的事件机制是事件方法会触发匹配元素的事件，或将函数绑定到所有匹配元素的某个事件。

7.1.1 什么是 jQuery 的事件机制

jQuery 的事件处理机制在 jQuery 框架中起着重要的作用。jQuery 的事件处理方法是 jQuery 中的核心函数。通过 jQuery 的事件处理机制，可以创造自定义的行为，比如说改变样式、效果显示、提交等，从而使网页效果更加丰富。

使用 jQuery 事件处理机制比直接使用 JavaScript 本身内置的一些事件响应方式更加灵活，且不容易暴露在外，并且有更加优雅的语法，大大减少了编写代码的工作量。

jQuery 的事件处理机制包括页面加载、事件绑定、事件委派、事件切换 4 种机制。

7.1.2 事件切换

事件切换是指在一个元素上绑定了两个以上的事件，在各个事件之间进行的切换动作。例如，当鼠标放在图片上时触发一个事件，当鼠标单击后又触发一个事件，可以用事件切换来实现。

在 jQuery 中，有两个方法用于事件的切换，一个方法是 hover()，另一个是 toggle()。

当需要设置在鼠标悬停和鼠标移出的事件中进行切换时，使用 hover()方法。下面的例子中，当鼠标悬停在文字上时，显示一段文字的效果。

【例 7.1】(示例文件 ch07\7.1.html)

事件切换：

```
<!DOCTYPE html>
<html>
<head>
<meta http-equiv="Content-Type" content="text/html; charset=gb2312" />
<title>hover()事件切换</title>
<script type="text/javascript" src="jquery.min.js"></script>
<script type="text/javascript">
$(document).ready(function(){
$(".clsContent").hide();
});
$(function(){
    $(".clsTitle").hover(function(){
        $(".clsContent").show();
    },
    function(){
        $(".clsContent").hide();
    })
})
```

```
</script>
</head>
<body>
<div class="clsTitle">石灰吟</div>
<div class="clsContent">千锤万凿出深山，烈火焚烧若等闲。粉身碎骨全不怕，要留清白在人间。</div>
</body>
</html>
```

在 IE 11.0 中浏览页面，效果如图 7-1 所示。将鼠标放在【石灰吟】文字上，最终结果如图 7-2 所示。

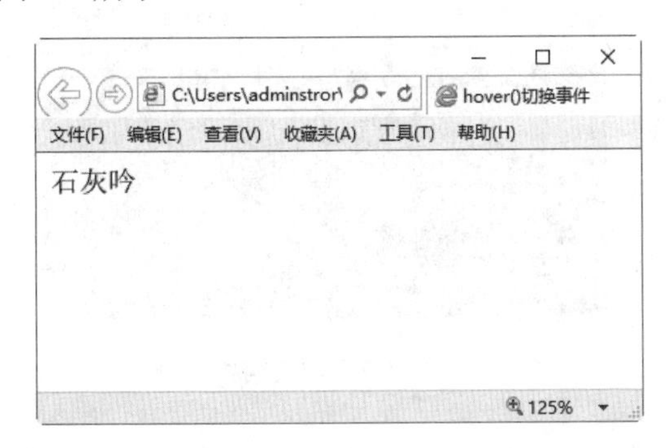

图 7-1　程序初始结果

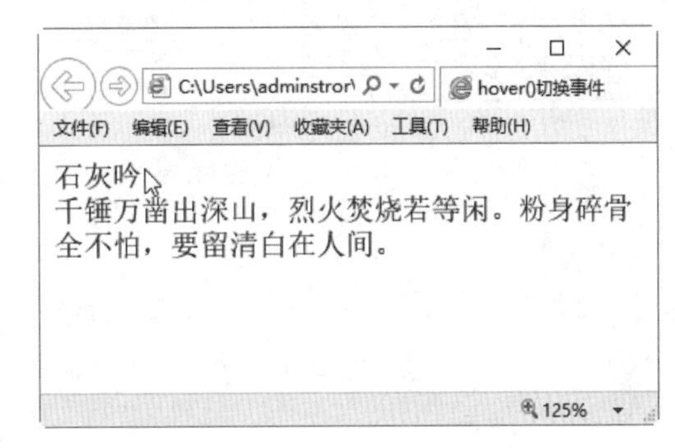

图 7-2　鼠标悬停后的结果

事件切换可以绑定两个或更多函数。当指定元素被点击时，在两个或多个函数之间轮流切换。

如果规定了两个以上的函数，则 toggle()方法将切换所有函数。例如，如果存在三个函数，则第一次点击将调用第一个函数，第二次点击调用第二个函数，第三次点击调用第三个函数，第四次点击再次调用第一个函数，以此类推。

【例 7.2】(示例文件 ch07\7.2.html)

在多个函数之间轮流切换：

```
<!DOCTYPE html>
<html>
<head>
<meta http-equiv="Content-Type" content="text/html; charset=gb2312" />
<title>toggle()切换事件</title>
<script type="text/javascript" src="jquery.min.js"></script>
<script type="text/javascript">
$(document).ready(function(){
$("button").toggle(function(){
$("body").css("background-color","red");},
function(){
$("body").css("background-color","yellow");},
function(){
$("body").css("background-color","green");}
);
});
</script>
```

```
</head>
<body>
<button>切换背景颜色</button>
</body>
</html>
```

在 IE 11.0 中浏览页面，效果如图 7-3 所示。单击【切换背景颜色】按钮，最终的结果如图 7-4 所示。通过不停地单击按钮，背景即可在指定的 3 个颜色之间转换。

图 7-3　程序初始结果

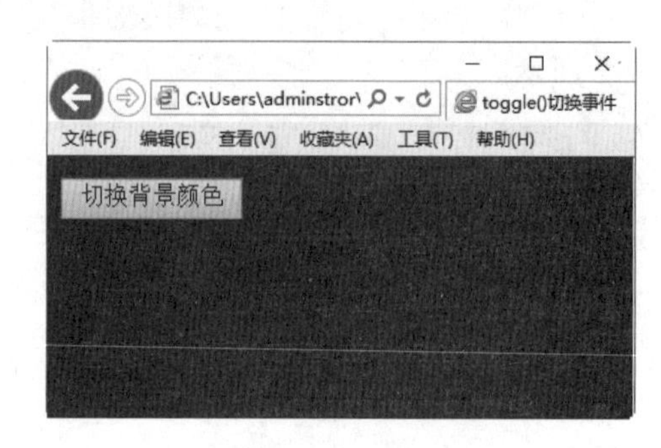

图 7-4　切换结果

7.1.3　事件冒泡

在一个对象上触发某类事件(如单击 onclick 事件)，如果此对象定义了此事件的处理程序，那么此事件就会调用这个处理程序，如果没有定义此事件处理程序或者事件返回 true，那么这个事件会向这个对象的父级对象传播，从里到外，直至它被处理(父级对象的所有同类事件都将被激活)，或者它到达了对象层次的最顶层，即 document 对象(有些浏览器是 window 对象)。

【例 7.3】(示例文件 ch07\7.3.html)

事件冒泡：

```
<!DOCTYPE html>
<html>
<head>
<meta http-equiv="Content-Type" content="text/html; charset=gb2312" />
<script type="text/javascript" src="jquery.min.js"></script>
<script type="text/javascript">
function add(Text){
   var Div = document.getElementById("display");
   Div.innerHTML += Text;  //输出点击顺序
}
</script>
</head>
<body onclick="add('第三层事件<br>');">
   <div onclick="add('第二层事件<br>');">
        <p onclick="add('第一层事件<br>');">事件冒泡</p>
```

```
    </div>
    <div id="display"></div>
</body>
</html>
```

在 IE 11.0 中浏览页面，效果如图 7-5 所示。单击【事件冒泡】文字，最终结果如图 7-6 所示。代码为 p、div、body 都添加了 onclick()函数，当单击 p 的文字时，触发事件，并且触发顺序是由最底层依次向上触发。

图 7-5 程序初始结果

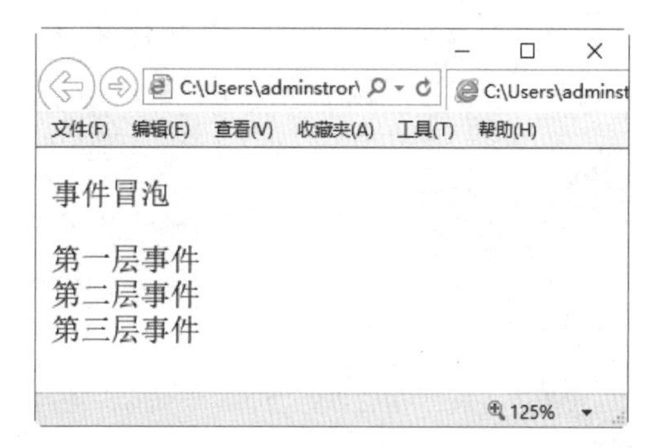

图 7-6 单击【事件冒泡】文字后

7.2 页面加载响应事件

jQuery 中的$(document).ready()事件是页面加载响应事件，ready()是 jQuery 事件模块中最重要的一个函数。这个方法可以看作是对 window.onload 注册事件的替代方法。通过使用这个方法，可以在 DOM 载入就绪时立刻调用所绑定的函数，而几乎所有的 JavaScript 函数都需要在那一刻执行。ready()函数仅能用于当前文档，因此无须选择器。

ready()函数的语法格式有如下 3 种。

- 语法 1：$(document).ready(function);
- 语法 2：$().ready(function);
- 语法 3：$(function);

其中参数 function 是必选项，规定当文档加载后要运行的函数。

【例 7.4】(示例文件 ch07\7.4.html)

使用 ready()函数：

```
<!DOCTYPE html>
<html>
<head>
<meta http-equiv="Content-Type" content="text/html; charset=gb2312" />
<script type="text/javascript" src="jquery.min.js"></script>
<script type="text/javascript">
$(document).ready(function(){
$(".btn1").click(function(){
$("p").slideToggle();
});
```

```
});
</script>
</head>
<body>
<p>此去经年，应是良辰好景虚设。便纵有千种风情，更与何人说？</p>
<button class="btn1">隐藏</button>
</body>
</html>
```

在 IE 11.0 中浏览页面，效果如图 7-7 所示。单击【隐藏】按钮，最终结果如图 7-8 所示。可见在文档加载后激活了函数。

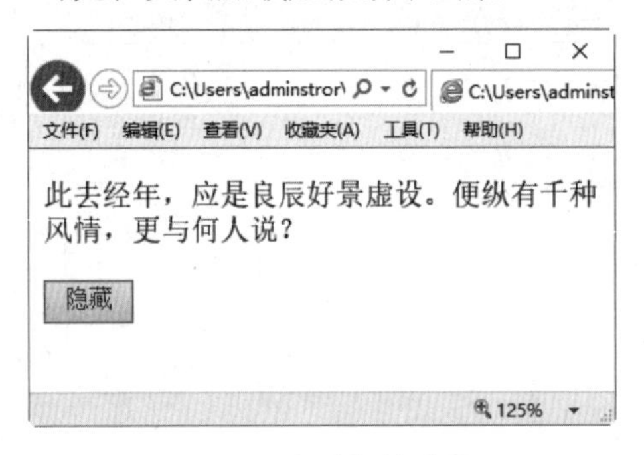

图 7-7　程序初始结果

图 7-8　单击按钮后的结果

7.3　jQuery 中的事件函数

在网站开发过程中，经常使用的事件函数包括键盘操作、鼠标操作、表单提交、焦点触发等事件。

7.3.1　键盘操作事件

日常开发中常见的键盘操作包括 keydown()、keypress()和 keyup()，如表 7-1 所示。

表 7-1　键盘操作事件

方　法	含　义
keydown()	触发或将函数绑定到指定元素的 key down 事件(按下键盘上某个按键时触发)
keypress()	触发或将函数绑定到指定元素的 key press 事件(按下某个按键并产生字符时触发)
keyup()	触发或将函数绑定到指定元素的 key up 事件(释放某个按键时触发)

完整的按键过程应该分为两步，按键被按下，然后按键被释放并复位。这里就触发了 keydown()和 keyup()事件函数。

下面通过例子来讲解 keydown()和 keyup()事件函数的使用方法。

【例 7.5】(示例文件 ch07\7.5.html)

使用 keydown()和 keyup()事件函数：

```
<!DOCTYPE html>
<html>
<head>
<meta http-equiv="Content-Type" content="text/html; charset=gb2312" />
<script type="text/javascript" src="jquery.min.js"></script>
<script type="text/javascript">
$(document).ready(function(){
$("input").keydown(function(){
$("input").css("background-color","yellow");
});
$("input").keyup(function(){
$("input").css("background-color","red");
});
});
</script>
</head>
<body>
Enter your name: <input type="text" />
<p>当发生 keydown 和 keyup 事件时，输入域会改变颜色。</p>
</body>
</html>
```

在 IE 11.0 中浏览页面，当按下按键时，输入域的背景色为黄色，效果如图 7-9 所示。当释放按键时，输入域的背景色为红色，效果如图 7-10 所示。

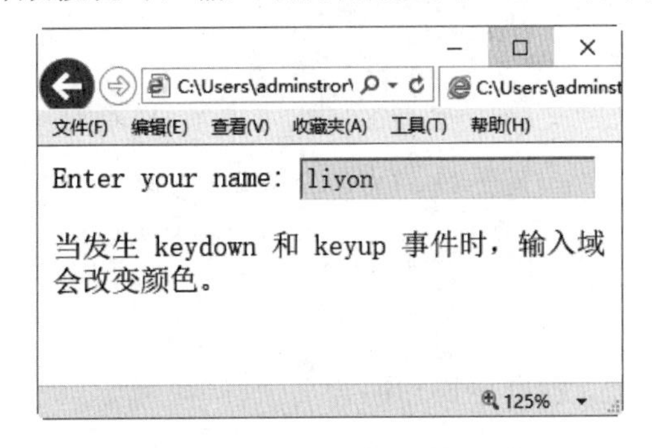

图 7-9 按下按键时输入域的背景色

图 7-10 释放按键时输入域的背景色

keypress 事件与 keydown 事件类似。当按键被按下时，会发生该事件。它发生在当前获得焦点的元素上。不过，与 keydown 事件不同，每插入一个字符，就会发生 keypress 事件。keypress()方法触发 keypress 事件，或规定当发生 keypress 事件时运行的函数。

下面通过例子来讲解 keypress()事件函数的使用方法。

【例 7.6】(示例文件 ch07\7.6.html)

使用 keypress()事件函数：

```
<!DOCTYPE html>
<html>
<head>
<meta http-equiv="Content-Type" content="text/html; charset=gb2312" />
<script type="text/javascript" src="jquery.min.js"></script>
```

```
<script type="text/javascript">
i = 0;
$(document).ready(function(){
$("input").keypress(function(){
$("span").text(i+=1);
});
});
</script>
</head>
<body>
Enter your name: <input type="text" />
<p>Keypresses:<span>0</span></p>
</body>
</html>
```

在 IE 11.0 中浏览页面，按下按键输入内容时，即可看到显示的按键次数，效果如图 7-11 所示。继续输入内容，则按下按键数发生相应的变化，效果如图 7-12 所示。

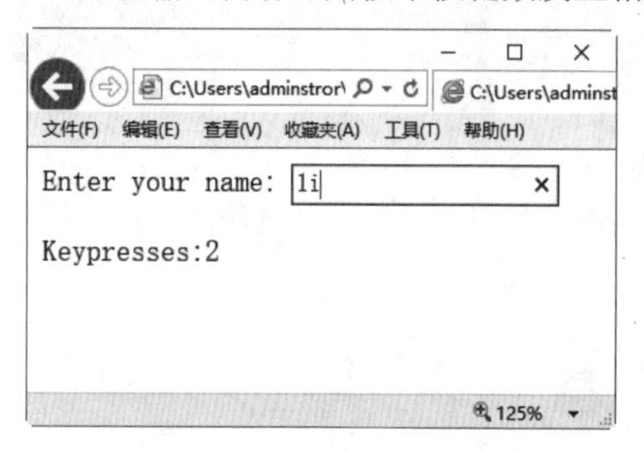

图 7-11 输入 2 个字母的效果

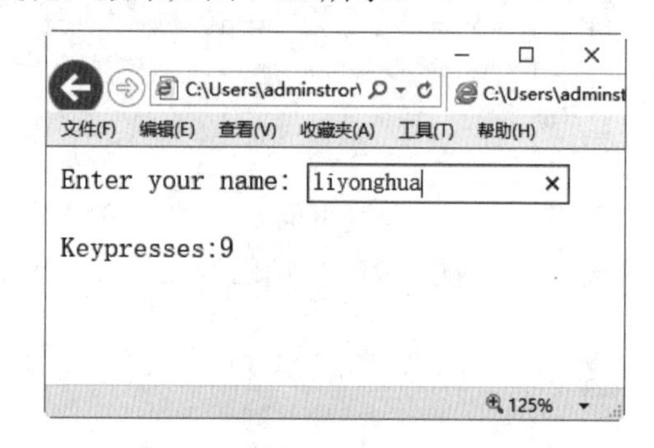

图 7-12 输入 9 个字母的效果

7.3.2 鼠标操作事件

与键盘操作事件相比，鼠标操作事件比较多，常见的鼠标操作的含义如表 7-2 所示。

表 7-2 鼠标操作事件

方 法	含 义
mousedown()	触发或将函数绑定到指定元素的 mouse down 事件(鼠标的按键被按下)
mouseenter()	触发或将函数绑定到指定元素的 mouse enter 事件(当鼠标指针进入或穿过目标时)
mouseleave()	触发或将函数绑定到指定元素的 mouse leave 事件(当鼠标指针离开目标时)
mousemove()	触发或将函数绑定到指定元素的 mouse move 事件(鼠标在目标的上方移动)
mouseout()	触发或将函数绑定到指定元素的 mouse out 事件(鼠标移出目标的上方)
mouseover()	触发或将函数绑定到指定元素的 mouse over 事件(鼠标移到目标的上方)
mouseup()	触发或将函数绑定到指定元素的 mouse up 事件(鼠标的按键被释放弹起)
click()	触发或将函数绑定到指定元素的 click 事件(单击鼠标的按键)
dblclick()	触发或将函数绑定到指定元素的 double click 事件(双击鼠标的按键)

下面通过使用 mousemove 事件函数实现鼠标定位的效果。

【例 7.7】(示例文件 ch07\7.7.html)

使用 mousemove 事件函数：

```
<!DOCTYPE html>
<html>
<head>
<meta http-equiv="Content-Type" content="text/html; charset=gb2312" />
<script type="text/javascript" src="jquery.min.js"></script>
<script type="text/javascript">
$(document).ready(function(){
$(document).mousemove(function(e){
$("span").text(e.pageX + ", " + e.pageY);
});
});
</script>
</head>
<body>
<p>鼠标位于坐标：<span></span>.</p>
</body>
</html>
```

在 IE 11.0 中浏览页面，效果如图 7-13 所示。可以看到，随着鼠标指针的移动，将显示鼠标指针的坐标。

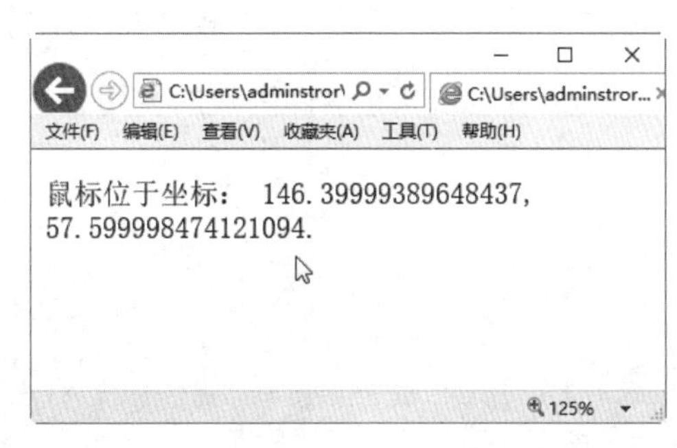

图 7-13　使用 mousemove 事件函数

下面通过例子来讲解鼠标 mouseover 和 mouseout 事件函数的使用方法。

【例 7.8】(示例文件 ch07\7.8.html)

使用 mouseover 和 mouseout 事件函数：

```
<!DOCTYPE html>
<html>
<head>
<meta http-equiv="Content-Type" content="text/html; charset=gb2312" />
<script type="text/javascript" src="jquery.min.js"></script>
<script type="text/javascript">
$(document).ready(function(){
$("p").mouseover(function(){
$("p").css("background-color","yellow");
});
```

```
$("p").mouseout(function(){
$("p").css("background-color","#E9E9E4");
});
});
</script>
</head>
<body>
<p style="background-color:#E9E9E4">请把鼠标指针移动到这个段落上。</p>
</body>
</html>
```

在 IE 11.0 中浏览页面，效果如图 7-14 所示。将鼠标放在段落上的效果如图 7-15 所示。该案例实现了当鼠标指针从元素上移入移出时改变元素的背景色。

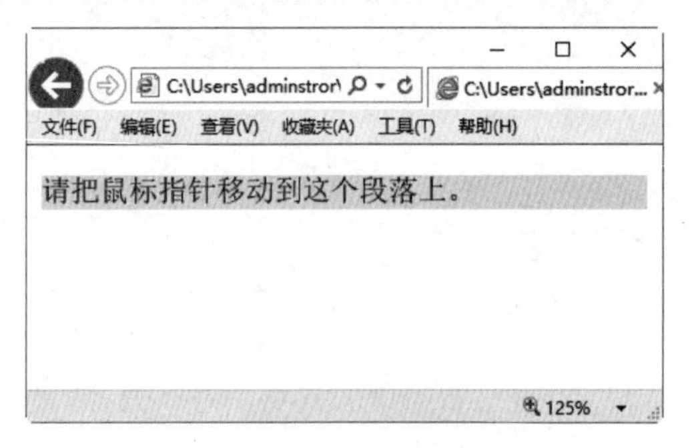

图 7-14　初始效果

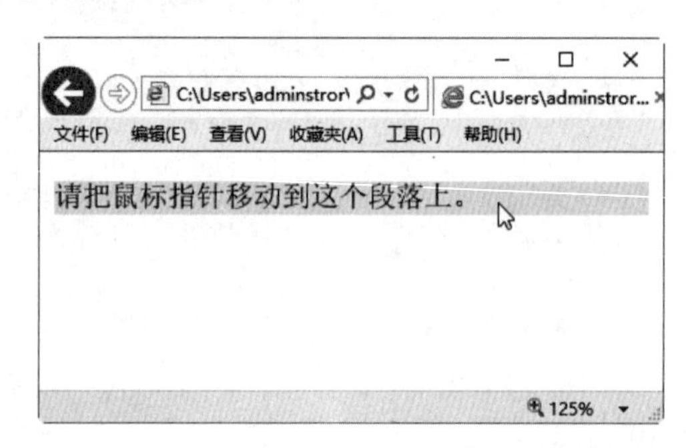

图 7-15　鼠标指针放在段落上的效果

下面通过例子来讲解鼠标 click 和 dblclick 事件函数的使用方法。

【例 7.9】(示例文件 ch07\7.9.html)

使用 click 和 dblclick 事件函数：

```
<!DOCTYPE html>
<html>
<head>
<meta http-equiv="Content-Type" content="text/html; charset=gb2312" />
<script type="text/javascript" src="jquery.min.js"></script>
<script type="text/javascript">
$(document).ready(function(){
$("#btn1").click(function(){
$("#id1").slideToggle();
});
$("#btn2").dblclick(function(){
$("#id2").slideToggle();
});
});
</script>
</head>
<body>
<div id="id1">墙角数枝梅，凌寒独自开。</div></p>
<button id="btn1">单击隐藏</button></p>
<div id="id2">遥知不是雪，为有暗香来。</div></p>
<button id="btn2">双击隐藏</button></p>
</body>
</html>
```

在 IE 11.0 中浏览页面，效果如图 7-16 所示。单击【单击隐藏】按钮，效果如图 7-17 所示。双击【双击隐藏】按钮，效果如图 7-18 所示。

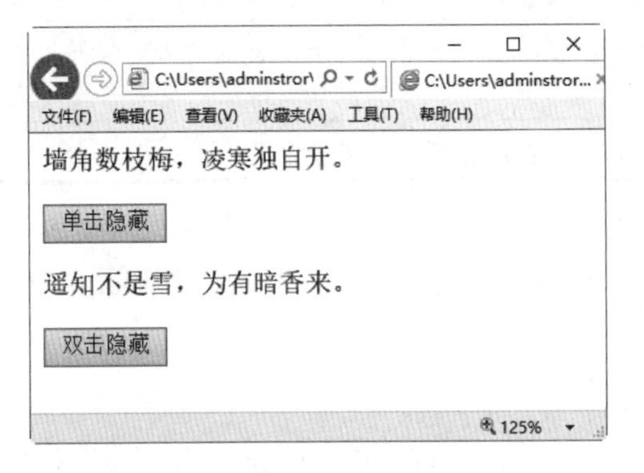

图 7-16　初始效果

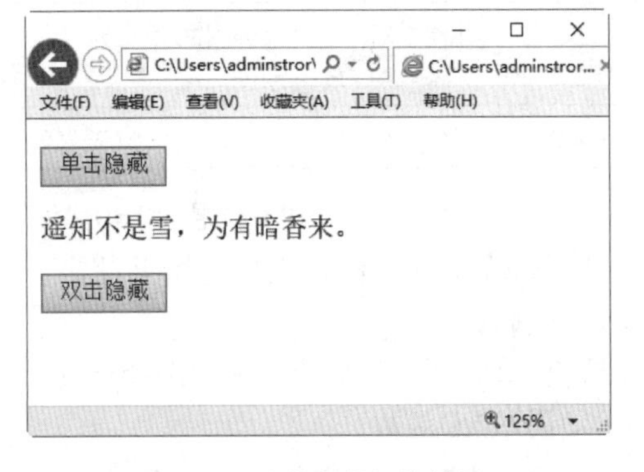

图 7-17　单击鼠标的效果

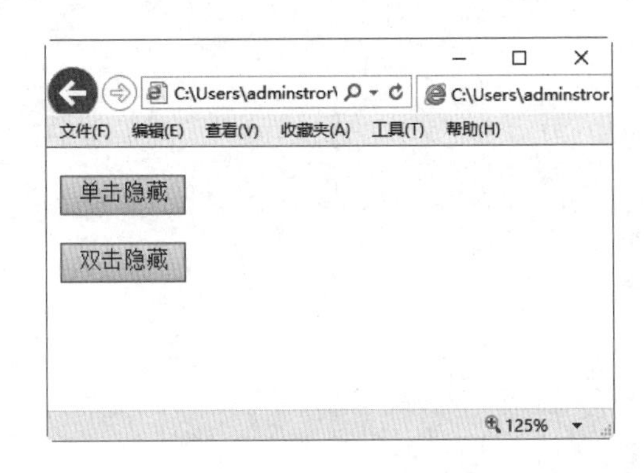

图 7-18　双击鼠标的效果

7.3.3　其他的常用事件

除了上面讲述的常用事件外，还有一些如表单提交、焦点触发等事件，如表 7-3 所示。

表 7-3　其他常用的事件

方　法	描　述
blur()	触发或将函数绑定到指定元素的 blur 事件(有元素或者窗口失去焦点时触发事件)
change()	触发或将函数绑定到指定元素的 change 事件(文本框内容改变时触发事件)
error()	触发或将函数绑定到指定元素的 error 事件(脚本或者图片加载错误、失败后触发事件)
resize()	触发或将函数绑定到指定元素的 resize 事件
scroll()	触发或将函数绑定到指定元素的 scroll 事件
focus()	触发或将函数绑定到指定元素的 focus 事件(有元素或者窗口获取焦点时触发事件)
select()	触发或将函数绑定到指定元素的 select 事件(文本框中的字符被选择之后触发事件)
submit()	触发或将函数绑定到指定元素的 submit 事件(表单“提交”之后触发事件)

续表

方　法	描　述
load()	触发或将函数绑定到指定元素的 load 事件(页面加载完成后在 window 上触发，图片加载完在自身触发)
unload()	触发或将函数绑定到指定元素的 unload 事件(与 load 相反，即卸载完成后触发)

下面挑选几个事件来讲解其使用方法。

blur()函数触发 blur 事件，如果设置了 function 参数，该函数也可规定当发生 blur 事件时执行的代码。

【例 7.10】(示例文件 ch07\7.10.html)

使用 blur()函数：

```
<!DOCTYPE html>
<html>
<head>
<meta http-equiv="Content-Type" content="text/html; charset=gb2312" />
<script type="text/javascript" src="jquery.min.js"></script>
<script type="text/javascript">
$(document).ready(function(){
$("input").focus(function(){
$("input").css("background-color","#FFFFCC");
});
$("input").blur(function(){
$("input").css("background-color","#D6D6FF");
});
});
</script>
</head>
<body>
Enter your name: <input type="text" />
<p>请在上面的输入域中点击，使其获得焦点，然后在输入域外面点击，使其失去焦点。</p>
</body>
</html>
```

在 IE 11.0 中浏览页面，在输入框中输入【洗衣机】文字，效果如图 7-19 所示。当鼠标单击文本框以外的空白处时，效果如图 7-20 所示。

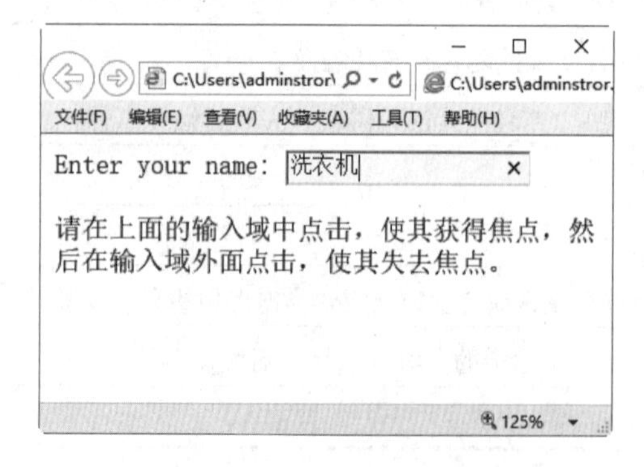

图 7-19　获得焦点后的效果

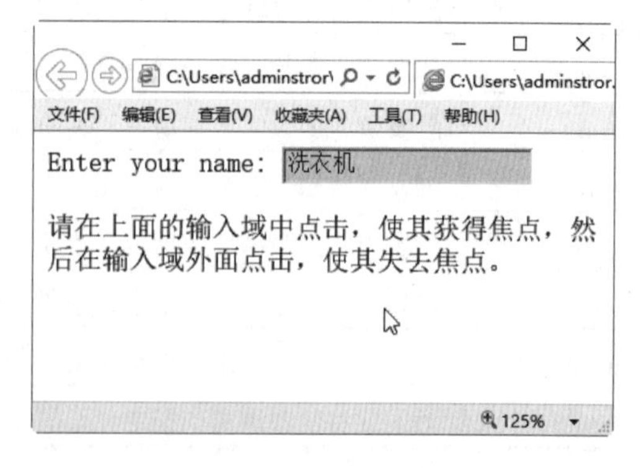

图 7-20　失去焦点后的效果

当元素的值发生改变时，可以使用 change 事件。该事件仅适用于文本域，以及 textarea 和 select 元素。change()函数触发 change 事件，或规定当发生 change 事件时运行的函数。

【例 7.11】(示例文件 ch07\7.11.html)

```
<!DOCTYPE html>
<html><head>
<meta http-equiv="Content-Type" content="text/html; charset=gb2312" />
<script type="text/javascript" src="jquery.min.js"></script>
<script type="text/javascript">
$(document).ready(function(){
$(".field").change(function(){
$(this).css("background-color","#FFFFCC");
});
});
</script>
</head><body>
<p>在某个域被使用或改变时，它会改变颜色。</p>
输入客户姓名：<input class="field" type="text" />
<p>汽车品牌：
<select class="field" name="cars">
<option value="volvo">Volvo</option>
<option value="saab">Saab</option>
<option value="fiat">Fiat</option>
<option value="audi">Audi</option>
</select></p>
</body></html>
```

在 IE 11.0 中浏览页面，效果如图 7-21 所示。输入客户的名称和选择汽车品牌后，即可看到文本框的底纹发生了变化，效果如图 7-22 所示。

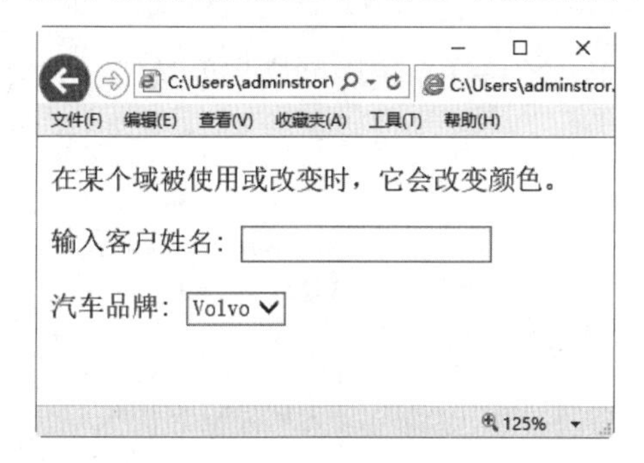

图 7-21 初始效果

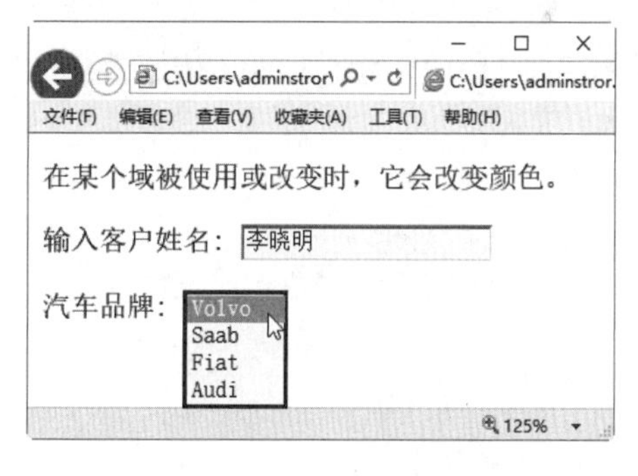

图 7-22 修改元素值后的效果

7.4 事件的基本操作

7.4.1 绑定事件

在 jQuery 中，可以用 bind()函数给 DOM 对象绑定一个事件。bind()函数为被选元素添加一个或多个事件处理程序，并规定事件发生时运行的函数。

规定向被选元素添加的一个或多个事件处理程序，以及当事件发生时运行的函数时，使用的语法格式如下：

```
$(selector).bind(event,data,function)
```

其中 event 为必需，规定添加到元素的一个或多个事件，由空格分隔多个事件，必须是有效的事件。data 为可选，规定传递到函数的额外数据。function 为必需，规定当事件发生时运行的函数。

【例 7.12】(示例文件 ch07\7.12.html)

用 bind()函数绑定事件：

```
<!DOCTYPE html>
<html>
<head>
<meta http-equiv="Content-Type" content="text/html; charset=gb2312" />
<script type="text/javascript" src="jquery.min.js"></script>
<script type="text/javascript">
$(document).ready(function(){
$("button").bind("click",function(){
$("p").slideToggle();
});
});
</script>
</head>
<body>
<p>寒雨连江夜入吴，平明送客楚山孤。洛阳亲友如相问，一片冰心在玉壶。</p>
<button>单击隐藏文字</button>
</body>
</html>
```

在 IE 11.0 中浏览页面，初始效果如图 7-23 所示。单击【单击隐藏文字】按钮，效果如图 7-24 所示。

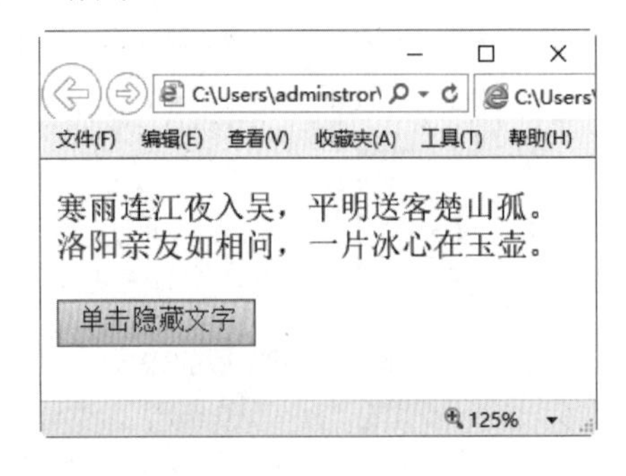

图 7-23　初始效果

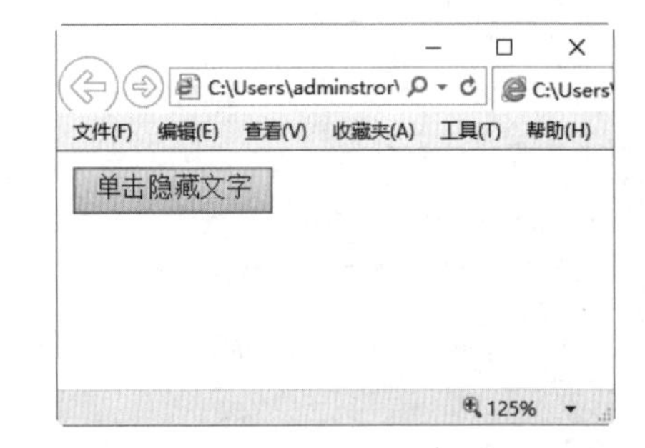

图 7-24　单击按钮后的效果

7.4.2　触发事件

事件绑定后，可用 trigger()方法进行触发操作。trigger()方法规定被选元素要触发的事件。trigger()函数的语法格式如下：

```
$(selector).trigger(event,[param1,param2,...])
```

其中 event 为触发事件的动作，如 click、dblclick。

【例 7.13】(示例文件 ch07\7.13.html)

使用 trigger()函数来触发事件：

```
<!DOCTYPE html>
<html>
<head>
<meta http-equiv="Content-Type" content="text/html; charset=gb2312" />
<script type="text/javascript" src="jquery.min.js"></script>
<script type="text/javascript">
$(document).ready(function(){
$("input").select(function(){
$("input").after("文本被选中！");
});
$("button").click(function(){
$("input").trigger("select");
});
});
</script>
</head>
<body>
<input type="text" name="FirstName" value="春花秋月何时了" />
<br />
<button>激活事件</button>
</body>
</html>
```

在 IE 11.0 中浏览页面，效果如图 7-25 所示。选择文本框中的文字或者单击【激活事件】按钮，效果如图 7-26 所示。

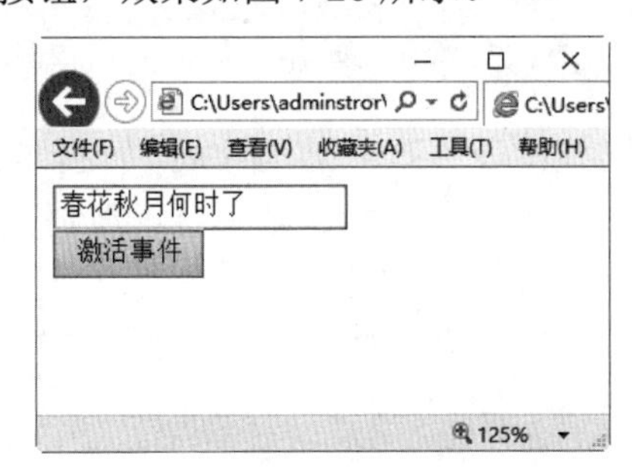

图 7-25 初始效果

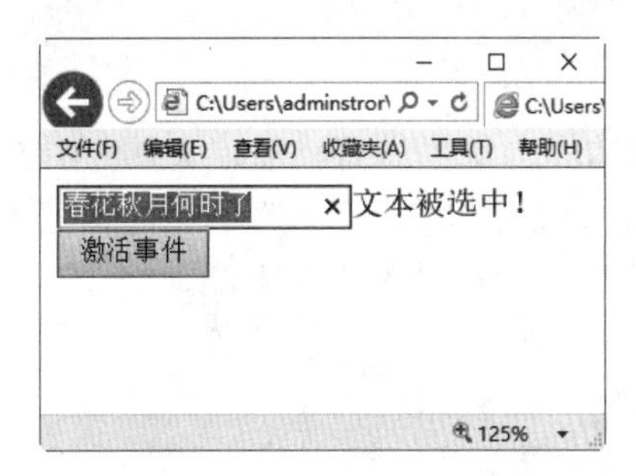

图 7-26 激活事件后的效果

7.4.3 移除事件

unbind()方法移除被选元素的事件处理程序。该方法能够移除所有的或被选的事件处理程序，或者当事件发生时终止指定函数的运行。unbind()适用于任何通过 jQuery 附加的事件处理程序。

unbind()方法使用的语法格式如下：

```
$(selector).unbind(event,function)
```

其中 event 是可选参数。规定删除元素的一个或多个事件，由空格分隔多个事件值。function 是可选参数，规定从元素的指定事件取消绑定的函数名。如果没有规定参数，unbind()方法会删除指定元素的所有事件处理程序。

【**例 7.14**】(示例文件 ch07\7.14.html)

使用 unbind()方法：

```
<!DOCTYPE html>
<html>
<head>
<meta http-equiv="Content-Type" content="text/html; charset=gb2312" />
<script type="text/javascript" src="jquery.min.js"></script>
<script type="text/javascript">
$(document).ready(function(){
$("p").click(function(){
$(this).slideToggle();
});
$("button").click(function(){
$("p").unbind();
});
});
</script>
</head>
<body>
<p>这是一个段落。</p>
<p>这是另一个段落。</p>
<p>点击任何段落可以令其消失。包括本段落。</p>
<button>删除 p 元素的事件处理器</button>
</body>
</html>
```

在 IE 11.0 中浏览页面，效果如图 7-27 所示。单击任意段落即可让其消失，如图 7-28 所示。单击【删除 p 元素的事件处理器】按钮后，再次单击任意段落，则不会出现消失的效果。可见此时已经移除了事件。

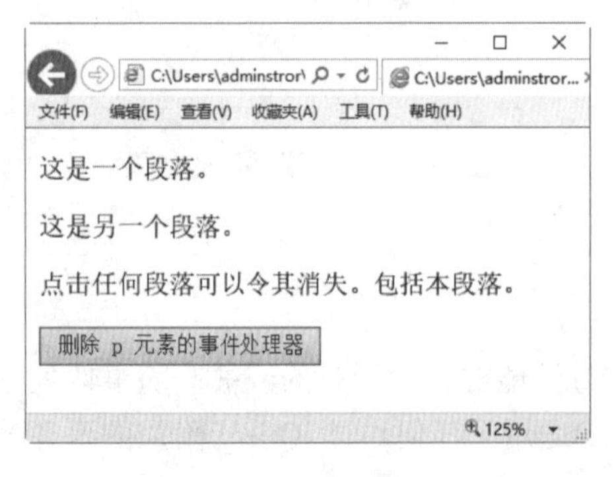

图 7-27　初始效果

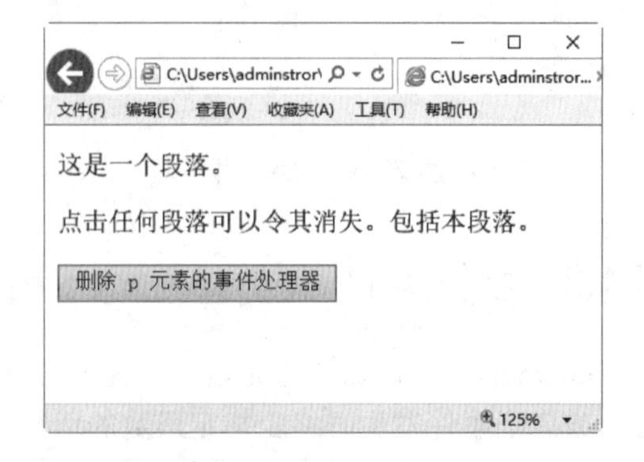

图 7-28　激活事件后的效果

7.5 实战演练——制作绚丽的多级动画菜单

本节主要制作绚丽的多级动画菜单效果。鼠标经过菜单区域时动画式展开大幅的下拉菜单，具有动态效果，显得更加生动活泼。具体操作步骤如下。

step 01 设计基本的网页框架。代码如下：

```
<!DOCTYPE html>
<html>
<head>
<meta http-equiv="Content-Type" content="text/html; charset=gb2312" />
</head>
<body>
<div class="box">
<ul id="veryhuo_menu" class="veryhuo_menu">
<li>
<span>淘宝特色服务</span><!-- Increases to 510px in width-->
<div class="ldd_submenu">
<ul>
<li class="ldd_heading">主题市场</li>
<li><a href="#">运动派</a></li>
<li><a href="#">情侣</a></li>
<li><a href="#">家具</a></li>
<li><a href="#">美食</a></li>
<li><a href="#">有车族</a></li>
</ul>
<ul>
<li class="ldd_heading">特色购物</li>
<li><a href="#">全球购</a></li>
<li><a href="#">淘女郎</a></li>
<li><a href="#">挑食</a></li>
<li><a href="#">搭配</a></li>
<li><a href="#">同城便民</a></li>
<li><a href="#">淘宝同学</a></li>
</ul>
<ul>
<li class="ldd_heading">优惠促销</li>
<li><a href="#">天天特价</a></li>
<li><a href="#">免费试用</a></li>
<li><a href="#">清仓</a></li>
<li><a href="#">一元起拍</a></li>
<li><a href="#">淘金币</a></li>
<li><a href="#t">聚划算</a></li>
</ul>
</div>
</body>
</html>
```

step 02 运行上述代码，效果如图 7-29 所示。

图 7-29 程序运行效果

step 03 为各级菜单添加 CSS 样式风格。代码如下：

```
<style>
*{
padding:0;
margin:0;
}
body{
background:#f0f0f0;
font-family:"Helvetica Neue",Arial,Helvetica,Geneva,sans-serif;
overflow-x:hidden;
}
span.reference{
position:fixed;
left:10px;
bottom:10px;
font-size:11px;
}
span.reference a{
color:#DF7B61;
text-decoration:none;
text-transform:uppercase;
text-shadow:0 1px 0 #fff;
}
span.reference a:hover{
color:#000;
}
.box{
margin-top:129px;
```

```
height:460px;
width:100%;
position:relative;
background:#fff url(/uploads/allimg/1202/veryhuo_click.png) no-repeat 380px
180px;
-moz-box-shadow:0px 0px 10px #aaa;
-webkit-box-shadow:0px 0px 10px #aaa;
-box-shadow:0px 0px 10px #aaa;
}
.box h2{
color:#f0f0f0;
padding:40px 10px;
text-shadow:1px 1px 1px #ccc;
}
ul.veryhuo_menu{
margin:0px;
padding:0;
display:block;
height:50px;
background-color:#D04528;
list-style:none;
font-family:"Trebuchet MS", sans-serif;
border-top:1px solid #EF593B;
border-bottom:1px solid #EF593B;
border-left:10px solid #D04528;
-moz-box-shadow:0px 3px 4px #591E12;
-webkit-box-shadow:0px 3px 4px #591E12;
-box-shadow:0px 3px 4px #591E12;
}
ul.veryhuo_menu a{
text-decoration:none;
}
ul.veryhuo_menu > li{
float:left;
position:relative;
}
ul.veryhuo_menu > li > span{
float:left;
color:#fff;
background-color:#D04528;
height:50px;
line-height:50px;
cursor:default;
padding:0px 20px;
text-shadow:0px 0px 1px #fff;
border-right:1px solid #DF7B61;
border-left:1px solid #C44D37;
}
ul.veryhuo_menu .ldd_submenu{
position:absolute;
```

```
top:50px;
width:550px;
display:none;
opacity:0.95;
left:0px;
font-size:10px;
background: #C34328;
border-top:1px solid #EF593B;
-moz-box-shadow:0px 3px 4px #591E12 inset;
-webkit-box-shadow:0px 3px 4px #591E12 inset;
-box-shadow:0px 3px 4px #591E12 inset;
}
a.ldd_subfoot{
background-color:#f0f0f0;
color:#444;
display:block;
clear:both;
padding:15px 20px;
text-transform:uppercase;
font-family: Arial, serif;
font-size:12px;
text-shadow:0px 0px 1px #fff;
-moz-box-shadow:0px 0px 2px #777 inset;
-webkit-box-shadow:0px 0px 2px #777 inset;
-box-shadow:0px 0px 2px #777 inset;
}
ul.veryhuo_menu ul{
list-style:none;
float:left;
border-left:1px solid #DF7B61;
margin:20px 0px 10px 30px;
padding:10px;
}
li.ldd_heading{
font-family: Georgia, serif;
font-size: 13px;
font-style: italic;
color:#FFB39F;
text-shadow:0px 0px 1px #B03E23;
padding:0px 0px 10px 0px;
}
ul.veryhuo_menu ul li a{
font-family: Arial, serif;
font-size:10px;
line-height:20px;
color:#fff;
padding:1px 3px;
}
ul.veryhuo_menu ul li a:hover{
-moz-box-shadow:0px 0px 2px #333;
```

```
-webkit-box-shadow:0px 0px 2px #333;
box-shadow:0px 0px 2px #333;
background:#AF412B;
}
</style>
```

step 04 添加实现多级动态菜单的代码，确保子菜单随着需求隐藏或者显现：

```
<!-- The JavaScript -->
<script type="text/javascript" src="jquery.min.js"></script>
<script type="text/javascript">
$(function() {
var $menu = $('#veryhuo_menu');
$menu.children('li').each(function(){
var $this = $(this);
var $span = $this.children('span');
$span.data('width',$span.width());
$this.bind('mouseenter',function(){
$menu.find('.ldd_submenu').stop(true,true).hide();
$span.stop().animate({'width':'510px'},300,function(){
$this.find('.ldd_submenu').slideDown(300);
});
}).bind('mouseleave',function(){
$this.find('.ldd_submenu').stop(true,true).hide();
$span.stop().animate({'width':$span.data('width')+'px'},300);
});
});
});
</script>
```

step 05 运行最终的案例代码，效果如图 7-30 所示。

图 7-30　程序运行初始效果

step 06 将鼠标放在【淘宝特色服务】链接文字上，动态显示多级菜单，效果如图 7-31 所示。

图 7-31　展开菜单的效果

7.6 疑难解惑

疑问 1：如何屏蔽鼠标的右键？

答：有些网站为了提高网页的安全性，屏蔽了鼠标右键。使用鼠标事件函数即可轻松地实现此功能。具体的功能代码如下：

```
<script language="javascript">
function block(Event){
    if(window.event)
        Event = window.event;
    if(Event.button == 2)
        alert("右键被屏蔽");
}
document.onmousedown = block;
</script>
```

疑问 2：mouseover 和 mouseenter 的区别是什么？

答：jQuery 中，mouseover 和 mouseenter 都在鼠标进入元素时触发，但是它们有所不同。具体说明如下。

(1) 如果元素内置有子元素，不论鼠标指针穿过被选元素还是其子元素，都会触发 mouseover()事件。而只有在鼠标指针穿过被选元素时，才会触发 mouseenter()事件，mouseenter 子元素不会反复触发事件，否则在 IE 中经常有闪烁情况发生。

(2) 在没有子元素时，mouseover()和 mouseenter()事件结果一致。

第 8 章 jQuery 的功能函数

jQuery 提供了很多功能函数。通过使用功能函数，用户可以轻松地实现需要的功能。本章主要讲述功能函数的基本概念、常用功能函数的使用方法、调用外部代码的方法等。

8.1 功能函数概述

jQuery 将常用功能的函数进行了总结和封装，这样用户在使用时，直接调用即可，不仅方便了开发者的使用，而且大大提高了开发者的效率。jQuery 提供的这些实现常用功能的函数，被称作功能函数。

例如，开发人员经常需要对数组和对象进行操作，jQuery 就提供了对元素进行遍历、筛选、合并等操作的函数。下面通过一个例子来理解。

【例 8.1】(示例文件 ch08\8.1.html)

对数组和对象进行操作：

```
<!DOCTYPE html>
<html>
<head>
<meta http-equiv="Content-Type" content="text/html; charset=gb2312" />
<title>合并数组 </title>
<script type="text/javascript" src="jquery.min.js"></script>
<script type="text/javascript">
$(function(){
    var first = ['A','B','C','D'];
    var second = ['E','F','G','H'];
    $("p:eq(0)").text("数组 a: " + first.join());
    $("p:eq(1)").text("数组 b: " + second.join());
    $("p:eq(2)").text("合并数组: "
      + ($.merge($.merge([],first), second)).join());
});
</script>
</head>
<body>
<p></p><p></p><p></p>
</body>
<html>
```

在 IE 11.0 中浏览，效果如图 8-1 所示。

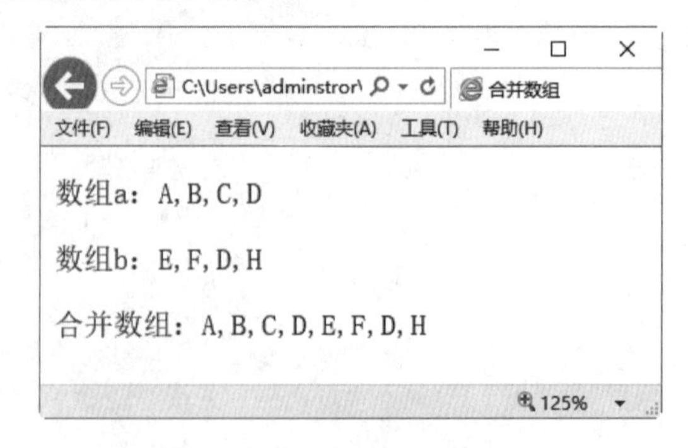

图 8-1 对数组和对象进行操作

8.2 常用的功能函数

了解功能函数的概念后，下面讲述常用功能函数的使用方法。

8.2.1 操作数组和对象

对于数组和对象的操作，主要包括元素的遍历、筛选和合并等。

(1) jQuery 提供的 each()方法用于为每个匹配元素规定运行的函数。可以使用 each()方法来遍历数组和对象。其语法格式如下：

```
$.each(object,fn);
```

其中，object 是需要遍历的对象。fn 是一个函数，这个函数是所遍历的对象都需要执行的，它可以接收两个参数：一个是数组对象的属性或者元素的序号；另一个是属性或者元素的值。这里需要注意的是：jQuery 还提供了$.each()方法，可以获取一些不熟悉对象的属性值。例如，不清楚一个对象包含什么属性，就可以使用$.each()方法进行遍历。

【例 8.2】(示例文件 ch08\8.2.html)

使用 each()方法：

```
<!DOCTYPE html>
<html>
<head>
<meta http-equiv="Content-Type" content="text/html; charset=gb2312" />
<title>each()方法</title>
<script type="text/javascript" src="jquery.min.js"></script>
<script type="text/javascript">
$(document).ready(function(){
$("button").click(function(){
$("li").each(function(){
alert($(this).text())
});
});
});
</script>
</head>
<body>
<button>输出每个列表项的值</button>
<ul>
<li>野径云俱黑</li>
<li>江船火独明</li>
<li>晓看红湿处</li>
<li>花重锦官城</li>
</ul>
</body>
</html>
```

在 IE 11.0 中浏览，单击【输出每个列表项的值】按钮，弹出每个列表中的值，依次单击

【确定】按钮，即可显示每个列表项的值，效果如图 8-2 所示。

图 8-2　显示每个列表项的值

(2)　jQuery 提供的 grep()方法用于数组元素过滤筛选。其语法格式如下：

```
grep(array,fn,invert)
```

其中，array 指待过滤数组；fn 是过滤函数，对于数组中的对象，如果返回值是 true，就保留，返回值是 false 就去除；invert 是可选项，当设置为 true 时 fn 函数取反，即满足条件的被剔除出去。

【例 8.3】(示例文件 ch08\8.3.html)

使用 grep()方法：

```
<!DOCTYPE html>
<html>
<head>
<meta http-equiv="Content-Type" content="text/html; charset=gb2312" />
<script type="text/javascript" src="jquery.min.js"></script>
<script type="text/javascript">
var Array = [1,2,3,4,5,6,7];
var Result = $.grep(Array,function(value){
    return (value > 2);
});
document.write("原数组: " + Array.join() + "<br>");
document.write("筛选大于 2 的结果为: " + Result.join());
</script>
</head>
<body>
</body>
</html>
```

在 IE 11.0 中浏览页面，效果如图 8-3 所示。

(3) jQuery 提供的 map()方法用于把每个元素通过函数传递到当前匹配集合中，生成包含返回值的新的 jQuery 对象。通过使用 map()方法，可以统一转换数组中的每一个元素值。其语法格式如下：

```
$.map(array,fn)
```

其中，array 是需要转化的目标数组。fn 显然就是转化函数，这个 fn 的作用就是对数组中的每一项都执行转化函数，它接收两个可选参数：一个是元素的值；另一个是元素的序号。

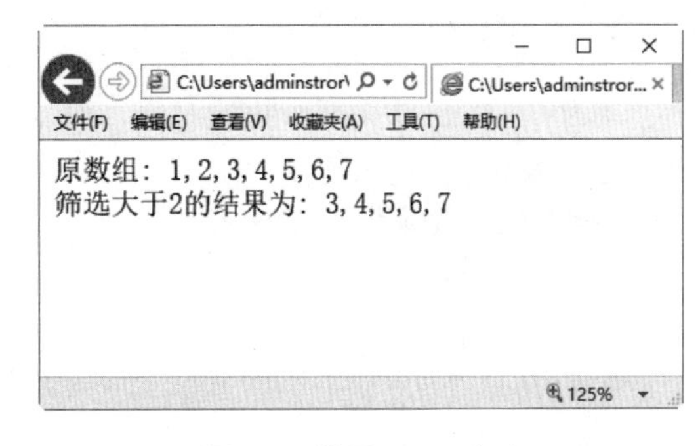

图 8-3　使用 grep()方法

【例 8.4】(示例文件 ch08\8.4.html)

使用 map()方法：

```
<!DOCTYPE html>
<html>
<head>
<meta http-equiv="Content-Type" content="text/html; charset=gb2312" />
<script type="text/javascript" src="jquery.min.js"></script>
<script type="text/javascript">
$(function(){
   var arr1 = ["apple", "apricot", "chestnut", "pear ","banana"];
   arr2 = $.map(arr1,function(value,index){
       return (value.toUpperCase());
   });
   $("p:eq(0)").text("原数组值: " + arr1.join());
   $("p:eq(1)").text("统一转化大写: " + arr2.join());
});
</script>
</head>
<body>
</body>
</html>
```

在 IE 11.0 中浏览，效果如图 8-4 所示。

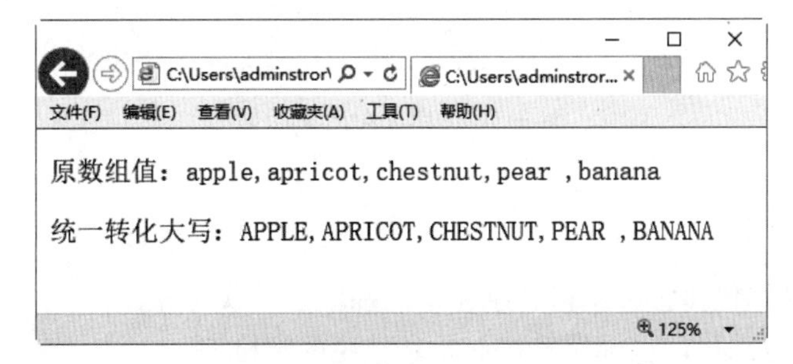

图 8-4　使用 map()方法

(4) jQuery 提供的$.inArray()函数很好地实现了数组元素的搜索功能。其语法格式如下：

```
$.inArray(value,array)
```

其中，value 是需要查找的对象，而 array 是数组本身，如果找到目标元素，就返回第一个元素所在位置，否则返回-1。

【例 8.5】(示例文件 ch08\8.5.html)

使用 inArray()函数：

```
<!DOCTYPE html>
<html>
<head>
<meta http-equiv="Content-Type" content="text/html; charset=gb2312" />
<script type="text/javascript" src="jquery.min.js">
</script>
<script type="text/javascript">
$(function(){
    var arr = ["This", "is", "an", "apple"];
    var add1 = $.inArray("apple",arr);
    var add2 = $.inArray("are",arr);
    $("p:eq(0)").text("数组：" + arr.join());
    $("p:eq(1)").text("“apple”的位置：" + add1);
    $("p:eq(2)").text("“are”的位置：" + add2);
});
</script>
</head>
<body></body>
</html>
```

在 IE 11.0 中浏览，效果如图 8-5 所示。

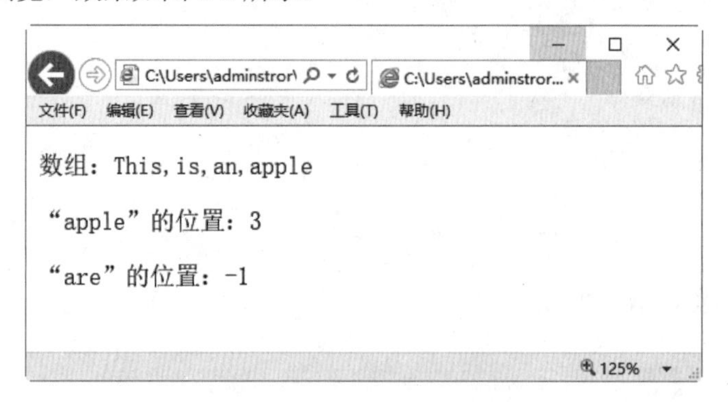

图 8-5　使用 inArray()函数

8.2.2　操作字符串

常用的字符串操作包括去除空格、字符串的抽取、替换等操作。

(1)　使用 trim()方法可以去掉字符串起始和结尾的空格。

【例 8.6】(示例文件 ch08\8.6.html)

使用 trim()方法：

```
<!DOCTYPE html>
<html>
<head>
<meta http-equiv="Content-Type" content="text/html; charset=gb2312" />
<script type="text/javascript" src="jquery.min.js"></script>
</head>
<body>
<pre id="original"></pre>
<pre id="trimmed"></pre>
<script>
  var str = "          此生此夜不长好，明月明年何处看        ";
  $("#original").html("原始字符串：/" + str + "/");
  $("#trimmed").html("去掉首尾空格：/" + $.trim(str) + "/");
</script>
</body>
</html>
```

在 IE 11.0 中浏览，效果如图 8-6 所示。

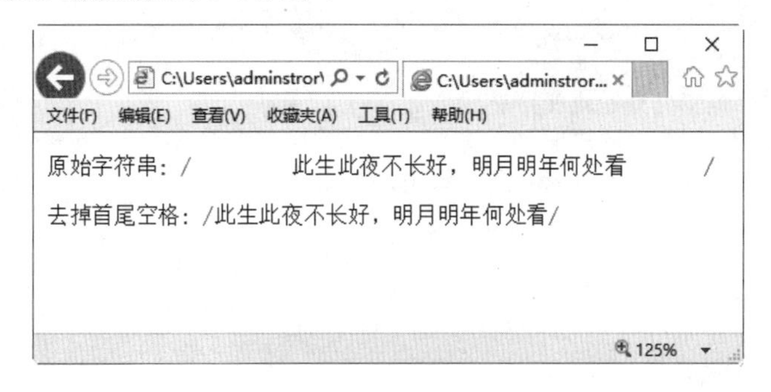

图 8-6　使用 trim()方法

(2)　使用 substr()方法可以在字符串中抽取指定下标的字符串片段。

【例 8.7】(示例文件 ch08\8.7.html)

使用 substr()方法：

```
<!DOCTYPE html>
<html>
<head>
<meta http-equiv="Content-Type" content="text/html; charset=gb2312" />
<script type="text/javascript" src="jquery.min.js"></script>
<script type="text/javascript">
  var str = "此生此夜不长好，明月明年何处看";
  document.write("原始内容：" + str);
  document.write("截取内容：" + str.substr(0,9));
</script>
</head>
<body>
</body>
</html>
```

在 IE 11.0 中浏览，效果如图 8-7 所示。

(3) 使用 replace()方法在字符串中用一些字符替换另一些字符，或替换一个与正则表达式匹配的子串，结果返回一个字符串。其语法格式如下：

```
replace(m,n);
```

其中，m 是要替换的目标；n 是替换后的新值。

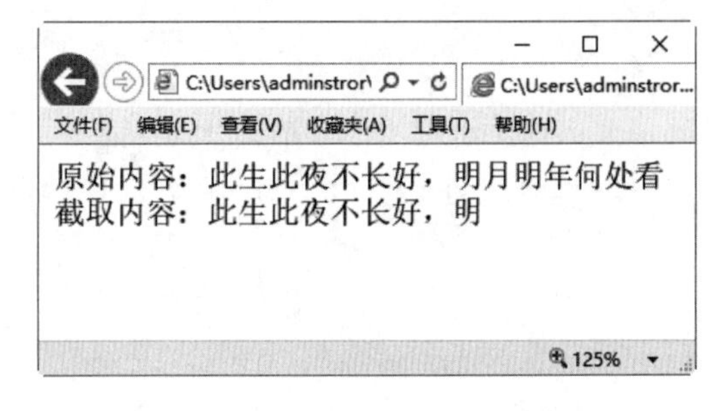

图 8-7 使用 substr()方法

【例 8.8】(示例文件 ch08\8.8.html)

使用 replace()方法：

```
<!DOCTYPE html>
<html>
<head>
<meta http-equiv="Content-Type" content="text/html; charset=gb2312" />
<script type="text/javascript" src="jquery.min.js"></script>
<script type="text/javascript">
  var str = "含苞待放的玫瑰！ ";
  str = str + "五彩盛开的玫瑰！";
  str = str + "香气扑鼻的玫瑰！";
  document.write(str.replace(/玫瑰/g, "玉兰"));
</script>
</head>
<body>
</body>
</html>
```

在 IE 11.0 中浏览，效果如图 8-8 所示。

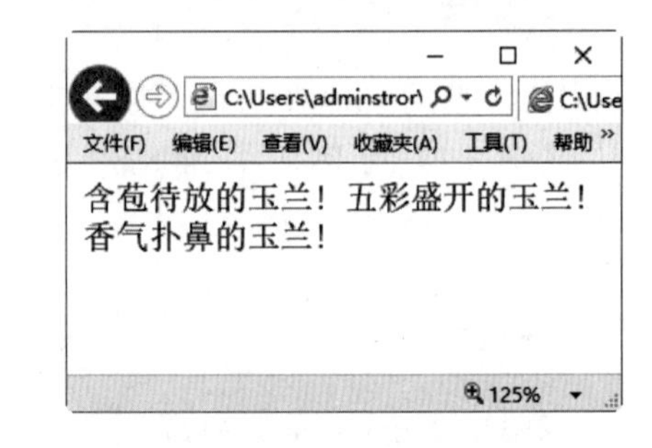

图 8-8 使用 replace()方法

8.2.3 序列化操作

jQuery 提供的 param(object)方法用于将表单元素数组或者对象序列化，返回值是 string。其中，数组或者 jQuery 对象会按照 name、value 进行序列化；普通对象会按照 key、value 进行序列化。

【例 8.9】(示例文件 ch08\8.9.html)

使用 param(object)方法：

```
<!DOCTYPE html>
<html>
<head>
<meta http-equiv="Content-Type" content="text/html; charset=gb2312" />
<script type="text/javascript" src="jquery.min.js"></script>
<script type="text/javascript">
$(document).ready(function(){
  personObj = new Object();
```

```
  personObj.firstname = "Bill";
  personObj.lastname = "Gates";
  personObj.age = 60;
  personObj.eyecolor = "blue";
  $("button").click(function(){
     $("div").text($.param(personObj));
  });
});
</script>
</head>
<body>
<button>序列化对象</button>
<div></div>
</body>
</html>
```

在 IE 11.0 中浏览，单击【序列化对象】按钮，效果如图 8-9 所示。

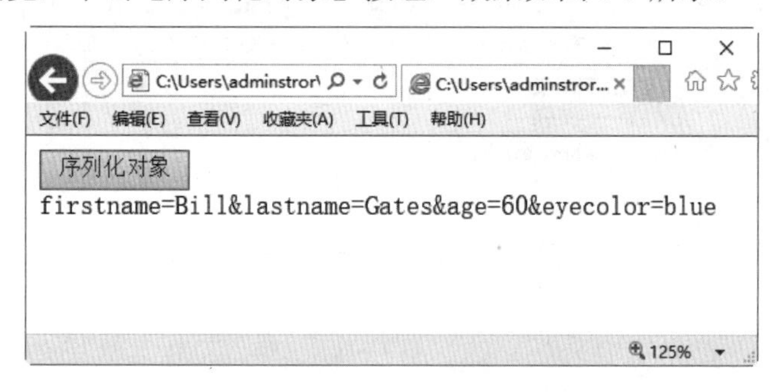

图 8-9　使用 param(object)方法

8.3　调用外部代码

通过使用 jQuery 提供的 getScript()方法，用户可以加载外部的代码，从而实现操作加载、运行不同代码的目的。其语法格式如下：

```
$.getScript(url,callback)
```

其中，url 是外部代码的地址，这里可以是相对地址，也可以是绝对地址；callback 是可选项，是获取外部代码之后需要运行的回调函数。

在调用代码前，先编写一个 text.js 代码文件，代码如下：

```
alert("滚滚长江东逝水，浪花淘尽英雄。");
```

【例 8.10】(示例文件 ch08\8.10.html)

使用 getScript()方法：

```
<!DOCTYPE html>
<html>
<head>
<meta http-equiv="Content-Type" content="text/html; charset=gb2312" />
```

```
<script type="text/javascript" src="jquery.min.js"></script>
<script type="text/javascript">
$(document).ready(function(){
$("button").click(function(){
$.getScript("text.js");
});
});
</script>
</head>
<body>
<button>调用外部代码</button>
</body>
</html>
```

在 IE 11.0 中浏览，单击【调用外部代码】按钮，最终效果如图 8-10 所示。

图 8-10　使用 getScript()方法

8.4　疑难解惑

疑问 1：如何加载外部文本文件的内容？

答：在 jQuery 中，load()方法是简单而强大的 Ajax 方法。用户可以使用 load()方法从服务器加载数据，并把返回的数据放入被选元素中。其语法格式如下：

```
$(selector).load(URL,data,callback);
```

其中，URL 是必需的参数，表示希望加载的文件路径。data 参数是可选的，规定与请求一同发送的查询字符串键值对集合。callback 也是可选的参数，是 load()方法完成后所执行的函数名称。

例如，用户想加载 test.txt 文件的内容到指定的<div>元素中，使用的代码如下：

```
$("#div1").load("test.txt");
```

疑问 2：jQuery 中的测试函数有哪些？

答：在 JavaScript 中，有自带的测试操作函数 isNaN()和 isFinite()。其中，isNaN()函数用于判断函数是否是非数值，如果是数值就返回 false；isFinite()函数是检查其参数是否是无穷大，如果参数是 NaN(非数值)，或者是正、负无穷大的数值时，就返回 false，否则返回 true。而在 jQuery 发展中，测试工具函数主要有下面两种，用于判断对象是否是某一种类型，返回值都是 boolean 值。

- $.isArray(object)：返回一个布尔值，指明对象是否是一个 JavaScript 数组(而不是类似数组的对象，如一个 jQuery 对象)。
- $.isFunction(object)：用于测试是否为函数的对象。

第 9 章 jQuery 与 Ajax 技术的应用

Ajax 是目前很新的一项网络技术。确切地说，Ajax 是一种用于创建更好、更快以及交互性更强的 Web 应用程序的技术。它能使浏览器为用户提供更为自然的浏览体验，就像在使用桌面应用程序一样。

9.1 Ajax 快速入门

Ajax 是一项很有生命力的技术，它的出现引发了 Web 应用的新革命。目前，网络上的许多站点中，使用 Ajax 技术的还非常有限。但是，可以预见在不远的将来，Ajax 技术会成为整个网络的主流。

9.1.1 什么是 Ajax

Ajax 的全称为 Asynchronous JavaScript And XML，是一种 Web 应用程序客户机技术，它结合了 JavaScript、层叠样式表(Cascading Style Sheets，CSS)、HTML、XMLHttpRequest 对象和文档对象模型(Document Object Model，DOM)多种技术。运行在浏览器上的 Ajax 应用程序，以一种异步的方式与 Web 服务器通信，并且只更新页面的一部分。通过利用 Ajax 技术，可以提供丰富的、基于浏览器的用户体验。

Ajax 让开发者在浏览器端更新被显示的 HTML 内容而不必刷新页面。换句话说，Ajax 可以使基于浏览器的应用程序更具交互性，而且更类似于传统型桌面应用程序。Google 的 Gmail 和 Outlook Express 就是两个使用 Ajax 技术的例子。而且，Ajax 可以用于任何客户端脚本语言中，这包括 JavaScript、JScript 和 VBScript。

下面给出一个简单的例子，来具体了解什么是 Ajax。

【例 9.1】(示例文件 ch09\HelloAjax.jsp)

本例从简单的角度入手，实现客户端与服务器异步通信，获取“你好，Ajax”的数据，并在不刷新页面的情况下将获得的“你好，Ajax”数据显示到页面上。

具体操作步骤如下。

step 01 使用记事本创建 HelloAjax.jsp 文件。代码如下：

```
<%@ page language="java" pageEncoding="gb2312"%>

<html>
  <head>
    <title>第一个 Ajax 实例</title>
    <style type="text/css">
      <!--
       body {
         background-image: url(images/img.jpg);
       }
      -->
    </style>
  </head>
  <script type="text/javascript">
   ...//省略了 script 代码
  </script>
  <body>
  <br>
```

```
    <center>
      <button onclick="hello()">Ajax</button>
      <P id="p">
         单击按钮后你会有惊奇的发现哟！
      </P>
    </center>
  </body>
</html>
```

JavaScript 代码嵌入在标签<script></script>之内，这里定义了一个函数 hello()，这个函数是通过一个按钮来驱动的。

step 02 在步骤 1 中省略的代码部分创建 XML Http Request 对象，创建完成后，把此对象赋值给 xmlHttp 变量。为了获得多种浏览器支持，应使用 createXMLHttpRequest()函数试着为多种浏览器创建 XMLHttpRequest 对象。代码如下：

```
var xmlHttp = false;

function createXMLHttpRequest()
{
    if (window.ActiveXObject)              //在 IE 浏览器中创建 XMLHttpRequest 对象
    {
        try{
            xmlHttp = new ActiveXObject("Msxml2.XMLHTTP");
        }
        catch(e){
            try{
                xmlHttp = new ActiveXObject("Microsoft.XMLHTTP");
            }
            catch(ee){
                xmlHttp = false;
            }
        }
    }
    else if (window.XMLHttpRequest)    //在非 IE 浏览器中创建 XMLHttpRequest 对象
    {
        try{
            xmlHttp = new XMLHttpRequest();
        }
        catch(e){
            xmlHttp = false;
        }
    }
}
```

step 03 在步骤 1 省略的代码部分再定义 hello()函数，为要与之通信的服务器资源创建一个 URL。xmlHttp.onreadystatechange=callback 与 xmlHttp.open("post", "HelloAjaxDo.jsp",true)定义了 JavaScript 回调函数，一旦响应它就自动执行，而 open 函数中所指定的 true 标志说明想要异步执行该请求，没有指定的情况下默认为 true。代码如下：

```
function hello()
{
    createXMLHttpRequest();    //调用创建 XMLHttpRequest 对象的方法
    xmlHttp.onreadystatechange = callback;    //设置回调函数

    //向服务器端 HelloAjaxDo.jsp 发送请求
    xmlHttp.open("post","HelloAjaxDo.jsp",true);
    xmlHttp.setRequestHeader("Content-Type",
      "application/x-www-form-urlencoded;charset=gb2312");
    xmlHttp.send(null);

    function callback()
    {
        if(xmlHttp.readyState==4)
        {
            if(xmlHttp.status==200)
            {
                var data = xmlHttp.responseText;
                var pNode = document.getElementById("p");
                pNode.innerHTML = data;
            }
        }
    }
}
```

函数 callback()是回调函数，它首先检查 XMLHttpRequest 对象的整体状态以保证它已经完成(readyStatus==4)，然后根据服务器的设定询问请求状态。如果一切正常(status==200)，就使用 var data = xmlHttp.responseText;来取得返回的数据，并且用 innerHTML 属性重写 DOM 的 pNode 节点的内容。

JavaScript 的变量类型使用的是弱类型，都使用 var 来声明。document 对象就是文档对应的 DOM 树。通过 document.getElementById(“p”);可以从标签的 id 值来取得此标签的一个引用(树的节点)；而 pNode.innerHTML=str;是为节点添加内容，这样就覆盖了节点的原有内容，如果不想覆盖，可以使用 pNode.innerHTML+=str;来追加内容。

step 04 通过步骤 3 可以知道，要异步请求的是 HelloAjaxDo.jsp，下面创建此文件：

```
<%@ page language="java" pageEncoding="gb2312"%>
<%
 out.println("你好，Ajax");
%>
```

step 05 将上述文件保存在 Ajax 站点下，启动 Tomcat 服务器打开浏览器，在地址栏中输入 http://localhost/Ajax/HelloAjax.jsp，然后单击转到按钮，看到的结果如图 9-1 所示。

step 06 单击 Ajax 按钮，结果变为如图 9-2 所示，注意按钮下内容的变化，这个变化没有看到刷新页面的过程。

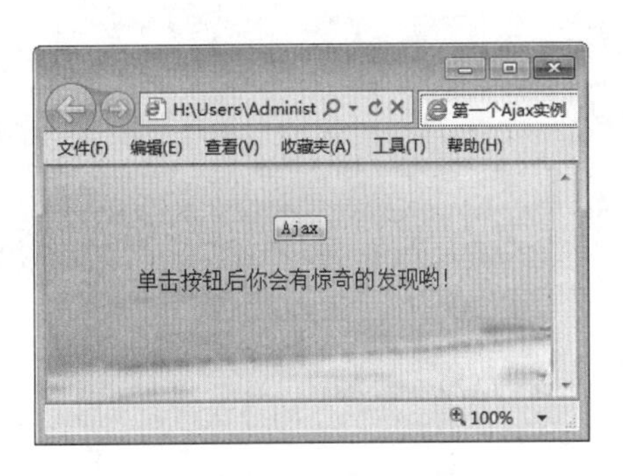

图 9-1 会变的页面

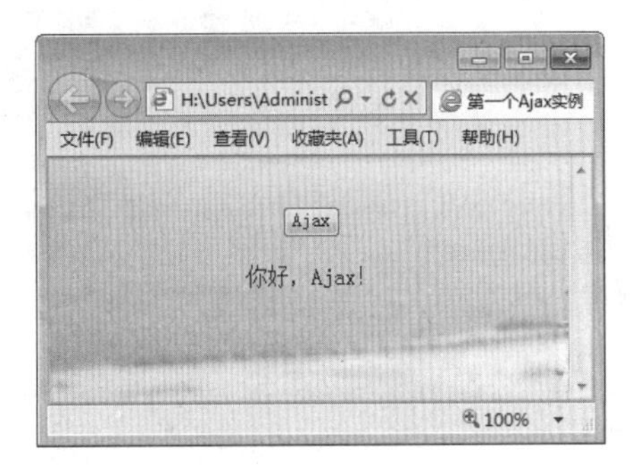

图 9-2 动态改变页面

9.1.2 Ajax 的关键元素

Ajax 不是单一的技术，而是 4 种技术的结合。要灵活地运用 Ajax，必须深入了解这些不同的技术。下面列出这些技术，并说明它们在 Ajax 中所扮演的角色。

- JavaScript：是通用的脚本语言，用来嵌入在某种应用之中。Web 浏览器中嵌入的 JavaScript 解释器允许通过程序与浏览器的很多内建功能进行交互。Ajax 应用程序是使用 JavaScript 编写的。
- CSS：为 Web 页面元素提供了一种可重用的可视化样式的定义方法。它提供了简单而又强大的方法，以一致的方式定义和使用可视化样式。在 Ajax 应用中，用户界面的样式可以通过 CSS 独立修改。
- DOM：以一组可以使用 JavaScript 操作的可编程对象展现出 Web 页面的结构。通过使用脚本修改 DOM，Ajax 应用程序可以在运行时改变用户界面，或者高效地重绘页面中的某个部分。
- XMLHttpRequest：该对象允许 Web 程序员从 Web 服务器以后台活动的方式获取数据。数据格式通常是 XML，但是也可以很好地支持任何基于文本的数据格式。

在 Ajax 的 4 种技术中，CSS、DOM 和 JavaScript 都是很早就出现的技术，它们以前结合在一起，称为动态 HTML，即 DHTML。

Ajax 的核心是 JavaScript 对象 XMLHttpRequest。该对象在 Internet Explorer 5 中首次引入，是一种支持异步请求的技术。简而言之，XMLHttpRequest 让我们可以使用 JavaScript 向服务器提出请求并处理响应，而不阻塞用户。

9.1.3 CSS 在 Ajax 应用中的地位

CSS 在 Ajax 中主要用于美化网页，是 Ajax 的美术师。无论 Ajax 的核心技术采用什么形式，任何时候显示在用户面前的都是一个页面，是页面就需要美化，那么就需要用 CSS 对显示在用户浏览器上的界面进行美化。

如果用户在浏览器中查看页面的源代码，就可以看到众多的<div>块以及 CSS 属性占据了源代码的很多部分。图 9-3 中也表明页面引用了外部的 CSS 样式文件。由此可见 CSS 在页面美化方面的重要性。

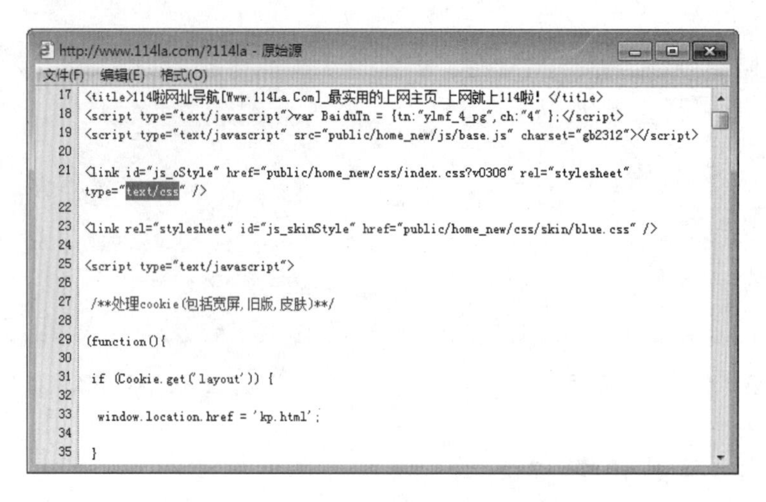

图 9-3 源文件中引用了外部 CSS 文件

9.2 Ajax 的核心技术

Ajax 作为一个新技术，结合了 4 种不同的技术，实现了客户端与服务器端的异步通信，并且对页面实现局部更新，大大提高了浏览器的工作速度。

9.2.1 全面剖析 XMLHttpRequest 对象

XMLHttpRequest 对象是当今所有 Ajax 和 Web 2.0 应用程序的技术基础。尽管软件经销商和开源社团现在都在提供各种 Ajax 框架以进一步简化 XMLHttpRequest 对象的使用，但是，我们仍然很有必要理解这个对象的详细工作机制。

1. XMLHttpRequest 概述

Ajax 利用一个构建到所有现代浏览器内部的 XMLHttpRequest 对象来实现发送和接收 HTTP 请求与响应信息。一个经由 XMLHttpRequest 对象发送的 HTTP 请求并不要求页面中拥有或回发一个<form>元素。

微软 Internet Explorer(IE) 5 中作为一个 ActiveX 对象形式引入了 XMLHttpRequest 对象。其他认识到这一对象重要性的浏览器制造商也都纷纷在其浏览器内实现了 XMLHttpRequest 对象，不过是作为一个本地 JavaScript 对象而不是作为一个 ActiveX 对象实现的。

如今，在认识到实现这一类型的价值及安全性特征之后，微软已经在其 IE 7 中把 XMLHttpRequest 实现为一个窗口对象属性。幸运的是，尽管其实现细节不同，但是所有的浏览器实现都具有类似的功能，并且实质上是相同的方法。目前，W3C 组织正在努力进行 XMLHttpRequest 对象的标准化。

2. XMLHttpRequest 对象的属性和事件

XMLHttpRequest 对象暴露各种属性、方法和事件，以便于脚本处理和控制 HTTP 请求与响应。下面进行详细的讨论。

1) readyState 属性

当 XMLHttpRequest 对象把一个 HTTP 请求发送到服务器时，将经历若干种状态，一直等待直到请求被处理；然后，它才接收一个响应。这样一来，脚本才正确响应各种状态，XMLHttpRequest 对象暴露描述对象当前状态的 readyState 属性，如表 9-1 所示。

表 9-1 XMLHttpRequest 对象的 readyState 属性

readyState 取值	描　述
0	描述一种“未初始化”状态；此时，已经创建一个 XMLHttpRequest 对象，但是还没有初始化
1	XMLHttpRequest 已经准备好把一个请求发送到服务器
2	描述一种“发送”状态；此时，已经通过 send()方法把一个请求发送到服务器端，但是还没有收到一个响应
3	描述一种“正在接收”状态；此时，已经接收到 HTTP 响应头部信息，但是消息体部分还没有完全接收结束
4	描述一种“已加载”状态；此时，响应已经被完全接收

2) onreadystatechange 事件

无论 readyState 值何时发生改变，XMLHttpRequest 对象都会激发一个 readystatechange 事件。其中，onreadystatechange 属性接收一个 EventListener 值，该值向该方法指示无论 readyState 值何时发生改变，该对象都将激活。

3) responseText 属性

这个 responseText 属性包含客户端接收到的 HTTP 响应的文本内容。当 readyState 值为 0、1 或 2 时，responseText 包含一个空字符串。当 readyState 值为 3(正在接收)时，响应中包含客户端还未完成的响应信息。当 readyState 为 4(已加载)时，该 responseText 包含完整的响应信息。

4) responseXML 属性

responseXML 属性用于当接收到完整的 HTTP 响应时描述 XML 响应；此时，Content-Type 头部指定 MIME(媒体)类型为 text/xml、application/xml 或以+xml 结尾。如果 Content-Type 头部并不包含这些媒体类型之一，那么 responseXML 的值为 null。无论何时，只要 readyState 值不为 4，那么该 responseXML 的值也为 null。

其实，这个 responseXML 属性值是一个文档接口类型的对象，用来描述被分析的文档。如果文档不能被分析(例如，如果文档不支持相应的字符编码)，那么 responseXML 的值将为 null。

5) status 属性

status 属性描述了 HTTP 状态代码，其类型为 short。而且，仅当 readyState 值为 3(正在接收中)或 4(已加载)时，这个 status 属性才可用。当 readyState 的值小于 3 时，试图存取 status 的值将引发一个异常。

6) statusText 属性

statusText 属性描述了 HTTP 状态代码文本；并且仅当 readyState 值为 3 或 4 时才可用。当 readyState 为其他值时，试图存取 statusText 属性将引发一个异常。

3. 创建 XMLHttpRequest 对象的方法

XMLHttpRequest 对象提供了各种方法，用于初始化和处理 HTTP 请求，下面详细介绍。

1) abort()方法

用户可以使用 abort()方法来暂停与一个 XMLHttpRequest 对象相联系的 HTTP 请求，从而把该对象复位到未初始化状态。

2) open()方法

用户需要调用 open()方法来初始化一个 XMLHttpRequest 对象。其中，method 参数是必须提供的，用于指定我们想用来发送请求的 HTTP 方法。为了把数据发送到服务器，应该使用 POST 方法；为了从服务器端检索数据，应该使用 GET 方法。

3) send()方法

在通过调用 open()方法准备好一个请求之后，用户需要把该请求发送到服务器。仅当 readyState 值为 1 时，才可以调用 send()方法；否则 XMLHttpRequest 对象将引发一个异常。

4) setRequestHeader()方法

setRequestHeader()方法用来设置请求的头部信息。当 readyState 值为 1 时，用户可以在调用 open()方法后调用这个方法；否则，将得到一个异常。

5) getResponseHeader()方法

getResponseHeader()方法用于检索响应的头部值。仅当 readyState 值是 3 或 4(换句话说，在响应头部可用以后)时，才可以调用这个方法；否则，该方法返回一个空字符串。

6) getAllResponseHeaders()方法

getAllResponseHeaders()方法以一个字符串形式返回所有的响应头部(每一个头部占单独的一行)。如果 readyState 的值不是 3 或 4，则该方法返回 null。

9.2.2 发出 Ajax 请求

在 Ajax 中，许多使用 XMLHttpRequest 的请求都是从一个 HTML 事件(例如一个调用 JavaScript 函数的按钮点击(onclick)或一个按键(onkeypress))中被初始化的。Ajax 支持包括表单校验在内的各种应用程序。有时，在填充表单的其他内容之前要求校验一个唯一的表单域。例如，要求使用一个唯一的 UserID 来注册表单。如果不是使用 Ajax 技术来校验这个 UserID 域，那么整个表单都必须被填充和提交。如果该 UserID 不是有效的，这个表单必须被重新提交。例如，相应于一个要求必须在服务器端进行校验的 Catalog ID 的表单域可按下列形式来指定：

```
<form name="validationForm" action="validateForm" method="post">
<table>
<tr>
   <td>Catalog Id:</td>
   <td>
```

```
        <input type="text" size="20" id="catalogId" name="catalogId"
          autocomplete="off" onkeyup="sendRequest()">
    </td>
    <td><div id="validationMessage"></div></td>
</tr>
</table>
</form>
```

在HTML中使用validationMessage div来显示相应于这个输入域Catalog Id的一个校验消息。onkeyup 事件调用一个 JavaScript sendRequest()函数。这个 sendRequest()函数创建一个XMLHttpRequest对象。创建一个XMLHttpRequest对象的过程因浏览器实现的不同而不同。

如果浏览器支持 XMLHttpRequest 对象作为一个窗口属性，那么代码可以调用XMLHttpRequest的构造器。如果浏览器把XMLHttpRequest对象实现为一个ActiveXObject对象，那么代码可以使用ActiveXObject的构造器。下面的函数将调用一个init()函数：

```
<script type="text/javascript">
function sendRequest(){
    var xmlHttpReq = init();
    function init(){
        if (window.XMLHttpRequest) {
            return new XMLHttpRequest();
        }
        else if (window.ActiveXObject) {
            return new ActiveXObject("Microsoft.XMLHTTP");
        }
    }
}
</script>
```

接下来，用户需要使用open()方法初始化XMLHttpRequest对象，从而指定HTTP方法和要使用的服务器URL：

```
var catalogId = encodeURIComponent(document.getElementById("catalogId").value);
xmlHttpReq.open("GET", "validateForm?catalogId=" + catalogId, true);
```

在默认情况下，使用XMLHttpRequest发送的HTTP请求是异步进行的，但是用户可以显式地把async参数设置为true。在这种情况下，对URL validateForm的调用将激活服务器端的一个Servlet。但是用户应该能够注意到服务器端技术不是根本性的；实际上，该URL可能是一个 ASP、ASP.NET 或 PHP 页面或一个 Web 服务，只要该页面能够返回一个响应，指示catalogID 值是否是有效的即可。因为用户在做异步调用时，需要注册一个 XMLHttpRequest对象来调用回调事件处理器，当它的 readyState 值改变时调用。记住，readyState 值的改变将会激发一个 readystatechange 事件。这时可以使用 onreadystatechange 属性来注册该回调事件处理器：

```
xmlHttpReq.onreadystatechange = processRequest;
```

然后，需要使用 send()方法发送该请求。因为这个请求使用的是 HTTP GET 方法，所以用户可以在不指定参数或使用null参数的情况下调用send()方法：

```
xmlHttpReq.send(null);
```

9.2.3 处理服务器响应

在上述示例中，因为 HTTP 方法是 GET，所以在服务器端的接收 Servlet 将调用一个 doGet()方法，该方法将检索在 URL 中指定的 catalogId 参数值，并且从一个数据库中检查它的有效性。

该示例中的 Servlet 需要构造一个发送到客户端的响应；而且，这个示例返回的是 XML 类型。因此，它把响应的 HTTP 内容类型设置为 text/xml 并且把 Cache-Control 头部设置为 no-cache。设置 Cache-Control 头部可以阻止浏览器简单地从缓存中重载页面。

代码如下：

```
public void doGet(HttpServletRequest request,HttpServletResponse response)
  throws ServletException,IOException {
    ...
    response.setContentType("text/xml");
    response.setHeader("Cache-Control", "no-cache");
}
```

从上述代码中可以看出，来自服务器端的响应是一个 XML DOM 对象，此对象将创建一个 XML 字符串，其中包含要在客户端进行处理的指令。另外，该 XML 字符串必须有一个根元素。代码如下：

```
out.println("<catalogId>valid</catalogId>");
```

注意 XMLHttpRequest 对象设计的目的是处理由普通文本或 XML 组成的响应；但是，一个响应也可能是另外一种类型(如果用户代理支持这种内容类型的话)。

当请求状态改变时，XMLHttpRequest 对象调用使用 onreadystatechange 注册的事件处理器。因此，在处理该响应之前，用户的事件处理器应该首先检查 readyState 的值和 HTTP 状态。当请求完成加载(readyState 值为 4)并且响应已经完成(HTTP 状态为 OK)时，用户就可以调用一个 JavaScript 函数来处理该响应内容。下列脚本负责在响应完成时检查相应的值并调用一个 processResponse()方法：

```
function processRequest(){
    if(xmlHttpReq.readyState==4){
        if(xmlHttpReq.status==200){
            processResponse();
        }
    }
}
```

该 processResponse()方法使用 XMLHttpRequest 对象的 responseXML 和 responseText 属性来检索 HTTP 响应。如上面所解释的，仅当在响应的媒体类型是 text/xml、application/xml 或以+xml 结尾时，这个 responseXML 才可用。这个 responseText 属性将以普通文本形式返回响

应。对于一个 XML 响应，用户将按如下方式检索内容：

```
var msg = xmlHttpReq.responseXML;
```

借助于存储在 msg 变量中的 XML，用户可以使用 DOM 方法 getElementsByTagName()来检索该元素的值，代码如下：

```
var catalogId =
 msg.getElementsByTagName("catalogId")[0].firstChild.nodeValue;
```

最后，通过更新 Web 页面的 validationMessage div 中的 HTML 内容并借助于 innerHTML 属性，用户可以测试该元素值以创建一个要显示的消息，代码如下：

```
if(catalogId=="valid"){
   var validationMessage = document.getElementById("validationMessage");
   validationMessage.innerHTML = "Catalog Id is Valid";
}
else
{
   var validationMessage = document.getElementById("validationMessage");
   validationMessage.innerHTML = "Catalog Id is not Valid";
}
```

9.3 jQuery 中的 Ajax

jQuery 提供多个与 Ajax 有关的方法。通过这些方法，用户可以通过 HTTP 的 Post 或 Get 方式从远程服务器上请求文本、HTML 或 XML 数据，然后把这些数据直接载入网页的被选元素上。

9.3.1 load()方法

jQuery 提供了一个简单但强大的方法 load()，其主要功能是从服务器加载数据，并把返回的数据放入被选元素中。

load()方法的语法格式如下：

```
$(selector).load(URL,data,callback);
```

其中，参数 URL 用于规定需要加载数据的 URL；参数 data 为可选参数，规定与请求一同发送的数据；参数 callback 为可选参数，规定参数是 load()方法完成后所执行的函数名称。

【例 9.2】(示例文件 ch09\9.1.html)

```
<!DOCTYPE html>
<html>
<head>
<meta charset="utf-8">
<title>使用 load()方法</title>
<script type="text/javascript" src="jquery.min.js"></script>
```

```
</script>
<script>
$(document).ready(function(){
    $("button").click(function(){
        $("#div1").load("test.txt");
    });
});
</script>
</head>
<body>

<div id="div1"><h2>使用 load()方法获取文本的内容</h2></div>
<button>更新页面</button>

</body>
</html>
```

其中加载文件 test.txt 文件的内容如图 9-4 所示。

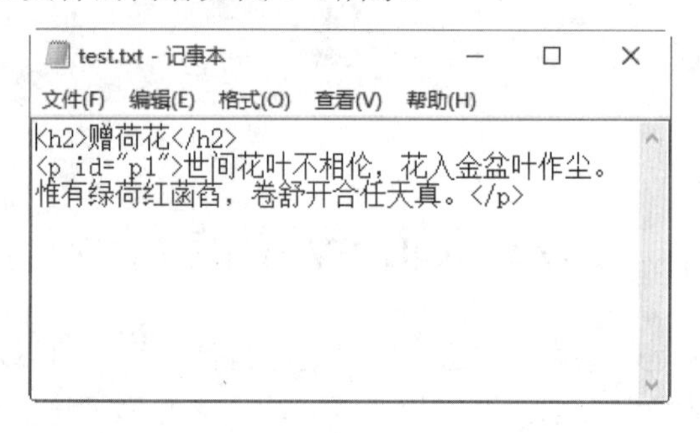

图 9-4　加载文件的内容

在 Firefox 53.0 中查看文件：http://localhost/code/ch09/9.1.html，效果如图 9-5 所示。单击【更新页面】按钮，即可加载文件的内容，如图 9-6 所示。

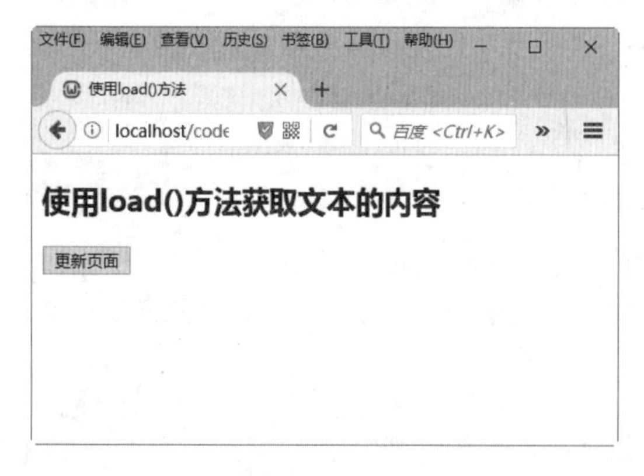

图 9-5　查看文件效果

图 9-6　加载文件后的效果

读者还可以把 jQuery 选择器添加到 URL 参数。下面的例子把 "test.txt" 文件中 id="p1" 的元素的内容，加载到指定的<div>元素中。

【例 9.3】(示例文件 ch09\9.2.html)

```
<!DOCTYPE html>
<html>
<head>
<meta charset="utf-8">
<title>使用 load()方法</title>
<script type="text/javascript" src="jquery.min.js"></script>
</script>
<script>
$(document).ready(function(){
    $("button").click(function(){
        $("#div1").load("test.txt #p1");
    });
});
</script>
</head>
<body>

<div id="div1"><h2>使用 load()方法获取文本的内容</h2></div>
<button>更新页面</button>

</body>
</html>
```

在 Firefox 53.0 中查看文件：http://localhost/code/ch09/9.2.html，效果如图 9-7 所示。单击【更新页面】按钮，即可加载文件的内容，如图 9-8 所示。

图 9-7　查看文件效果

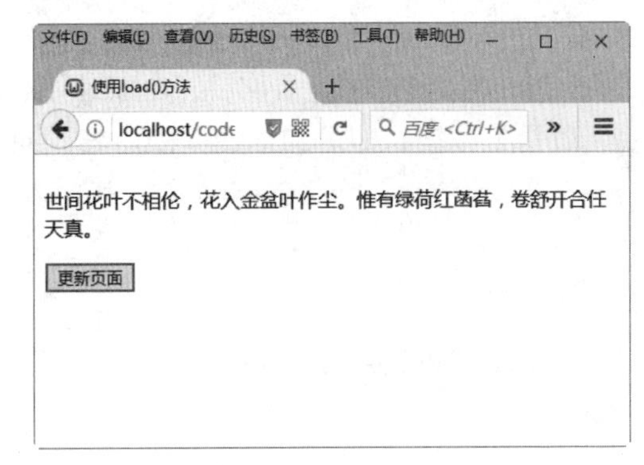

图 9-8　加载文件后的效果

load()方法的可选参数 callback 规定 load()方法完成后调用的函数。该调用函数的可以设置的参数如下。

(1) responseTxt：包含调用成功时的结果内容。

(2) statusTxt ：包含调用的状态。

(3) xhr：包含 XMLHttpRequest 对象。

下面的案例将在 load()方法完成后显示一个提示框。如果 load()方法已成功，则显示【加载内容已经成功了!】，而如果失败，则显示错误消息。

【例 9.4】(示例文件 ch09\9.3.html)

```
<!DOCTYPE html>
<html>
<head>
<meta charset="utf-8">
<title>使用 load()方法</title>
<script type="text/javascript" src="jquery.min.js"></script>
</script>
<script>
$(document).ready(function(){
  $("button").click(function(){
    $("#div1").load("test.txt",function(responseTxt,statusTxt,xhr){
      if(statusTxt=="success")
        alert("加载内容已经成功了!");
      if(statusTxt=="error")
        alert("Error: "+xhr.status+": "+xhr.statusText);
    });
  });
});
</script>
</head>
<body>

<div id="div1"><h2>检验 load()方法是否执行成功</h2></div>
<button>更新页面</button>

</body>
</html>
```

在 Firefox 53.0 中查看文件：http://localhost/code/ch09/9.3.html，效果如图 9-9 所示。单击【更新页面】按钮，即可加载文件的内容，同时打开信息提示对话框，如图 9-10 所示。

图 9-9　查看文件效果

图 9-10　信息提示对话框

9.3.2　$.get()方法和$.post()方法

jQuery 的$.get()和$.post()方法用于通过 HTTP GET 或 POST 方式从服务器获取数据。

1. $.get()方法

$.get()方法通过 HTTP GET 方式从服务器上获取数据。$.get()方法的语法格式如下：

```
$.get(URL,callback);
```

其中，参数 URL 用于规定需要加载数据的 URL；参数 callback 为可选参数，规定参数是$.get()方法完成后所执行的函数名称。

【例 9.5】(示例文件 ch09\9.4.html)

使用$.get()方法：

```
<!DOCTYPE html>
<html>
<head>
<meta charset="utf-8">
<title>使用$.get()方法</title>
<script type="text/javascript" src="jquery.min.js"></script>
</script>
<script>
$(document).ready(function(){
    $("button").click(function(){
        $.get("test.txt",function(data,status){
            alert("数据: " + data + "\n状态: " + status);
        });
    });
});
</script>
</head>
<body>

<button>通过$.get()方法请求并获取结果</button>

</body>
</html>
```

在 Firefox 53.0 中查看文件：http://localhost/code/ch09/9.4.html，效果如图 9-11 所示。单击【通过$.get()方法请求并获取结果】按钮，即可打开信息提示对话框，如图 9-12 所示。

图 9-11　查看文件效果

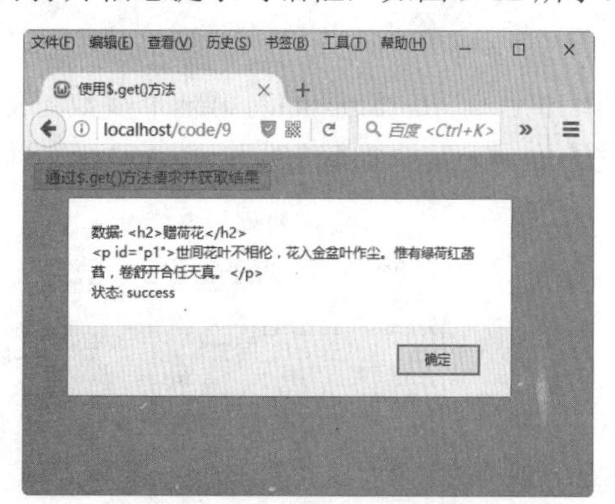

图 9-12　信息提示对话框

2. $.post()方法

$.post()方法通过 HTTP POST 方式从服务器上获取数据。$.post()方法的语法格式如下：

```
$.post(URL,data,callback);
```

其中，参数 URL 用于规定需要加载数据的 URL；参数 data 为可选参数，规定与请求一同发送的数据；参数 callback 为可选参数，规定参数是 load() 方法完成后所执行的函数名称。

【例 9.6】(示例文件 ch09\9.5.html)

使用$.post()方法：

```
<!DOCTYPE html>
<html>
<head>
<meta charset="utf-8">
<title>使用$.post()方法</title>
<script type="text/javascript" src="jquery.min.js"></script>
</script>
<script>
$(document).ready(function(){
    $("button").click(function(){
        $.post("mytest.php",{
            name:"zhangxiaoming",
            age:"26"
        },
        function(data,status){
            alert("数据: \n" + data + "\n 状态: " + status);
        });
    });
});
</script>
</head>
<body>

<button>通过$.post()方法请求并获取结果</button>

</body>
</html
```

在上述代码中，$.post()的第一个参数是 URL ("mytest.php")，然后连同请求(name 和 age)一起发送数据。其中加载文件 mytest.php 文件的内容如图 9-13 所示。

mytest.php - 记事本

文件(F) 编辑(E) 格式(O) 查看(V) 帮助(H)

```
<?php
$name = isset($_POST['name']) ? htmlspecialchars($_POST['name']) : '';
$url = isset($_POST['age']) ? htmlspecialchars($_POST['age']) : '';
echo '员工姓名：' . $name;
echo "\n";
echo '员工年龄：' .$age;
?>
```

图 9-13　加载文件的内容

在 Firefox 53.0 中查看文件：http://localhost/code/ch09/9.5.html，效果如图 9-14 所示。单击【通过$.post()方法请求并获取结果】按钮，即可打开信息提示对话框，如图 9-15 所示。

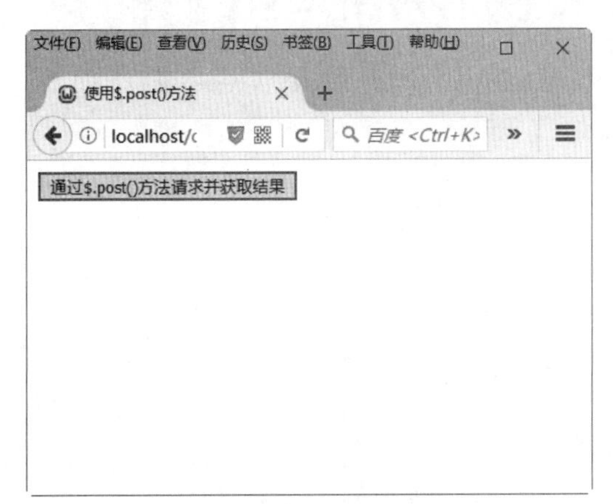

图 9-14　查看文件效果

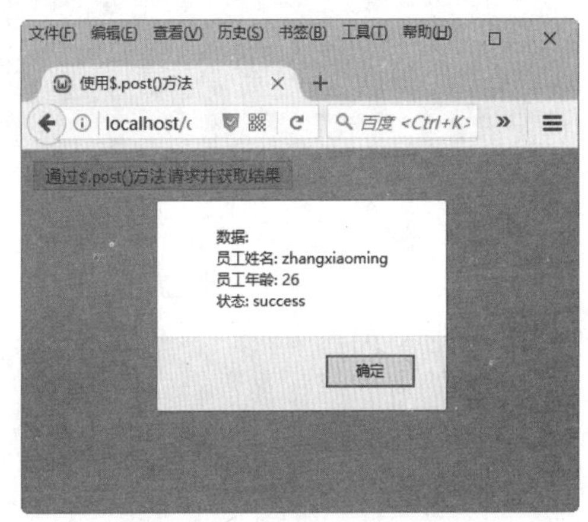

图 9-15　信息提示对话框

9.3.3　$.getScript()方法和$.getJson()方法

下面分别来介绍如何使用$.getScript()方法和$.getJson()方法。

1. $.getScript()方法

$.getScript()方法通过 HTTP GET 的方式载入并执行 JavaScript 文件。其语法格式如下：

```
$.getScript(URL,success(response,status))
```

其中，参数 URL 用于规定需要加载数据的 URL；如果请求成功，则返回结果 response 和求助状态 status。

【例 9.7】(示例文件 ch09\9.6.html)

使用$.getScript()方法：

```
<!DOCTYPE html>
<html>
<head>
<meta charset="utf-8">
<title>使用$.getScript()方法</title>
<script type="text/javascript" src="jquery.min.js"></script>
</script>
<script type="text/javascript">
$(document).ready(function(){
  $("button").click(function(){
    $.getScript("myscript.js");
  });
});
</script>
</head>
```

```
<body>

<button>通过$.getScript()方法请求并执行一个 JavaScript 文件</button>

</body>
</html
```

其中加载的 JavaScript 文件的内容如图 9-16 所示。

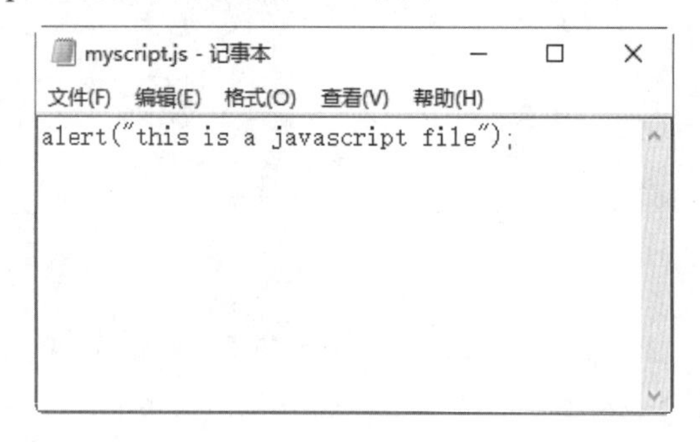

图 9-16　加载文件的内容

在 Firefox 53.0 中查看文件：http://localhost/code/ch09/9.6.html，效果如图 9-17 所示。单击【通过$.getScript()方法请求并执行一个 JavaScript 文件】按钮，即可打开信息提示对话框，如图 9-18 所示。

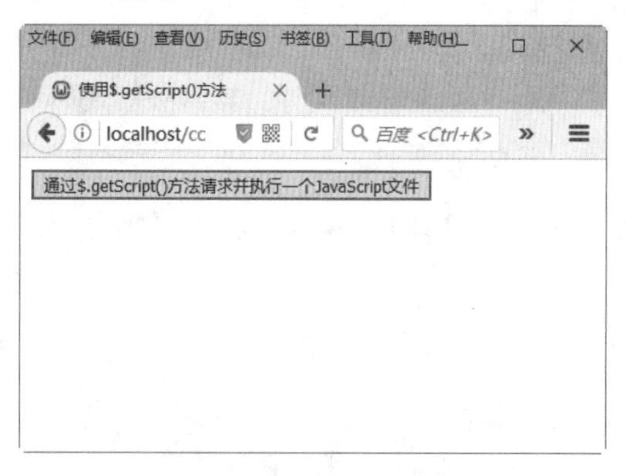

图 9-17　查看文件效果

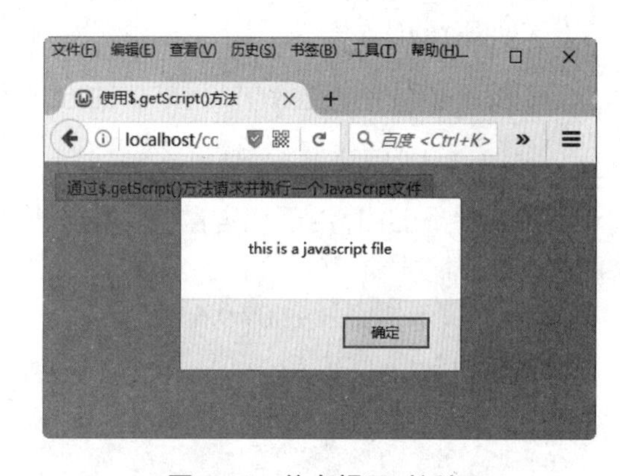

图 9-18　信息提示对话框

2. $.getJson()方法

$.getJson()方法主要是通过 HTTP GET 方式获取 JSON 数据。其语法格式如下：

```
$.getJSON(URL,data,success(data,status,xhr))
```

其中参数 URL 用于规定需要加载数据的 URL。

【例 9.8】(示例文件 ch09\9.7.html)

使用$.getJson()方法：

```
<!DOCTYPE html>
<html>
```

```
<head>
<meta charset="utf-8">
<title>使用$.getJson()方法</title>
<script type="text/javascript" src="jquery.min.js"></script>
</script>
<script type="text/javascript">
$(document).ready(function(){
  $("button").click(function(){
    $.getJSON("myjson.js",function(result){
      $.each(result, function(i, field){
        $("p").append(field + " ");
      });
    });
  });
});
</script>
</script>
</head>
<body>

<button>获取 JSON 数据</button>

</body>
</html
```

其中加载的 JSON 数据文件的内容如图 9-19 所示。

```
myjson.js - 记事本
文件(F) 编辑(E) 格式(O) 查看(V) 帮助(H)
{
  "汽车品牌": "BUICK",
  "汽车车型": "VELITE 5",
   "汽车价格": 1600000
}
```

图 9-19　加载文件的内容

在 Firefox 53.0 浏览器中查看文件：http://localhost/code/ch09/9.7.html，效果如图 9-20 所示。单击【获取 JSON 数据】按钮，即可加载 JSON 数据，如图 9-21 所示。

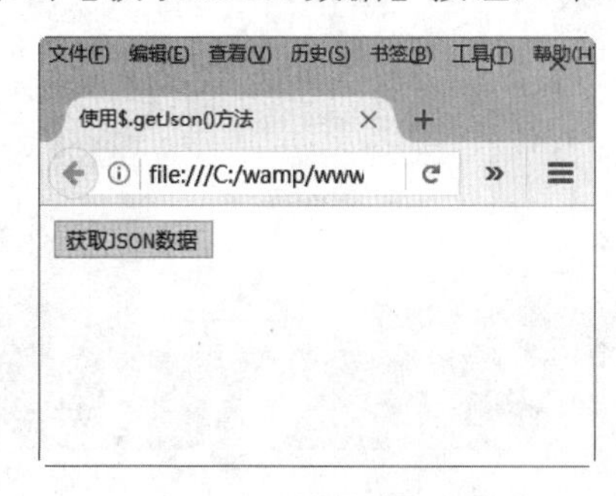

图 9-20　查看文件效果

图 9-21　加载 JSON 数据

9.3.4 $.ajax()方法

$.ajax()方法用于执行异步 HTTP 请求。所有的 jQuery Ajax 方法都使用 ajax() 方法。该方法通常用于其他方法不能完成的请求。其语法格式如下：

```
$.ajax({name:value, name:value, ... })
```

其中的参数规定 Ajax 请求的一个或多个名称。

【例 9.9】(示例文件 ch09\9.8.html)

使用$.ajax()方法：

```
<!DOCTYPE html>
<html>
<head>
<meta charset="utf-8">
<title>使用$.ajax()</title>
<script type="text/javascript" src="jquery.min.js"></script>
</script>
<script>
$(document).ready(function(){
    $("button").click(function(){
        $.ajax({url:"myscript.js",dataType:"script"});
    });
});
</script>
</head>
<body>

<div id="div1"><h2>使用$.ajax()方法执行 JavaScript 文件</h2></div>
<button>更新部分页面内容</button>

</body>
</html>
```

使用 Firefox 53.0 浏览器中查看文件：http://localhost/code/ch09/9.8.html，效果如图 9-22 所示。单击“开始执行”按钮，弹出信息提示对话框，如图 9-23 所示。

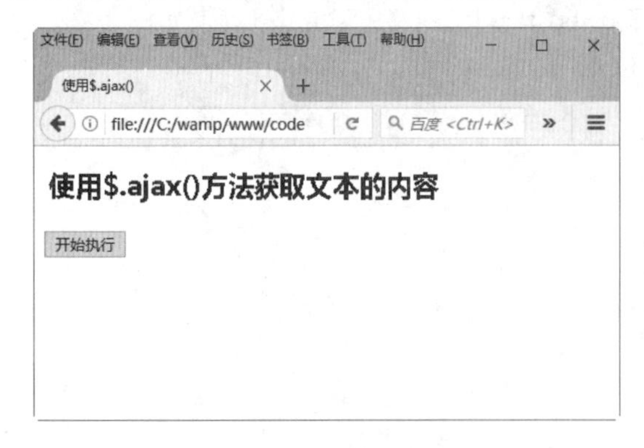

图 9-22 查看文件效果

图 9-23 执行 JavaScript 文件

9.4 疑难解惑

疑问 1：jQuery 对 Ajax 技术有什么好处？

答：编写常规的 Ajax 代码比较困难，因为不同的浏览器对 Ajax 的实现并不相同。这样就必须编写额外的代码对浏览器进行测试。jQuery 为大家解决了这一难题，只需要引用 jQuery 库函数，就可以实现 Ajax 功能，更可以实现浏览器的兼容问题。

疑问 2：使用 load()方法加载中文内容时出现乱码怎么办？

答：如果用 jQuery 的 load()方法加载的文档中包含中文字符，可能会引起乱码问题。要解决这个问题，需要注意以下两点。

(1) 在源文件的 head 中加入编码方式如下：

```
<meta charset="utf-8">
```

(2) 修改加载文档的编码方式为 UTF-8 格式编码，例如在记事本文件中，选择【文件】→【另存为】命令，如图 9-24 所示。打开【另存为】对话框，在【编码】下拉列表框中选择 UTF-8 选项，然后单击【保存】按钮即可，如图 9-25 所示。

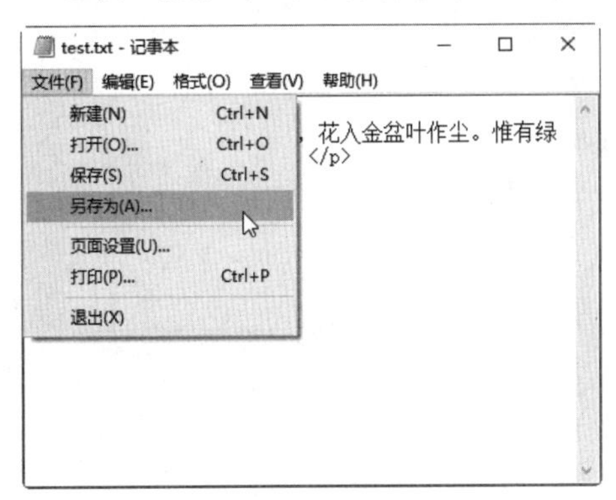

图 9-24　选择【另存为】命令

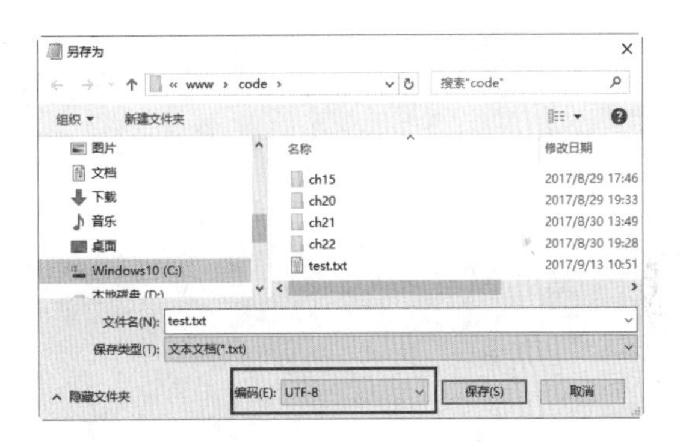

图 9-25　【另存为】对话框

第 10 章 jQuery 插件的开发与使用

jQuery 具有强大的扩展功能，允许开发人员使用或自己创建 jQuery 插件来扩充 jQuery 的功能。使用插件可以提高项目的开发效率，解决人力成本问题。特别是一些比较著名的插件，受到了开发者的追捧。插件又将 jQuery 的功能提升到了一个新的层次。

10.1 理 解 插 件

在学习插件之前，用户需要了解插件的基本概念。

10.1.1 什么是插件

编写插件的目的是给已有的一系列方法或函数做一个封装，以便在其他地方重复使用，同时方便后期维护。随着 jQuery 的广泛使用，已经出现了大量的 jQuery 插件，如 thickbox、iFX、jQuery-googleMap 等，简单地引用这些源文件就可以方便地使用这些插件。

jQuery 除了提供一个简单、有效的方式来管理元素及脚本外，还提供了添加方法和额外功能到核心模块的机制。通过这种机制，jQuery 允许用户自己创建属于自己的插件，提高开发过程中的效率。

10.1.2 如何使用插件

由于 jQuery 插件其实就是 JS 包，所以使用方法比较简单，基本步骤如下。

(1) 将下载的插件或者自定义的插件放在主 jQuery 源文件下，然后在<head>标记中引用插件的 JS 文件和 jQuery 库文件。

(2) 包含一个自定义的 JavaScript 文件，并在其中使用插件创建的方法。

下面通过一个例子来讲解具体的使用方法。

【例 10.1】 使用 jQuery 插件。

(1) 用户可以从官方网站下载 jquery.form.js 文件，然后放在网站目录下。

(2) 创建服务器端处理文件 10.1.aspx，然后放在网站目录下。具体代码如下：

```
<%@ Page Language="C#" ContentType="text/html" ResponseEncoding="gb2312" %>
<%@ Import Namespace="System.Data" %>
<%
Response.CacheControl = "no-cache";
Response.AddHeader("Pragma","no-cache");
string back = "";
back += "用户: " + Request["name"];
back += "<br>";
back += "评论: " + Request["comment"];
Response.Write(back);
%>
```

(3) 新建网页文件 10.1.html，在 head 部分引入 jQuery 库和 Form 插件库文件。具体代码如下：

```
<!DOCTYPE html>
<html>
<head>
```

```
<script src="jquery.min.js"></script>
<script src="jquery.form.js"></script>
<script>
   // 等待加载
   $(document).ready(function() {
       // 给 myForm 绑定一个回调函数
       $('#myForm').ajaxForm(function() {
           alert("恭喜，评论发表成功！");
       });
   });
</script>
</head>
<body>
<form id="myForm" action="10.1.aspx" method="post">
   用户名：<input type="text" name="name" />
   </br>
   评论内容：<textarea name="comment"></textarea>
   <input type="submit" value="发表评论" />
</form>
</body>
<html>
```

在 IE 11.0 中浏览，输入用户名和评论内容，单击【发表评论】按钮，结果如图 10-1 所示。

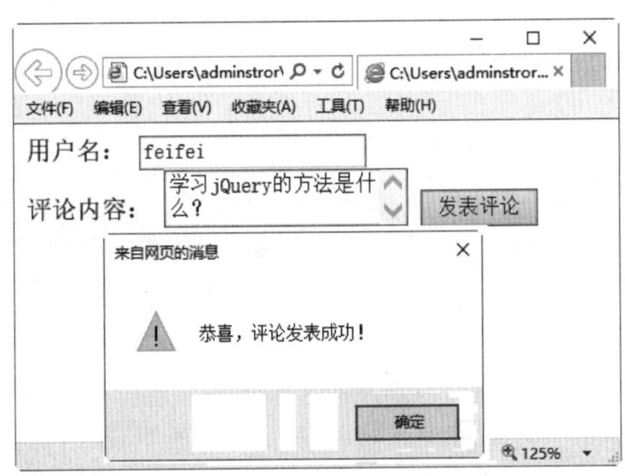

图 10-1 程序运行的结果

10.2 流行的插件

jQuery 官方网站中有很多现成的插件。在官方主页中单击 Plugins 超链接，即可在打开的页面中查看和下载 jQuery 提供的插件，如图 10-2 所示。本章主要介绍目前比较流行的插件。

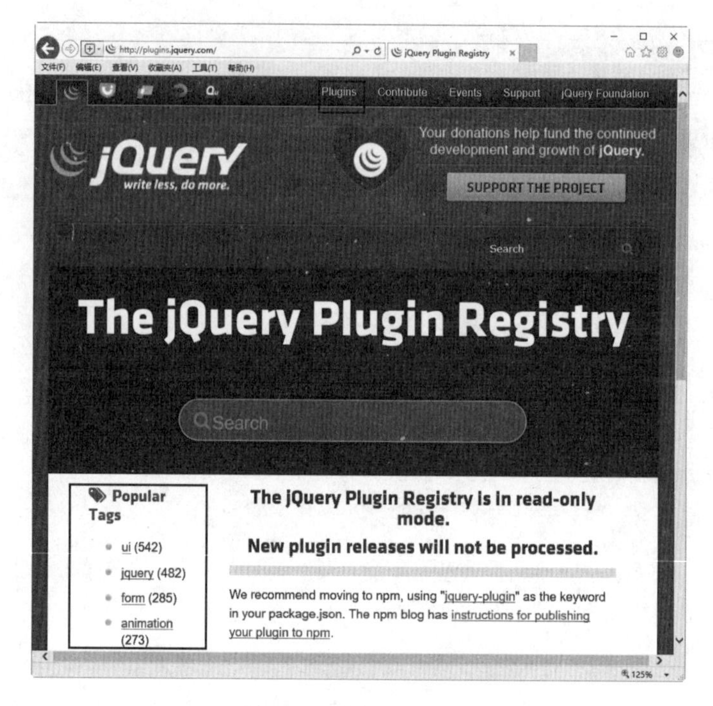

图 10-2　插件下载页面

10.2.1　jQueryUI 插件

jQueryUI 是一个基于 jQuery 的用户界面开发库，主要由 UI 小部件和 CSS 样式表集合而成，它们被打包到一起，以完成常用的任务。

在下载 jQueryUI 包时，还需要注意其他一些文件。development-bundle 目录下包含了 demonstrations 和 documentation，它们虽然有用，但不是产品环境下部署所必需的。但是，在 css 和 js 目录下的文件，必须部署到 Web 应用程序中。js 目录包含 jQuery 和 jQueryUI 库；而 css 目录包括 CSS 文件和所有生成小部件和样式表所需的图片。

UI 插件主要可以实现鼠标互动，包括拖曳、排序、选择、缩放等效果，另外还有折叠菜单、日历、对话框、滑动条、表格排序、页签、放大镜效果)和阴影效果等。

下面通过拖曳的示例来讲解具体的使用方法。

jQueryUI 提供的 API 极大地简化了拖曳功能的开发。只需要分别在拖曳源(source)和目标(target)上调用 draggable 函数即可。

【例 10.2】(示例文件 ch10\10.2.html)

使用 jQueryUI 提供的 API 实现拖曳功能：

```
<!DOCTYPE html>
<html>
<head>
<title>draggable()</title>
<style type="text/css">
<!--
.block{
```

```
border: 2px solid #760022;
background-color: #ffb5bb;
width: 80px; height: 25px;
margin: 5px; float: left;
padding: 20px; text-align: center;
font-size: 14px;
}
-->
</style>
<script language="javascript" src="jquery.ui/jquery.js"></script>
<script type="text/javascript" src="jquery.min.js"></script>
<script language="javascript" src="jquery.ui/ui.mouse.js"></script>
<script language="javascript" src="jquery.ui/ui.draggable.js"></script>
<script language="javascript">
$(function(){
for(var i=0; i<2; i++){  //添加两个<div>块
$(document.body).append($("<div class='block'>拖块"
  + i.toString() + "</div>").css("opacity",0.6));
}
$(".block").draggable();
});
</script>
</head>
<body>
</body>
</html>
```

在 IE 11.0 中浏览，按住拖块，即可拖曳到指定的位置，效果如图 10-3 所示。

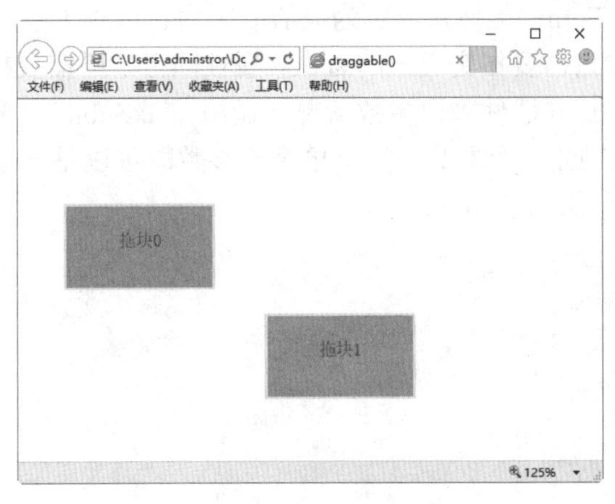

图 10-3 实现了拖曳功能

10.2.2 Form 插件

jQuery Form 插件是一个优秀的 Ajax 表单插件，可以非常容易地使 HTML 表单支持 Ajax。jQuery Form 有两个核心方法：ajaxForm()和 ajaxSubmit()，它们集合了从控制表单元素到决定如何管理提交进程的功能。另外，插件还包括其他一些方法，如 formToArray()、 formSerialize()、fieldSerialize()、fieldValue()、clearForm()、clearFields()、resetForm()等。

1. ajaxForm()

ajaxForm()方法适用于以提交表单方式处理数据。需要在表单中标明表单的 action、id、method 属性，最好在表单中提供 submit 按钮。此方式大大简化了使用 Ajax 提交表单时的数据传递问题，不需要逐个地以 JavaScript 的方式获取每个表单属性的值，并且也不需要通过 url 重写的方式传递数据。ajaxForm()会自动收集当前表单中每个属性的值，然后以表单提交的方式提交到目标 url。这种方式提交数据较安全，并且使用简单，不需要冗余的 JavaScript 代码。

使用时，需要在 document 的 ready 函数中使用 ajaxForm()来为 Ajax 提交表单进行准备。ajaxForm()接收 0 个或 1 个参数。这个单个的参数既可以是一个回调函数，也可以是一个 Options 对象。代码如下：

```
<script>
$(document).ready(function() {
// 给 myFormId 绑定一个回调函数
$('#myFormId').ajaxForm(function() {
alert("成功提交!");
});
});
</script>
```

2. ajaxSubmit()

ajaxSubmit()方法适用于以事件机制提交表单，如通过超链接、图片的 click 事件等提交表单。此方法的作用与 ajaxForm()类似，但更为灵活，因为它依赖于事件机制，只要有事件存在就能使用该方法。使用时只需要指定表单的 action 属性即可，不需要提供 submit 按钮。

在使用 jQuery 的 Form 插件时，多数情况下调用 ajaxSubmit()来对用户提交表单进行响应。ajaxSubmit()接收 0 个或 1 个参数。这个单个的参数既可以是一个回调函数，也可以是一个 options 对象。一个简单的例子如下：

```
$(document).ready(function(){
$('#btn').click(function(){
$('#registerForm').ajaxSubmit(function(data){
alert(data);
});
return false;
});
});
```

上述代码通过表单中 id 为 btn 的按钮的 click 事件触发，并通过 ajaxSubmit()方法以异步 Ajax 方式提交表单到表单的 action 所指路径。

简单地说，通过 Form 插件的这两个核心方法，都可以在不修改表单的 HTML 代码结构的情况下，轻易地将表单的提交方式升级为 Ajax 提交方式。当然，Form 插件还拥有很多方法，这些方法可以帮助用户很容易地管理表单数据和表单提交。

10.2.3 提示信息插件

在网站开发过程中，有时想要实现对于一篇文章的关键词部分的提示，也就是当鼠标移动到这个关键词时，弹出相关的一段文字或图片的介绍。这就需要使用到 jQuery 的 clueTip 插件来实现。

clueTip 是一个 jQuery 工具提示插件，可以方便地为链接或其他元素添加 Tooltip 功能。当链接包括 title 属性时，它的内容将变成 clueTip 的标题。clueTip 中显示的内容可以通过 Ajax 获取，也可以从当前页面的元素中获取。

使用 clueTip 的具体操作步骤如下。

(1) 引入 jQuery 库和 clueTip 插件的 js 文件。插件的下载地址为：

```
http://plugins.learningjquery.com/cluetip/demo/
```

引用插件的.js 文件如下：

```
<link rel="stylesheet" href="jquery.cluetip.css" type="text/css" />
<script src="jquery.min.js" type="text/javascript"></script>
<script src="jquery.cluetip.js" type="text/javascript"></script>
```

(2) 建立 HTML 结构，如下面的格式：

```
<!-- use ajax/ahah to pull content from fragment.html: -->
<p>
<a class="tips" href="fragment.html"
  rel="fragment.html">show me the cluetip!</a>
</p>
<!-- use title attribute for clueTip contents, but don't include anything
in the clueTip's heading -->
<p>
<a id="houdini" href="houdini.html"
  title="|Houdini was an escape artist.
  |He was also adept at prestidigitation.">Houdini</a>
</p>
```

(3) 初始化插件，代码如下：

```
$(document).ready(function() {
$('a.tips').cluetip();
$('#houdini').cluetip({
   //使用调用元素的 title 属性来填充 clueTip，在有“|”的地方将内容分裂成独立的 div
splitTitle: '|',
showTitle: false    //隐藏 clueTip 的标题
});
});
```

10.2.4 jcarousel 插件

jcarousel 是一款 jQuery 插件，用来控制水平或垂直排列的列表项。例如如图 10-4 所示的滚动切换效果。单击左右两侧的箭头，可以向左或者向右查看图片。当到达第一张图片时，左边的箭头变为不可用状态；当到达最后一张图片时，右边的箭头变为不可用状态。

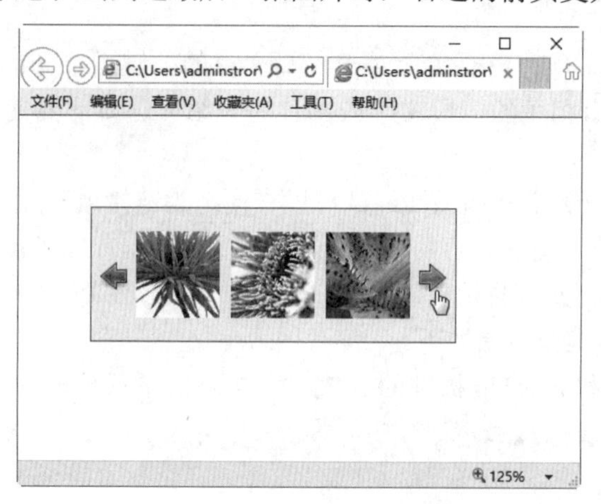

图 10-4　图片滚动切换效果

使用的相关代码如下：

```
<script type="text/javascript" src="../lib/jquery-3.0.6.pack.js"></script>
<script type="text/javascript"
  src="../lib/jquery.jcarousel.pack.js"></script>
<link rel="stylesheet" type="text/css"
  href="../lib/jquery.jcarousel.css" />
<link rel="stylesheet" type="text/css" href="../skins/tango/skin.css" />
<script type="text/javascript">
jQuery(document).ready(function() {
jQuery('#mycarousel').jcarousel();
});
```

10.3　定义自己的插件

除了可以使用现成的插件以外，用户还可以自定义插件。

10.3.1　插件的工作原理

jQuery 插件的机制很简单，就是利用 jQuery 提供的 jQuery.fn.extend()和 jQuery.extend()方法扩展 jQuery 的功能。知道了插件的机制之后，编写插件就容易了，只要按照插件的机制和功能要求编写代码，就可以实现自定义功能的插件。

而要按照机制编写插件，还需要了解插件的种类，插件一般分为 3 类：封装对象方法插件、封装全局函数插件和选择器插件。

1. 封装对象方法插件

这种插件是将对象方法封装起来，用于对通过选择器获取的 jQuery 对象进行操作，是最常见的一种插件。此类插件可以发挥出 jQuery 选择器的强大优势，有相当一部分的 jQuery 的方法都是在 jQuery 脚本库内部通过这种形式“插”在内核上的，如 parent()方法、appendTo()方法等。

2. 封装全局函数插件

可以将独立的函数加到 jQuery 命名空间下。添加一个全局函数，只需要做如下定义：

```
jQuery.foo = function() {
alert('这是函数的具体内容.');
};
```

当然，用户也可以添加多个全局函数：

```
jQuery.foo = function() {
alert('这是函数的具体内容.');
};
jQuery.bar = function(param) {
alert('这是另外一个函数的具体内容".');
};
```

调用时与函数是一样的：jQuery.foo()、jQuery.bar()或者$.foo()、$.bar(‘bar’)。

例如，常用的 jQuery.ajax()方法、去首尾空格的 jQuery.trim()方法都是 jQuery 内部作为全局函数的插件附加到内核上去的。

3. 选择器插件

虽然 jQuery 的选择器十分强大，但在少数情况下，还是会需要用到选择器插件来扩充一些自己喜欢的选择器。

jQuery.fn.extend()多用于扩展上面提到的 3 种类型中的第一种，而 jQuery.extend()则用于扩展后两种插件。这两个方法都接收一个类型为 Object 的参数。Object 对象的“名/值对”分别代表“函数或方法名/函数主体”。

10.3.2 自定义一个简单的插件

下面通过一个例子来讲解如何自定义一个插件。定义的插件功能是：在列表元素中，当鼠标在列表项上移动时，其背景颜色会根据设定的颜色而改变。

【例 10.3】(示例文件 10.3.html 和 10.3.js)

一个简单的插件示例 10.3.js：

```
/// <reference path="jquery.min.js"/>
/*-------------------------------------------------------------/
功能：设置列表中表项获取鼠标焦点时的背景色
参数：li_col【可选】 鼠标所在表项行的背景色
返回：原调用对象
示例：$("ul").focusColor("red");
/------------------------------------------------------------*/
; (function($) {
```

```
    $.fn.extend({
        "focusColor": function(li_col) {
            var def_col = "#ccc"; //默认获取焦点的色值
            var lst_col = "#fff"; //默认丢失焦点的色值
            //如果设置的颜色不为空，使用设置的颜色，否则为默认色
            li_col = (li_col == undefined) ? def_col : li_col;
            $(this).find("li").each(function() { //遍历表项<li>中的全部元素
                $(this).mouseover(function() { //获取鼠标焦点事件
                    $(this).css("background-color", li_col); //使用设置的颜色
                }).mouseout(function() { //鼠标焦点移出事件
                    $(this).css("background-color", "#fff"); //恢复原来的颜色
                })
            })
            return $(this); //返回 jQuery 对象，保持链式操作
        }
    });
})(jQuery);
```

不考虑实际的处理逻辑时，该插件的框架如下：

```
; (function($) {
    $.fn.extend({
        "focusColor": function(li_col) {
            //各种默认属性和参数的设置
            $(this).find("li").each(function() { //遍历表项<li>中的全部元素
            //插件的具体实现逻辑
            })
            return $(this); //返回 jQuery 对象，保持链式操作
        }
    });
})(jQuery);
```

各种默认属性和参数设置的处理中，创建颜色参数以允许用户设定自己的颜色值；并根据参数是否为空来设定不同的颜色值。代码如下：

```
var def_col = "#ccc"; //默认获取焦点的色值
var lst_col = "#fff"; //默认丢失焦点的色值
//如果设置的颜色不为空，使用设置的颜色，否则为默认色
li_col = (li_col == undefined)? def_col : li_col;
```

在遍历列表项时，针对鼠标移入事件 mouseover()设定对象的背景色，并且在鼠标移出事件 mouseout()中恢复原来的背景色。代码如下：

```
$(this).mouseover(function() { //获取鼠标焦点事件
    $(this).css("background-color", li_col); //使用设置的颜色
}).mouseout(function() { //鼠标焦点移出事件
    $(this).css("background-color", "#fff"); //恢复原来的颜色
})
```

当调用此插件时，需要先引入插件的.js 文件，然后调用该插件中的方法。

示例的 HTML 代码如下：

```
<!DOCTYPE html>
<html>
<head>
   <title>简单的插件示例</title>
   <script type="text/javascript"  src="jquery.min.js"></script>
   <script type="text/javascript" src="10.3.js"></script>
   <style type="text/css">
body{font-size:12px}
.divFrame{width:260px;border:solid 1px #666}
.divFrame .divTitle{
padding:5px;background-color:#eee;font-weight:bold}
.divFrame .divContent{padding:8px;line-height:1.6em}
.divFrame .divContent ul{padding:0px;margin:0px;
   list-style-type:none}
.divFrame .divContent ul li span{margin-right:20px}
   </style>
   <script type="text/javascript">
      $(function() {
         $("#u1").focusColor("red"); //调用自定义的插件
      })
   </script>
</head>
<body>
   <div class="divFrame">
       <div class="divTitle">对象级别的插件</div>
       <div class="divContent">
           <ul id="u1">
               <li><span>张三</span><span>男</span></li>
               <li><span>李四</span><span>女</span></li>
               <li><span>王五</span><span>男</span></li>
           </ul>
       </div>
    </div>
</body>
</html>
```

在 IE 11.0 中浏览，效果如图 10-5 所示。

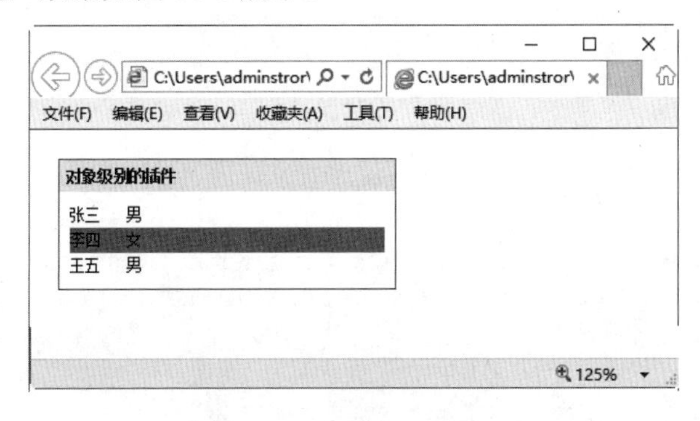

图 10-5 使用自定义插件

10.4 实战演练——创建拖曳购物车效果

jQueryUI 插件除了提供 draggable()来实现鼠标的拖曳功能外，还提供了一个 droppable()方法，实现接收容器。通过上述方法，可以实现购物的拖曳效果。

【例 10.4】(示例文件 ch10\10.4.html)

创建拖曳购物车效果：

```
<!DOCTYPE html>
<html>
<head>
<title>droppable()</title>
<style type="text/css">
<!--
.draggable{
   width:70px; height:40px;
   border:2px solid;
   padding:10px; margin:5px;
   text-align:center;
}
.green{
   background-color:#73d216;
   border-color:#4e9a06;
}
.red{
   background-color:#ef2929;
   border-color:#cc0000;
}
.droppable {
   position:absolute;
   right:20px; top:20px;
   width:400px; height:300px;
   background-color:#b3a233;
   border:3px double #c17d11;
   padding:5px;
   text-align:center;
}
-->
</style>
<script language="javascript" src="jquery.ui/jquery-1.2.4a.js"></script>
<script language="javascript" src="jquery.ui/ui.base.min.js"></script>
<script language="javascript" src="jquery.ui/ui.draggable.min.js"></script>
<script language="javascript" src="jquery.ui/ui.droppable.min.js"></script>
<script language="javascript">
$(function(){
   $(".draggable").draggable({helper:"clone"});
   $("#droppable-accept").droppable({
       accept: function(draggable){
```

```
            return $(draggable).hasClass("green");
        },
        drop: function(){
            $(this).append($("<div></div>").html("成功添加到购物车！"));
        }
    });
});
</script>
</head>
<body>
<div class="draggable red">冰箱</div>
<div class="draggable green">空调</div>
<div id="droppable-accept" class="droppable">购物车<br></div>
</body>
</html>
```

在 IE 11.0 中浏览，按住拖块，即可拖曳到指定的购物车中，效果如图 10-6 所示。

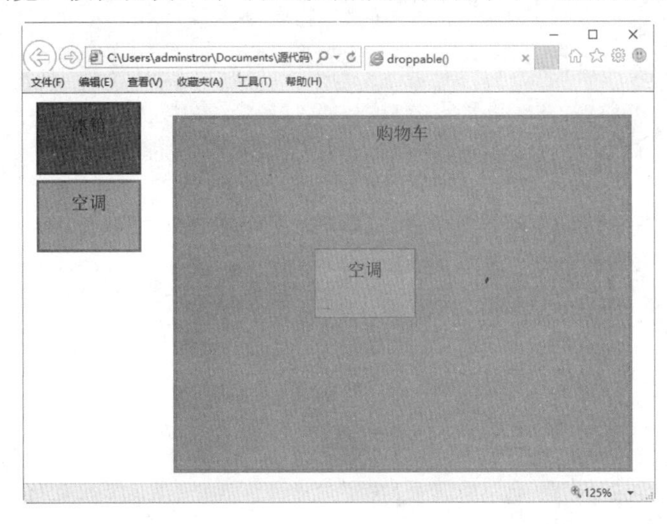

图 10-6　拖曳购物车效果

10.5　疑难解惑

疑问 1：编写 jQuery 插件时需要注意什么？

答：需要注意以下几点。

(1) 插件的推荐命名方法为 jquery.[插件名].js。

(2) 所有的对象方法都应当附加到 jQuery.fn 对象上面，而所有的全局函数都应当附加到 jQuery 对象本身上。

(3) 在插件内部，this 指向的是当前通过选择器获取的 jQuery 对象，而不像一般方法那样，内部的 this 指向的是 DOM 元素。

(4) 可以通过 this.each 来遍历所有的元素。

(5) 所有方法或函数插件，都应当以分号结尾，否则压缩的时候可能会出现问题。为了更加保险些，可以在插件头部添加一个分号(;)，以免它们的不规范代码给插件带来影响。

(6) 插件应该返回一个 jQuery 对象，以便保证插件的可链式操作。

(7) 避免在插件内部使用$作为 jQuery 对象的别名，而应当使用完整的 jQuery 来表示。这样可以避免冲突。

疑问 2：如何避免插件函数或变量名冲突？

答：虽然在 jQuery 命名空间中禁止使用了大量的 JavaScript 函数名和变量名，但是仍然不可避免某些函数或变量名将与其他 jQuery 插件冲突。因此，需要将一些方法封装到另一个自定义的命名空间。

例如下面的使用空间的例子：

```
jQuery.myPlugin = {
foo:function() {
alert('This is a test. This is only a test.');
},
bar:function(param) {
alert('This function takes a parameter, which is "' + param + '".');
}
};
```

采用命名空间的函数仍然是全局函数，调用时采用的代码如下：

```
$.myPlugin.foo();
$.myPlugin.bar('baz');
```

第 3 篇

移动网页开发

- 第 11 章　走进 jQuery Mobile
- 第 12 章　jQuery Mobile UI 组件
- 第 13 章　jQuery Mobile 事件

第 11 章
走进 jQuery Mobile

jQuery 函数库往往用于 PC 端网页的开发。为了实现跨平台设计网页，用户可以使用 jQuery Mobile 开发移动网站。HTML 5 和函数库 jQuery Mobile 一起开发动移动网站，可以解决不同移动设备上显示界面统一的问题。本章重点学习 jQuery Mobile 的基础知识。

11.1 认识 jQuery Mobile

jQuery Mobile 是 jQuery 在手机上和平板设备上的版本。jQuery Mobile 不仅会给主流移动平台带来 jQuery 核心库，而且会发布一个完整统一的 jQuery 移动 UI 框架。通过 jQuery Mobile 制作出来的网页能够支持全球主流的移动平台，而且在浏览网页时，能够拥有操作应用软件一样的触碰和滑动效果。

jQuery Mobile 的优势如下。

(1) 简单易用。jQuery Mobile 简单易用。页面开发主要使用标记，无须或仅需很少 JavaScript。jQuery Mobile 通过 HTML 5 标记和 CSS 3 规范来配置和美化页面，对于已经熟悉 HTML 5 和 CSS 3 的读者来说，上手非常容易，架构清晰。

(2) 跨平台。目前大部分的移动设备浏览器都支持 HTML 5 标准和 jQuery Mobile，所以可以实现跨平台的移动设备。例如 Android、Apple IOS、BlackBerry、Windows Phone、Symbian 和 MeeGo 等。

(3) 提供丰富的函数库。常见的键盘、触碰功能等，开发人员不用编写代码，只需要经过简单的设置，就可以实现需要的功能，大大减少了程序员开发的时间。

(4) 丰富的布景主题和 ThemeRoller 工具。jQuery Mobile 提供了布局主题，通过这些主题，可以轻轻松松地快速创建绚丽多彩的网页。通过使用 jQuery UT 的 ThemeRoller 在线工具，只需要在下拉菜单中进行简单的设置，就可以制作出丰富多彩的网页风格，并且可以将代码下载下来应用。

jQuery Mobile 的操作流程如下。

(1) 创建 HTML 5 文件。

(2) 载入 jQuery、jQuery Mobile 和 jQuery Mobile CSS 链接库。

(3) 使用 jQuery Mobile 定义的 HTML 标准，编写网页架构和内容。

11.2 跨平台移动设备网页 jQuery Mobile

学习移动设备的网页设计开发，遇到的最大难题是跨浏览器支持的问题。为了解决这个问题，jQuery 推出了新的函数库 jQuery Mobile，主要用于统一当前移动设备的用户界面。

11.2.1 移动设备模拟器

网页制作完成后，需要在移动设备上预览最终的效果。为了方便预览效果，用户可以使用移动设备模拟器，常见的移动设备模拟器是 Opera Mobile Emulator。

Opera Mobile Emulator 是一款针对电脑桌面开发的模拟移动设备的浏览器，几乎完全重现 Opera Mobile 手机浏览器的使用效果，可自行设置需要模拟的不同型号的手机和平板电脑配置，然后在电脑上模拟各类手机等移动设备访问网站。

Opera Mobile Emulator 的下载网址：http://www.opera.com/zh-cn/developer/mobile-emulator/，根据不同的系统选择不同的版本，这里选择 Windows 系统下的版本，如图 11-1 所示。

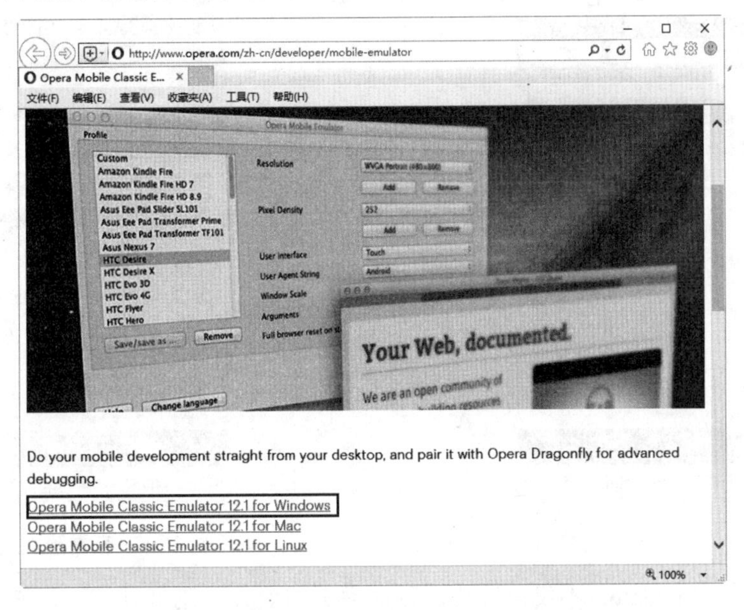

图 11-1　Opera Mobile Emulator 的下载页面

下载并安装之后启动 Opera Mobile Emulator，打开如图 11-2 所示的窗口，在【资料】列表中选择移动设备的类型，这里选择 LG Optimus 3D 选项，单击【启动】按钮。

打开欢迎界面，用户可以单击不同的链接，查看该软件的功能，如图 11-3 所示。

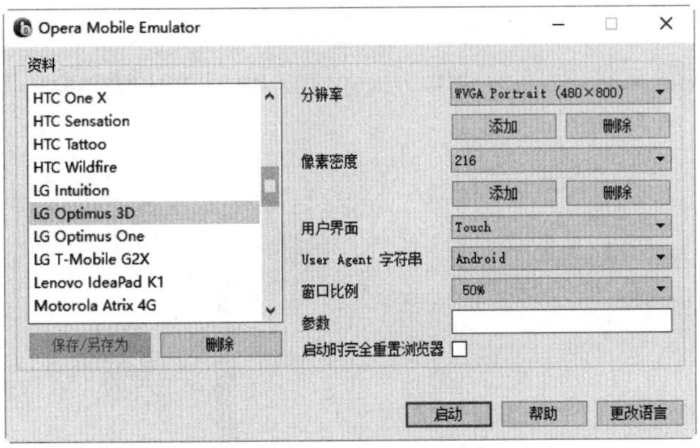

图 11-2　参数设置界面

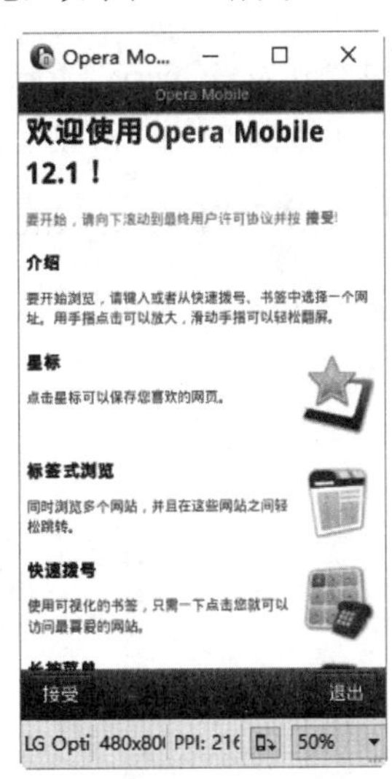

图 11-3　欢迎界面

单击【接受】按钮，打开手机模拟器窗口，在【输入网址】文本框中输入需要查看网页效果的地址即可，如图 11-4 所示。

例如，这里直接单击【当当网】图标，即可查看当当网在该移动设备模拟器中的效果，如图 11-5 所示。

图 11-4　手机模拟器窗口

图 11-5　查看预览效果

Opera Mobile Emulator 不仅可以查看移动网页的效果，还可以任意调整窗口的大小，从而可以查看不同屏幕尺寸的效果。这点也是 Opera Mobile Emulator 与其他移动设备模拟器相比最大的优势。

11.2.2　jQuery Mobile 的安装

想要开发 jQuery Mobile 网页，必须引用 JavaScript 函数库(.js)、CSS 样式表和配套的 jQuery 函数库文件。常见的引用方法有以下两种。

1. 直接引用 jQuery Mobile 库文件

从 jQuery Mobile 的官网下载该库文件(网址是 http://jquerymobile.com/download/)，如图 11-6 所示。

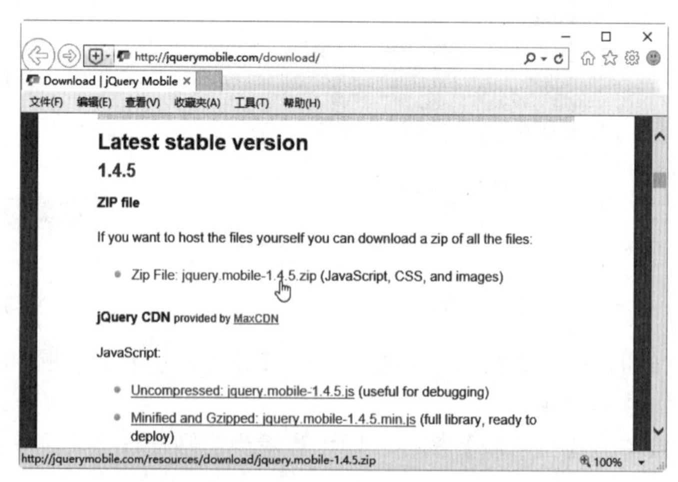

图 11-6　下载 jQuery Mobile 库文件

下载完成即可解压，然后直接引用文件即可，代码如下：

```
<head>
<meta name="viewport" content="width=device-width, initial-scale=1">
<link rel="stylesheet" href="jquery.mobile-1.4.5.css">
<script src="jquery.js"></script>
<script src="jquery.mobile-1.4.5.js"></script>
</head>
```

将下载的文件解压到和网页位于同一目录下，否则会无法引用而报错。

细心的读者会发现，在<script>标签中没有插入 type="text/javascript"，这是什么原因呢？因为所有浏览器中 HTML 5 的默认脚本语言都是 JavaScript，所以在 HTML 5 中已经不再需要该属性。

2. 从 CDN 中加载 jQuery Mobile

CDN 的全称是 Content Delivery Network，即内容分发网络。其基本思路是尽可能避开互联网上有可能影响数据传输速度和稳定性的瓶颈和环节，使内容传输得更快、更稳定。

从 CDN 中加载 jQuery Mobile, 用户不需要在电脑上安装任何东西。用户仅仅需要在网页中加载层叠样式(.css)和 JavaScript 库 (.js) 就能够使用 jQuery Mobile 了。

用户可以从 jQuery Mobile 官网中查找引用路径，网址是：http://jquerymobile.com/download/，进入该网站后，找到 jQuery Mobile 的引用链接，然后将其复制后添加到 HTML 文件<head>标记中即可，如图 11-7 所示。

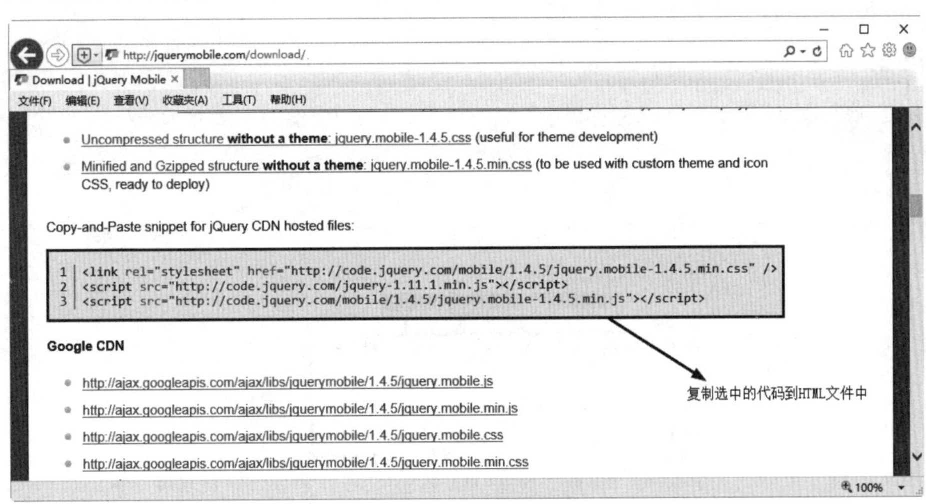

图 11-7　复制代码

将代码复制到<head>标记块内，代码如下：

```
<head>
<!-- meta 使用 viewport 以确保页面可自由缩放 -->
<meta name="viewport" content="width=device-width, initial-scale=1">
```

```
<!-- 引入 jQuery Mobile 样式 -->
 <link rel="stylesheet"
href="http://code.jquery.com/mobile/1.4.5/jquery.mobile-1.4.5.min.css"/>
<!-- 引入 jQuery 库 -->
 <script src="http://code.jquery.com/jquery-1.11.3.min.js"></script>
<!-- 引入 jQuery Mobile 库 -->
 <script src="http://code.jquery.com/mobile/1.4.5/jquery.mobile-
1.4.5.min.js"></script>
 </head>
```

由于 jQuery Mobile 函数库仍然在开发中，所以引用的链接中的版本号可能会与本书不同，请使用官方提供的最新版本，只要按照上述方法将代码复制下来引用即可。

11.2.3 jQuery Mobile 网页的架构

jQuery Mobile 网页是由 header、content 与 footer 3 个区域组成的架构。利用<div>标记加上 HTML 5 自定义属性 data-*来定义移动设备网页组件样式。最基本的属性 data-role 可以用来定义移动设备的页面架构，语法格式如下：

```
<div data-role="page">
<!—开始一个 page-->
  <div data-role="header">
    <h1>这个是标题</h1>
  </div>
  <div data-role="main" class="ui-content">
    <p>这里是内容</p>
  </div>
  <div data-role="footer">
    <h1>底部文本</h1>
  </div>
</div>
```

在模拟器中的预览效果如图 11-8 所示。

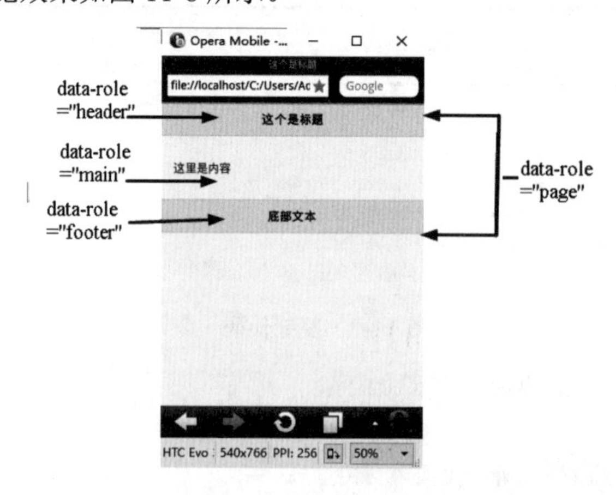

图 11-8　程序预览效果

从结果可以看出，jQuery Mobile 网页以页(page)为单位，一个 HTML 页面可以放一个页面，也可以放多个页面，只是浏览器每次只会显示一页。如果有多个页面，需要在页面中添加超链接，从而实现多个页面的切换。

【案例分析】

(1) data-role="page" 是在浏览器中显示的页面。

(2) data-role="header" 是在页面顶部创建的工具条，通常用于标题或者搜索按钮。

(3) data-role="main" 定义了页面的内容，如文本、图片、表单、按钮等。

(4) "ui-content" 类用于在页面添加内边距和外边距。

(5) data-role="footer" 用于创建页面底部工具条。

11.3 创建多页面的 jQuery Mobile 网页

本案例将使用 jQuery Mobile 制作一个多页面的 jQuery Mobile 网页，并创建多个页面。使用不同的 id 属性来区分不同的页面。

【例 11.1】(示例文件 ch11\11.1.html)

```
<!DOCTYPE html>
<html>
<head>
<meta name="viewport" content="width=device-width, initial-scale=1">
<link rel="stylesheet"
href="http://code.jquery.com/mobile/1.4.5/jquery.mobile-1.4.5.min.css">
<script src="http://code.jquery.com/jquery-1.11.3.min.js"></script>
 <script src="http://code.jquery.com/mobile/1.4.5/jquery.mobile-
1.4.5.min.js"></script>
 </head>
<body>
<div data-role="page" id="first">
  <div data-role="header">
    <h1>古诗欣赏</h1>
  </div>
  <div data-role="main" class="ui-content">
    <p>几回花下坐吹箫，银汉红墙入望遥。</p>
    <a href="#second">下一页</a>
  </div>
  <div data-role="footer">
    <h1>清代诗人</h1>
  </div>
</div>
<div data-role="page" id="second">
  <div data-role="header">
    <h1>古诗欣赏</h1>
  </div>
  <div data-role="main" class="ui-content">
    <p>似此星辰非昨夜，为谁风露立中宵。</p>
```

```
    <a href="#first">上一页</a>
  </div>
  <div data-role="footer">
    <h1>清代诗人</h1>
  </div>
</div>
</body>
</html>
```

在模拟器中的预览效果如图 11-9 所示。单击【下一页】超链接，即可进入第二页，如图 11-10 所示。单击【上一页】超链接，即可返回到第一页中。

图 11-9　程序预览效果

图 11-10　第二页预览效果

11.4　将页面作为对话框使用

对话框是用于页面信息显示或者表单信息的输入。jQuery Mobile 通过在链接中添加如下属性，即可将页面作为对话框使用：

```
data-dialog="true"
```

【例 11.2】(示例文件 ch11\11.2.html)

```
<!DOCTYPE html>
<html>
<head>
<meta name="viewport" content="width=device-width, initial-scale=1">
<link rel="stylesheet"
href="http://code.jquery.com/mobile/1.4.5/jquery.mobile-1.4.5.min.css">
<script src="http://code.jquery.com/jquery-1.11.3.min.js"></script>
 <script src="http://code.jquery.com/mobile/1.4.5/jquery.mobile-
1.4.5.min.js"></script>
</head>
<body>
<div data-role="page" id="first">
```

```
  <div data-role="header">
    <h1>古诗鉴赏</h1>
  </div>
  <div data-role="main" class="ui-content">
    <p>浩荡离愁白日斜，吟鞭东指即天涯。落红不是无情物，化作春泥更护花。</p>
    <a href="#second">查看详情</a>
  </div>
  <div data-role="footer">
    <h1>清代诗词</h1>
  </div>
</div>
<div data-role="page" data-dialog="true" id="second">
  <div data-role="header">
    <h1>诗词鉴赏</h1>
  </div>
  <div data-role="main" class="ui-content">
    <p>这首诗是《己亥杂诗》的第五首，写诗人离京的感受。虽然载着"浩荡离愁"，却表示仍然要
为国为民尽自己最后一份心力。</p>
    <a href="#first">上一页</a>
  </div>
  <div data-role="footer">
    <h1>清代诗词</h1>
  </div>
</div>
</body>
</html>
```

在模拟器中的预览效果如图 11-11 所示。单击【查看详情】超链接，即可打开一个对话框，如图 11-12 所示。

图 11-11　程序预览效果

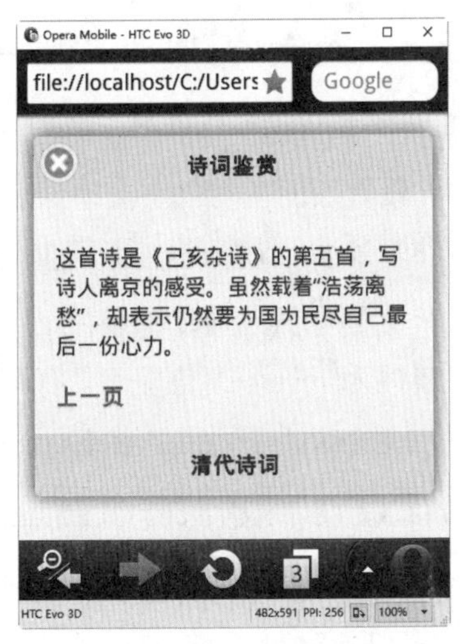

图 11-12　对话框预览效果

从结果可以看出，对话框与普通页面不同，它显示在当前页面上，但又不会填充完整的页面，顶部图标用于关闭对话框，单击【上一页】链接，也可以关闭对话框。

11.5　绚丽多彩的页面切换效果

jQuery Mobile 提供了各种页面切换到下一个页面的效果。主要通过设置 data-transition 属性来完成各种页面切换效果，语法格式如下：

```
<a href="#link" data-transition="切换效果">切换下一页</a>
```

其中切换效果有很多，如表 11-1 所示。

表 11-1　页面切换效果

页面效果参数	含　义
fade	默认的切换效果。淡入到下一页
none	无过渡效果
flip	从后向前翻转到下一页
flow	抛出当前页，进入下一页
pop	像弹出窗口那样转到下一页
slide	从右向左滑动到下一页
slidefade	从右向左滑动并淡入到下一页
slideup	从下到上滑动到下一页
slidedown	从上到下滑动到下一页
turn	转向下一页

在 jQuery Mobile 的所有链接上，默认使用淡入淡出的效果。

例如，设置页面从右向左滑动到下一页，代码如下：

```
<a href="#second" data-transition="slide">切换下一页</a>
```

上面的所有效果支持后退行为。例如，用户想让页面从左向右滑动，可以添加 data-direction 属性为"reverse" 值即可，代码如下：

```
<a href="#second" data-transition="slide" data-direction="reverse">切换下一页</a>
```

【例 11.3】(示例文件 ch11\11.3.html)

```
<!DOCTYPE html>
<html>
<head>
<meta name="viewport" content="width=device-width, initial-scale=1">
<link rel="stylesheet"
href="http://code.jquery.com/mobile/1.4.5/jquery.mobile-1.4.5.min.css">
<script src="http://code.jquery.com/jquery-1.11.3.min.js"></script>
 <script src="http://code.jquery.com/mobile/1.4.5/jquery.mobile-
```

```
1.4.5.min.js"></script>
 </head>
<body>
<div data-role="page" id="first">
  <div data-role="header">
    <h1>古诗欣赏</h1>
  </div>
  <div data-role="main" class="ui-content">
    <p>老农家贫在山住，耕种山田三四亩。</p>
<!—实现从右到左切换到下一页 -->
    <a href="#second" data-transition="slide" >下一页</a>
  </div>
  <div data-role="footer">
    <h1>野老歌</h1>
  </div>
</div>
<div data-role="page" id="second">
  <div data-role="header">
    <h1>古诗欣赏</h1>
  </div>
  <div data-role="main" class="ui-content">
    <p>岁暮锄犁傍空室，呼儿登山收橡实。</p>
<!—实现从左到右切换到下一页 -->
    <a href="#first" data-transition="slide" data-direction="reverse">上一页</a>
  </div>
  <div data-role="footer">
    <h1>野老歌</h1>
  </div>
</div>
</body>
</html>
```

在模拟器中的预览效果如图 11-13 所示。单击【下一页】超链接，即可从右到左滑动进入第二页，如图 11-14 所示。单击【上一页】超链接，即可从左到右滑动返回到第一页中。

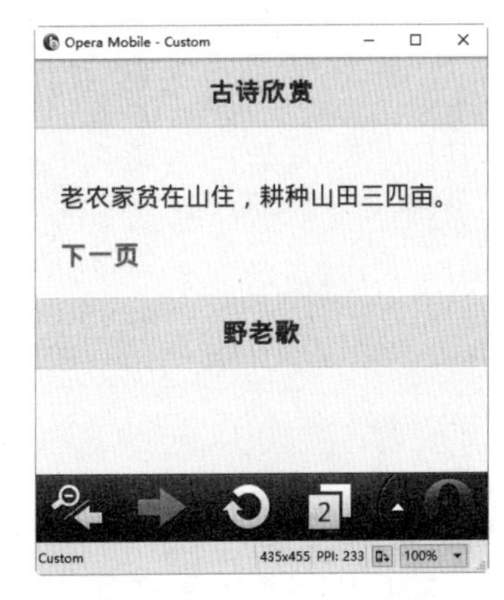

图 11-13 程序预览效果

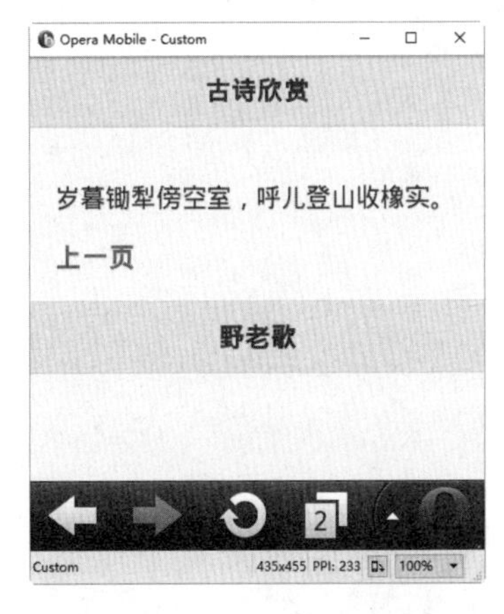

图 11-14 第二页预览效果

11.6 疑难解惑

疑问 1：如何在模拟器中查看做好的网页效果？

答：HTML 文件制作完成后，要想在模拟器中测试，可以在地址栏中输入文件的路径，例如输入如下：

```
file://localhost/ch16/16.2.html
```

为了防止输入错误，可以直接将文件拖曳到地址栏中，模拟器会自动帮助用户添加完整路径。

疑问 2：jQuery Mobile 支持哪些移动设备？

答：目前市面上移动设备非常多，如果想查询 jQuery Mobile 支持哪些移动设备，可以参照 jQuery Mobile 网站的各厂商支持表，还可以参考维基百科网站对 jQuery Mobile 说明中提供的 Mobile browser support 一览表。

疑问 3：我的浏览器为什么不支持页面切换效果？

答：要想实现页面切换效果，浏览器必须支持 CSS3 3D 切换功能。目前，支持 CSS3 3D 切换功能的浏览器最小版本包括 IE 10.0、Chrome 12.0、Firefox 16.0、Safari 4.0 和 Opera 15.0 等。

第 12 章 jQuery Mobile UI 组件

jQuery Mobile 针对用户界面提供了各种可视化的标签，包括按钮、复选框、选择菜单、列表、弹窗、工具栏、面板、导航和布局等。这些可视化标签与 HTML 5 标记一起使用，即可轻轻松松地开发出绚丽多彩的移动网页。本章重点学习这些标签的使用方法和技巧。

12.1 套用 UI 组件

jQuery Mobile 提供很多可视化的 UI 组件，只要套用之后，就可以生成绚丽并且适合移动设备使用的组件。jQuery Mobile 中各种可视化的 UI 组件与 HTML 5 标记大同小异。下面介绍常用的组件的用法，其中按钮、列表等功能变化比较大的后面会做详细介绍。

12.1.1 表单组件

jQuery Mobile 使用 CSS 自动为 HTML 表单添加样式，让它们看起来更具吸引力，触摸起来更具友好性。

在 jQuery Mobile 中，经常使用的表单控件如下。

1. 文本框

文本框的语法规则如下：

```
<input type="text" name="fname" id="fname" value=" ">
```

其中 value 属性是文本框中显示的内容，也可以使用 placeholder 来指定一个简短的描述，用来描述输入内容的含义。

【例 12.1】(示例文件 ch12\12.1.html)

使用文本框：

```
<!DOCTYPE html>
<html>
<head>
<meta name="viewport" content="width=device-width, initial-scale=1">
<link rel="stylesheet"
href="http://code.jquery.com/mobile/1.4.5/jquery.mobile-1.4.5.min.css">
<script src="http://code.jquery.com/jquery-1.11.3.min.js"></script>
 <script src="http://code.jquery.com/mobile/1.4.5/jquery.mobile-
1.4.5.min.js"></script>
 </head>
<body>
<div data-role="first">
  <div data-role="header">
  <h1>输入会员信息</h1>
  </div>
  <div data-role="main" class="ui-content">
    <form>
      <div class="ui-field-contain">
        <label for="fullname">姓名：</label>
        <input type="text" name="fullname" id="fullname">
        <label for="bday">出生年月：</label>
        <input type="date" name="bday" id="bday">
        <label for="email">E-mail:</label>
```

```
      <input type="email" name="email" id="email" placeholder="输入您的电子邮箱">
    </div>
    <input type="submit" data-inline="true" value="注册">
   </form>
  </div>
</div>
</body>
</html>
```

在模拟器中的预览效果如图 12-1 所示。单击【出生年月】文本框时，会自动打开日期选择器，用户直接选择相应的日期即可，如图 12-2 所示。

图 12-1 程序预览效果

图 12-2 日期选择器

2. 文本域

使用<textarea>可以实现多行文本输入效果。

【例 12.2】(示例文件 ch12\12.2.html)

使用文本域：

```
<!DOCTYPE html>
<html>
<head>
<meta name="viewport" content="width=device-width, initial-scale=1">
<link rel="stylesheet"
href="http://code.jquery.com/mobile/1.4.5/jquery.mobile-1.4.5.min.css">
<script src="http://code.jquery.com/jquery-1.11.3.min.js"></script>
 <script src="http://code.jquery.com/mobile/1.4.5/jquery.mobile-
1.4.5.min.js"></script>
 </head>
<body>
<div data-role="first">
  <div data-role="header">
  <h1>多行文本域</h1>
  </div>
```

```
  <div data-role="main" class="ui-content">
    <form>
      <div class="ui-field-contain">
        <label for="info">输入最喜欢的一首古诗:</label>
        <textarea name="addinfo" id="info"></textarea>
      </div>
      <input type="submit" data-inline="true" value="提交">
    </form>
  </div>
</div>
</body>
</html>
```

在模拟器中的预览效果如图 12-3 所示。输入多行内容时，文本框会根据输入的内容，自动调整文本框的高度，如图 12-4 所示。

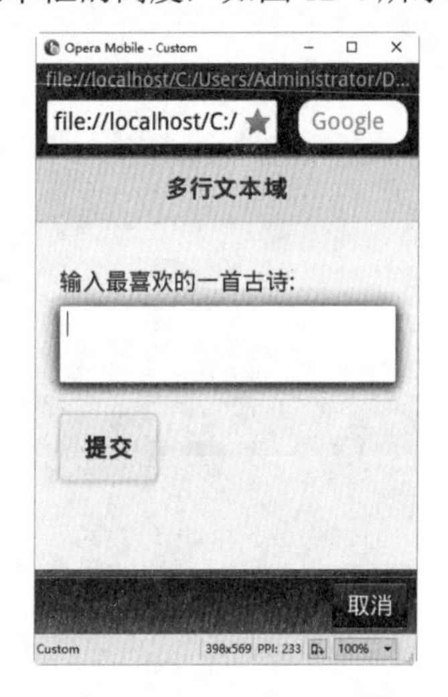

图 12-3　程序预览效果

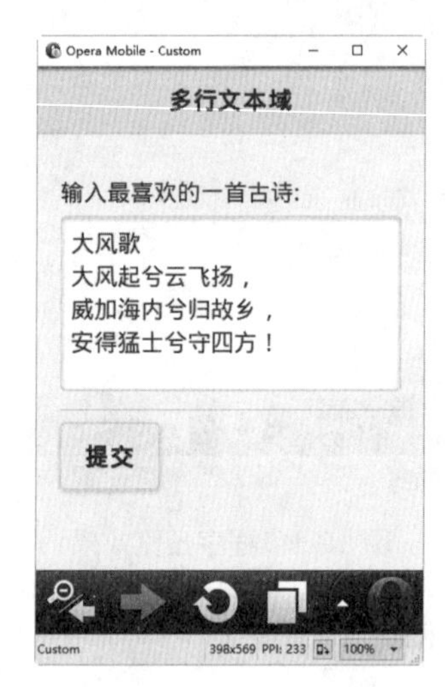

图 12-4　输入多行内容

3. 搜索输入框

HTML 5 中新增的 type="search"类型为搜索输入框，它是为输入搜索定义文本字段。

搜索输入框的语法格式如下：

```
<input type="search" name="search" id="search" placeholder="搜索内容">
```

搜索输入框的效果如图 12-5 所示。

图 12-5　搜索输入框

4. 范围滑动条

使用<input type="range">控件，即可创建范围滑动条，其语法格式如下：

```
<input type="range" name="points" id="points" value="50" min="0" max="100"
data-show-value="true">
```

其中，max 属性规定允许的最大值。min 属性规定允许的最小值。step 属性规定合法的数

字间隔。value 属性规定默认值。data-show-value 属性规定是否在按钮上显示进度的值，如果设置为 true，则表示显示进度的值。如果设置为 false，则表示不显示进度的值。

【例 12.3】(示例文件 ch12\12.3.html)

使用范围滑动条：

```
<!DOCTYPE html>
<html>
<head>
<meta name="viewport" content="width=device-width, initial-scale=1">
<link rel="stylesheet"
href="http://code.jquery.com/mobile/1.4.5/jquery.mobile-1.4.5.min.css">
<script src="http://code.jquery.com/jquery-1.11.3.min.js"></script>
 <script src="http://code.jquery.com/mobile/1.4.5/jquery.mobile-
1.4.5.min.js"></script>
 </head>
<body>
<div data-role="first">
  <div data-role="header">
    <h1>工作进度申报</h1>
  </div>
  <div data-role="main" class="ui-content">
    <form>
      <label for="points">工作完成的进度:</label>
      <input type="range" name="points" id="points" value="50" min="0"
max="100" data-show-value="true">
      <input type="submit" data-inline="true" value="提交">
    </form>
  </div>
</div>
</body>
</html>
```

在模拟器中的预览效果如图 12-6 所示。用户可以拖动滑块，选择需要的值。也可以通过加减按钮，精确选择进度的值。

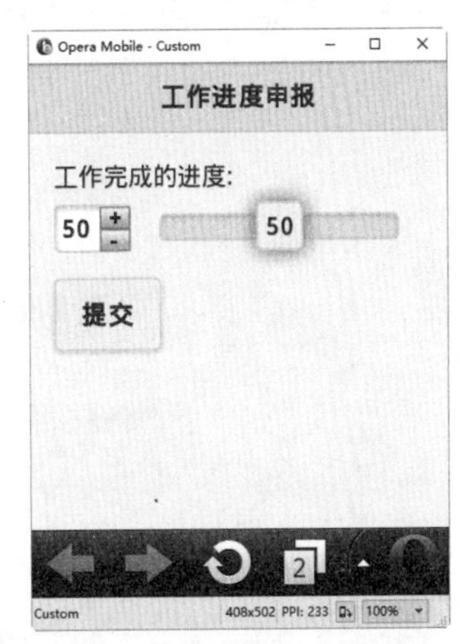

图 12-6　程序预览效果

使用 data-popup-enabled 属性可以设置小弹窗效果，代码如下：

```
<input type="range" name="points" id="points"
value="50" min="0" max="100" data-popup-
enabled="true">
```

添加后的效果如图 12-7 所示。

使用 data-highlight 属性可以高亮度显示滑动条的值，代码如下：

```
<input type="range" name="points" id="points"
value="50" min="0" max="100" data-
highlight="true">
```

添加后的效果如图 12-8 所示。

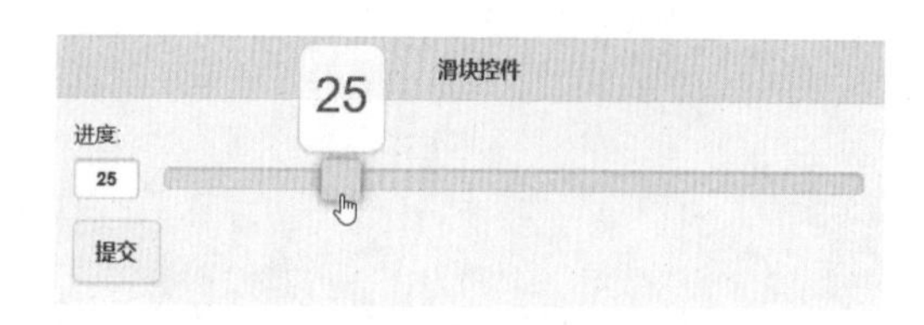

图 12-7　进度值显示效果

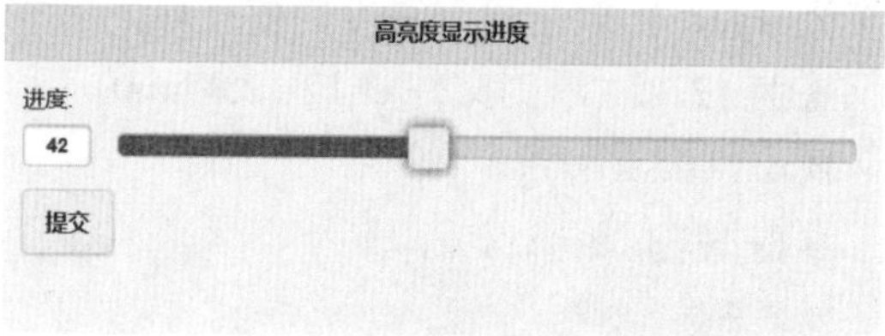

图 12-8　高亮度显示进度值效果

5. 表单按钮

表单按钮分为 3 种：普通按钮、提交按钮和取消按钮。只需要在 type 属性中设置表单的类型即可，代码如下：

```
<input type="submit" value="提交按钮">
<input type="reset" value="取消按钮">
<input type="button" value="普通按钮">
```

在模拟器中的预览效果如图 12-9 所示。

当用户在有限数量的选择中仅选取一个选项时，经常用到表单中的单选按钮。通过 type="radio"来创建一系列单选按钮，代码如下：

```
<fieldset data-role="controlgroup">
<legend>请选择您的年级：</legend>
  <label for="one">一年级</label>
  <input type="radio" name="grade" id="one" value="one">
  <label for="two">二年级</label>
  <input type="radio" name="grade" id="two" value="two">
  <label for="three">三年级</label>
  <input type="radio" name="grade" id=" three" value=" three">
</fieldset>
```

在模拟器中的预览效果如图 12-10 所示。

图 12-9　表单按钮预览效果

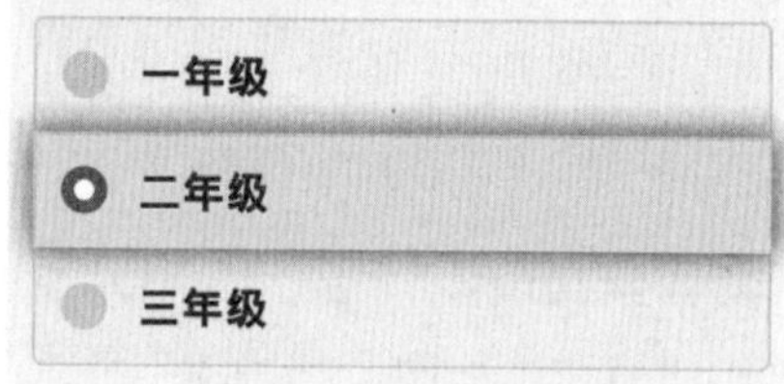

图 12-10　单选按钮

<fieldset>标记用来创建按钮组，组内各个组件保持自己的功能。在<fieldset>标记内添加 data-role="controlgroup"，这样这些单选按钮样式统一，看起来像一个组合。其中<legend>标签用来定义按钮组的标题。

6. 复选框

当用户在有限数量的选择中选取一个或多个选项时，需要使用复选框，代码如下：

```
<fieldset data-role="controlgroup">
  <legend>请选择您喜爱的季节：</legend>
  <label for="spring">春天</label>
  <input type="checkbox" name="season" id="spring" value="spring">
  <label for="summer">夏天</label>
  <input type="checkbox" name="season" id="summer" value="summer">
  <label for="fall">秋天</label>
  <input type="checkbox" name="season" id="fall" value="fall">
  <label for="winter">冬天</label>
  <input type="checkbox" name="season" id="winter" value="winter">
</fieldset>
```

在模拟器中的预览效果如图 12-11 所示。

7. 选择菜单

使用<select>标签可以创建带有若干选项的下拉列表。<select>标签内的<option>属性定义了列表中的可用选项，代码如下：

```
<fieldset data-role="fieldcontain">
      <label for="day">选择值日时间：</label>
      <select name="day" id="day">
       <option value="mon">星期一</option>
       <option value="tue">星期二</option>
       <option value="wed">星期三</option>
       <option value="thu">星期四</option>
       <option value="fri">星期五</option>
       <option value="sat">星期六</option>
       <option value="sun">星期日</option>
      </select>
</fieldset>
```

在模拟器中的预览效果如图 12-12 所示。

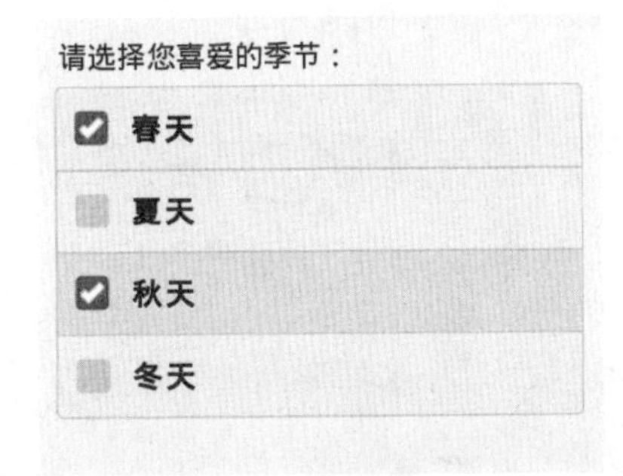

图 12-11　复选框

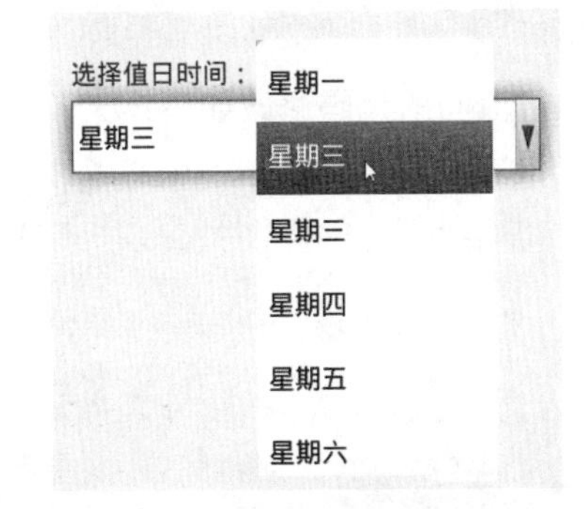

图 12-12　选择菜单

如果菜单中的选项还需要再次分组，可以在<select>内使用<optgroup>标签，添加后的代码如下：

```
<fieldset data-role="fieldcontain">
       <label for="day">选择值日时间：</label>
       <select name="day" id="day">
       <optgroup label="工作日">
        <option value="mon">星期一</option>
        <option value="tue">星期二</option>
        <option value="wed">星期三</option>
        <option value="thu">星期四</option>
        <option value="fri">星期五</option>
       </optgroup>
       <optgroup label="休息日">
        <option value="sat">星期六</option>
        <option value="sun">星期日</option>
       </optgroup>
       </select>
</fieldset>
```

在模拟器中的预览效果如图 12-13 所示。

图 12-13　菜单选项分组后的效果

如果想选择菜单中的多个选项，需要设置<select>标签的 multiple 属性，设置代码如下：

```
<select name="day" id="day" multiple data-native-menu="false">
```

例如把上面的代码修改如下：

```
<fieldset data-role="fieldcontain">
       <label for="day">选择值日时间：</label>
       <select name="day" id="day" multiple data-native-menu="false">
       <optgroup label="工作日">
        <option value="mon">星期一</option>
        <option value="tue">星期二</option>
        <option value="wed">星期三</option>
        <option value="thu">星期四</option>
        <option value="fri">星期五</option>
       </optgroup>
       <optgroup label="休息日">
```

```
            <option value="sat">星期六</option>
            <option value="sun">星期日</option>
        </optgroup>
        </select>
</fieldset>
```

在模拟器中预览，选择菜单时的效果如图 12-14 所示。选择完成后，即可看到多个菜单选项被选择，如图 12-15 所示。

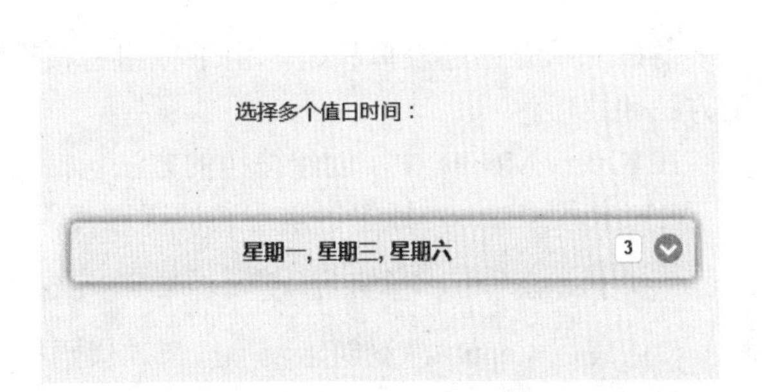

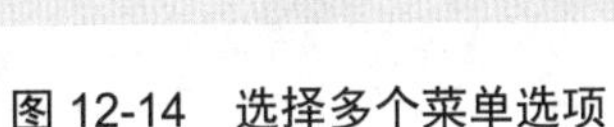

图 12-14　选择多个菜单选项

图 12-15　多个菜单选项被选择后的效果

8. 翻转波动开关

设置<input type="checkbox">标签的 data-role 为"flipswitch"时，可以创建翻转波动开关。代码如下：

```
<form>
  <label for="switch">切换开关：</label>
   <input type="checkbox" data-role="flipswitch" name="switch"
id="switch">
</form>
```

在模拟器中的预览效果如图 12-16 所示。

同时，用户还可以使用 checked 属性来设置默认的选项。代码如下：

```
<input type="checkbox" data-role="flipswitch" name="switch" id="switch"
checked>
```

修改后的预览效果如图 12-17 所示。

在默认情况下，开关切换的文本为 On 和 Off。可以使用 data-on-text 和 data-off-text 属性来修改。代码如下：

```
<input type="checkbox" data-role="flipswitch" name="switch" id="switch"
data-on-text="打开" data-off-text="关闭">
```

修改后的预览效果如图 12-18 所示。

切换开关:
Off

图 12-16 开关默认效果

切换开关:
On

图 12-17 修改默认选项后的效果

切换开关:
关闭

图 12-18 修改切换开关文本后的效果

12.1.2 按钮和按钮组

前面简单介绍过表单按钮，由于按钮和按钮组功能变化比较大，下面详细讲述它们的使用方法和技巧。

在 jQuery Mobile 中，创建按钮的方法包括以下 3 种。

(1) 使用<button>标签创建普通按钮。代码如下：

```
<button>按钮</button>
```

(2) 使用<input>标签创建表单按钮。代码如下：

```
<input type="button" value="按钮">
```

(3) 使用 data-role="button"属性创建链接按钮。代码如下：

```
<a href="#" data-role="button">按钮</a>
```

在 jQuery Mobile 中，按钮的样式会被自动添加上。为了让按钮在移动设备上更具吸引力和可用性，推荐在页面间进行链接时，使用第三种方法；在表单提交时，用第一种或第二种方法。

在默认情况下，按钮占满整个屏幕宽度。如果想要一个仅是与内容一样宽的按钮，或者需要并排显示两个或多个按钮，可以通过设置 data-inline="true"来完成。代码如下：

```
<a href="#pagetwo" data-role="button" data-inline="true">下一页</a>
```

下面通过一个案例来区别默认按钮和设置后按钮的区别。

【例 12.4】(示例文件 ch12\12.4.html)

```
<!DOCTYPE html>
<html>
<head>
<meta name="viewport" content="width=device-width, initial-scale=1">
<link rel="stylesheet"
href="http://code.jquery.com/mobile/1.4.5/jquery.mobile-1.4.5.min.css">
<script src="http://code.jquery.com/jquery-1.11.3.min.js"></script>
 <script src="http://code.jquery.com/mobile/1.4.5/jquery.mobile-
1.4.5.min.js"></script>
 </head>
<body>
<div data-role="page" id="first">
```

```
  <div data-role="header">
    <h1>按钮的区别</h1>
  </div>
  <div data-role="content" class="content">
    <p>普通 / 默认按钮:</p>
    <a href="#second" data-role="button">下一页</a>
    <p>设置后的按钮:</p>
    <a href="#second" data-inline="true">下一页</a>
    <a href="#first" data-inline="true">上一页</a>
  </div>
  <div data-role="footer">
    <h1>2 种按钮</h1>
  </div>
</div>
</body>
</html>
```

在模拟器中的预览效果如图 12-19 所示。

jQuery Mobile 提供了一个简单的方法来将按钮组合在一起。使用 data-role="controlgroup"属性即可通过按钮组来组合按钮。同时使用 data-type="horizontal/vertical"属性来设置按钮的排列方式是水平还是垂直。

【例 12.5】(示例文件 ch12\12.5.html)

```
<!DOCTYPE html>
<html>
<head>
<meta name="viewport" content="width=device-width, initial-scale=1">
<link rel="stylesheet"
href="http://code.jquery.com/mobile/1.4.5/jquery.mobile-1.4.5.min.css">
<script src="http://code.jquery.com/jquery-1.11.3.min.js"></script>
 <script src="http://code.jquery.com/mobile/1.4.5/jquery.mobile-
1.4.5.min.js"></script>
 </head>
<body>
<div data-role="page" id="first">
  <div data-role="header">
    <h1>组按钮的排列</h1>
  </div>
  <div data-role="content" class="content">
<div data-role="controlgroup" data-type="horizontal">
    <p>水平排列的按钮: </p>
    <a href="#" data-role="button">按钮 a</a>
    <a href="#" data-role="button">按钮 b</a>
    <a href="#" data-role="button">按钮 c</a>
</div><br>
    <div data-role="controlgroup" data-type="vertical"
    <p>垂直排列的按钮:</p>
     <a href="#" data-role="button">按钮 a</a>
    <a href="#" data-role="button">按钮 b </a>
    <a href="#" data-role="button">按钮 c</a>
```

```
</div>
  </div>
  <div data-role="footer">
    <h1>2 种排列方式</h1>
  </div>
</div>
</body>
</html>
```

在模拟器中的预览效果如图 12-20 所示。

图 12-19　不同按钮的效果

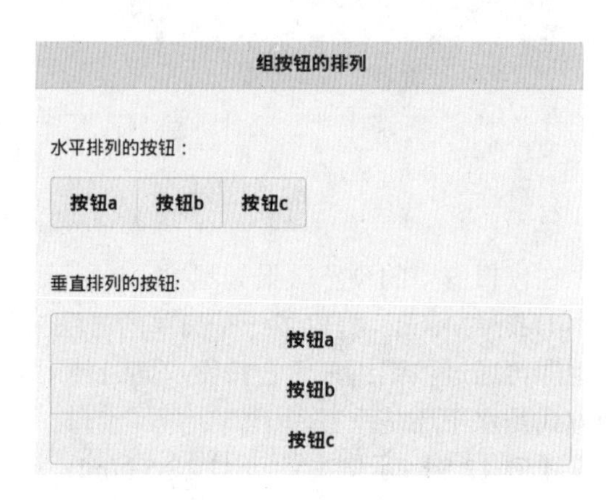

图 12-20　不同排列方式的按钮组

12.1.3　按钮图标

jQuery Mobile 提供了一套丰富多彩的按钮图标，用户只需要使用 data-icon 属性即可添加按钮图标。常用的图标样式如表 12-1 所示。

表 12-1　常用的按钮图标样式

图标参数	外观样式	说　明
data-icon="arrow-l"	左箭头	左箭头
data-icon="arrow-r"	右箭头	右箭头
data-icon="arrow-u"	上箭头	上箭头
data-icon="arrow-d"	下箭头	下箭头
data-icon="info"	信息	信息
data-icon="plus"	加号	加号
data-icon="minus"	减号	减号
data-icon="check"	复选	复选
data-icon="refresh"	重新整理	重新整理
data-icon="delete"	删除	删除

续表

图标参数	外观样式	说 明
data-icon="forward"	前进	前进
data-icon="back"	后退	后退
data-icon="star"	星形	星形
data-icon="audio"	扬声器	扬声器
data-icon="lock"	挂锁	挂锁
data-icon="search"	搜索	搜索
data-icon="alert"	警告	警告
data-icon="grid"	网格	网格
data-icon="home"	首页	首页

例如以下代码：

```
<a href="#" data-role="button" data-icon="lock">挂锁</a>
<a href="#" data-role="button" data-icon="check">复选</a>
<a href="#" data-role="button" data-icon="refresh">重新整理</a>
<a href="#" data-role="button" data-icon="delete">删除</a>
```

在模拟器中的预览效果如图 12-21 所示。

细心的读者会发现，按钮上的图标默认情况下会出现在按钮的左边。如果需要设置图标的位置，可以设置 data-iconpos 属性来指定位置，包括 top(顶部)、right(右侧)和 bottom(底部)。例如以下代码：

```
<a href="#" data-role="button" data-icon="refresh">重新整理</a>
<a href="#" data-role="button" data-icon="refresh" data-iconpos="top">重新整
理</a>
<a href="#" data-role="button" data-icon="refresh" data-iconpos="right">重
新整理</a>
<a href="#" data-role="button" data-icon="refresh" data-iconpos="bottom">重
新整理</a>
```

在模拟器中的预览效果如图 12-22 所示。

图 12-21　按钮图标效果

图 12-22　设置图标的位置

如果不想让按钮上出现文字，可以将 data-iconpos 属性设置为 notext，这样只会显示按钮而没有文字。

12.1.4 弹窗

弹窗是一个非常流行的对话框，弹窗可以覆盖在页面上展示。弹窗可用于显示一段文本、图片、地图或其他内容。创建一个弹窗，需要使用<a>和<div>标签。在<a>标签上添加data-rel="popup"属性，<div>标签添加 data-role="popup"属性。然后为<div>设置 id，设置<a>的 href 值为<div>指定的 id，其中<div>中的内容为弹窗显示的内容。代码如下：

```
<a href="#firstpp" data-rel="popup">显示弹窗</a>
<div data-role="popup" id="firstpp">
  <p>这是弹出窗口显示的内容</p>
</div>
```

在模拟器中的预览效果如图 12-23 所示。单击【显示弹窗】按钮即可显示弹出窗口的内容。

<div>弹窗与单击的<a>链接必须在同一个页面上。

在默认情况下，单击弹窗之外的区域或按 Esc 键即可关闭弹窗。用户也可以在弹窗上添加关闭按钮，只需要设置属性 data-rel="back"即可，结果如图 12-24 所示。

图 12-23 弹窗效果

这是弹出窗口显示的内容

图 12-24 带关闭按钮的弹窗效果

用户还可以在弹窗中显示图片，代码如下：

```
<div id="pageone" data-role="content" class="content" >
    <p>单击下面的小图片</p>
    <a href="#firstpp" data-rel="popup" >
    <img src="123.jpeg" style="width:200px;"></a>
    <div data-role="popup" id="firstpp">
    <p>这是我的图片！</p>
    </a><img src="123.jpeg" style="width:500px;height:500px;" >
    </div>
  </div>
```

在模拟器中的预览效果如图 12-25 所示。单击图片，即可弹出如图 12-26 所示的图片弹窗。

图 12-25 初始预览效果

图 12-26 图片弹窗效果

12.2 列 表

和电脑相比，移动设备屏幕比较小，所以常常以列表的形式显示数据。下面学习列表的使用方法和技巧。

12.2.1 列表视图

jQuery Mobile 中的列表视图是标准的 HTML 列表，包括有序列表<ol>和无序列表<ul>。列表视图是 jQuery Mobile 中功能强大的一个特性。它会使标准的无序或有序列表应用更广泛。

列表的使用方法非常简单，只需要在<ul>或<ol>标签中添加属性 data-role="listview"。每个项目(<li>)中可以添加链接。下面通过一个案例来学习。

【例 12.6】(示例文件 ch12\12.6.html)

使用列表视图：

```
<!DOCTYPE html>
<html>
<head>
<meta name="viewport" content="width=device-width, initial-scale=1">
<link rel="stylesheet"
href="http://code.jquery.com/mobile/1.4.5/jquery.mobile-1.4.5.min.css">
<script src="http://code.jquery.com/jquery-1.11.3.min.js"></script>
 <script src="http://code.jquery.com/mobile/1.4.5/jquery.mobile-
1.4.5.min.js"></script>
 </head>
<body>
<div data-role="page" id="first">
  <div data-role="header">
    <h1>列表视图</h1>
  </div>
```

```
  <div data-role="content" class="content">
  <h2>有序列表：</h2>
    <ol data-role="listview">
      <li><a href="#">香蕉</a></li>
      <li><a href="#">橘子</a></li>
      <li><a href="#">苹果</a></li>
    </ol>
    <h2>无序列表：</h2>
    <ul data-role="listview">
      <li><a href="#">芹菜</a></li>
      <li><a href="#">韭菜</a></li>
      <li><a href="#">菠菜</a></li>
    </ul>
  </div>
  </div>
  <div data-role="footer">
    <h1>有序列表和无序列表</h1>
  </div>
</div>
</body>
</html>
```

在模拟器中的预览效果如图 12-27 所示。

图 12-27　有序列表和无序列表

在默认情况下，列表项的链接会自动变成一个按钮，此时不再需要使用 data-role="button"属性。

从结果可以看出，列表样式中没有边缘和圆角效果，这里可以通过设置属性 data-inset="true"来完成，代码如下：

```
<ul data-role="listview" data-inset="true">
```

上面案例的部分代码修改如下：

```
 <div data-role="content" class="content">
 <h2>标准列表样式：</h2>
   <ol data-role="listview">
     <li><a href="#">香蕉</a></li>
     <li><a href="#">橘子</a></li>
     <li><a href="#">苹果</a></li>
   </ol>
   <h2>添加 data-inset="true"属性后的样式：</h2>
   <ul data-role="listview" data-inset="true">
     <li><a href="#">芹菜</a></li>
     <li><a href="#">韭菜</a></li>
     <li><a href="#">菠菜</a></li>
   </ul>
</div>
```

在模拟器中的预览效果如图 12-28 所示。

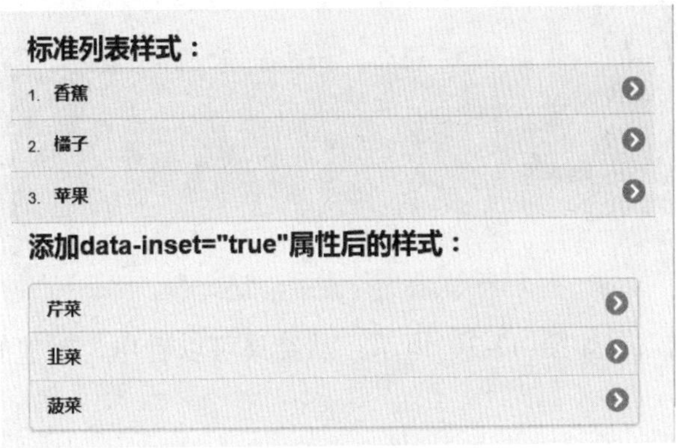

图 12-28　有边缘和圆角的列表效果

如果列表项比较多，用户可以使用列表分隔项对列表进行分组操作，这样使列表看起来更整齐。通过在列表项<li>标签中添加 data-role="list-divider" 属性即可指定列表分隔，例如以下代码：

```
<ul data-role="listview">
 <li data-role="list-divider">蔬菜</li>
  <li><a href="#">芹菜</a></li>
  <li><a href="#">韭菜</a></li>
<li data-role="list-divider">水果</li>
  <li><a href="#">苹果</a></li>
  <li><a href="#">橘子</a></li>
<li data-role="list-divider">乳制品</li>
  <li><a href="#">酸奶</a></li>
  <li><a href="#">奶酪</a></li>
 </ul>
```

在模拟器中的预览效果如图 12-29 所示。

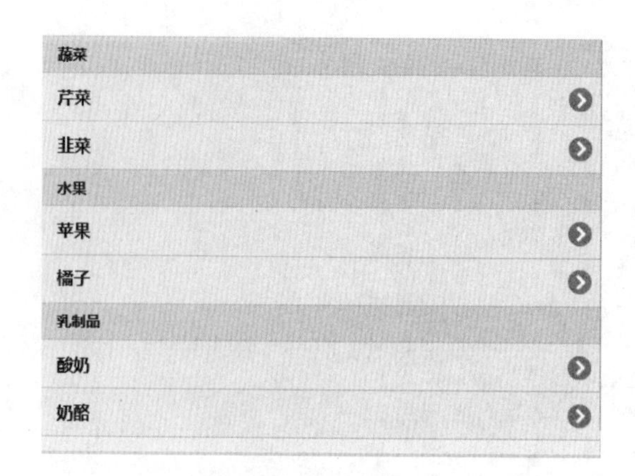

图 12-29　对项目进行分隔后的效果

如果项目列表是一个按字母顺序排列的列表，通过添加 data-autodividers="true"属性，可以自动生成项目的分隔，代码如下：

```
<ul data-role="listview" data-autodividers="true">
  <li><a href="#">Avocado</a></li>
  <li><a href="#"> Apricot</a></li>
  <li><a href="#">Banana</a></li>
  <li><a href="#">Bramley</a></li>
  <li><a href="#"> Cherry </a></li>
</ul>
```

在模拟器中的预览效果如图 12-30 所示。从结果可以看出，创建的分隔文本是列表项文本的第一个大写字母。

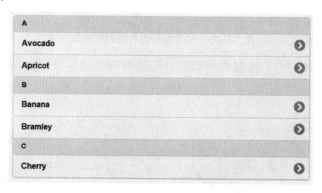

图 12-30　自动生成分隔后的效果

12.2.2　列表内容

在列表内容中，既可以添加图片和说明，也可以添加计数泡泡，同时还能拆分按钮和列表的链接。

1. 添加图片和说明

在前面做的案例中，列表项目前没有图片或说明。下面来讲述如何添加图片和说明，代码如下：

```
<li>
   <a href="#">
   <img src="124.jpg">
   <h3>香蕉</h3>
   <p>香蕉的原产地是东南亚</p>
   </a>
</li>
```

在模拟器中的预览效果如图 12-31 所示。

图 12-31　添加图片和说明

2. 加入计数泡泡

计数泡泡主要是在列表中显示数字时使用，只需要在<li>标签加入以下标签即可：

```
<span class="ui-li-count">数字</span>
```

例如下面的例子：

```
<li>
   <a href="#">
   <img src="124.jpg">
   <h3>香蕉</h3>
   <p>香蕉的原产地是东南亚</p>
   <span class="ui-li-count">111</span>
   </a>
</li>
```

在模拟器中的预览效果如图 12-32 所示。

图 12-32　添加计数泡泡

3. 拆分按钮和列表的链接

在默认情况下，单击列表项或按钮，都是转向同一个链接。用户也可以拆分按钮和列表项的链接，这样单击按钮和列表项时，会转向不同的链接。设置方法比较简单，只需要在<li>标签中加入两组<a>标签即可。

例如：

```
<li>
    <a href="122.html">
    <img src="124.jpg">
    <h3>香蕉</h3>
    <p>香蕉的原产地是东南亚</p>
    </a>
    <a href="123.html data-icon="star"></a>
</li>
```

在模拟器中的预览效果如图 12-33 所示。

图 12-33　拆分按钮和列表的链接

12.2.3　列表过滤

在 jQuery Mobile 中，用户可以对列表项目进行搜索过滤。添加过滤效果的思路如下：

(1)　创建一个表单，并添加类"ui-filterable"，该类的作用是自动调整搜索字段与过滤元素的外边距。代码如下：

```
<form class="ui-filterable">
</form>
```

(2)　在<form>标签内创建一个<input>标签，并添加 data-type="search"属性，并指定 id，从而创建基本的搜索字段。代码如下：

```
<form class="ui-filterable">
  <input id="myFilter" data-type="search">
</form>
```

(3)　为过滤的列表添加 data-input 属性，该值为<input>标签的 id。代码如下：

```
<ul data-role="listview" data-filter="true" data-input="#myFilter">
```

下面通过一个案例来理解列表是如何过滤的。

【例 12.7】(示例文件 ch12\12.7.html)

```
<!DOCTYPE html>
<html>
<head>
<meta name="viewport" content="width=device-width, initial-scale=1">
<link rel="stylesheet"
href="http://code.jquery.com/mobile/1.4.5/jquery.mobile-1.4.5.min.css">
<script src="http://code.jquery.com/jquery-1.11.3.min.js"></script>
 <script src="http://code.jquery.com/mobile/1.4.5/jquery.mobile-
1.4.5.min.js"></script>
```

```
</head>
<body>
<div data-role="page" id="first">
  <div data-role="content" class="content">
   <h2>进货商联系表</h2>
   <form>
    <input id="myFilter" data-type="search">
   </form>
    <ul data-role="listview" data-filter="true" data-input="#myFilter">
      <li><a href="#">张小名</a></li>
      <li><a href="#">刘名园</a></li>
      <li><a href="#">刘鲲鹏</a></li>
     <li><a href="#">张鹏举</a></li>
     <li><a href="#">张鹏远</a></li>
    </ul>
   </div>
</div>
</body>
</html>
```

在模拟器中的预览效果如图 12-34 所示。输入需要过滤的关键字，例如这里搜索姓张的进货商，结果如图 12-35 所示。

进货商联系表

张小名

刘名园

刘鲲鹏

张鹏举

张鹏远

图 12-34　程序预览效果

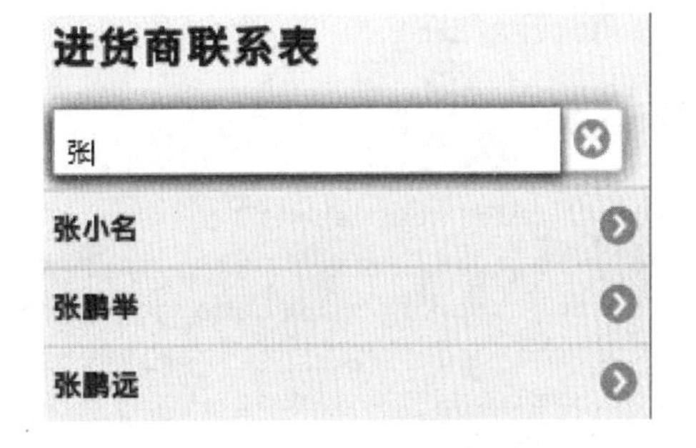

图 12-35　列表过滤后的效果

如果需要在搜索框内添加提示信息，可以通过设置 placeholder 属性来完成。代码如下：

```
<input id="myFilter" data-type="search" placeholder="请输入联系人的姓">
```

12.3　面板和可折叠块

在 jQuery Mobile 中，可以通过面板或可折叠块来隐藏或显示指定的内容。下面重点学习面板和可折叠块的使用方法和技巧。

12.3.1 面板

在 jQuery Mobile 中可以添加面板，面板会在屏幕上从左到右划出。通过为<div>标签添加 data-role="panel"属性来创建面板。具体思路如下。

(1) 通过<div>标签来定义面板的内容，并定义 id 属性。例如以下代码：

```
<div data-role="panel" id="myPanel">
    <h2>长恨歌</h2>
    <p>天生丽质难自弃，一朝选在君王侧。回眸一笑百媚生，六宫粉黛无颜色。</p>
</div>
```

注意

定义的面板内容必须置于头部、内容和底部组成的页面之前或之后。

(2) 要访问面板，需要创建一个指向面板<div>的链接，单击该链接即可打开面板。例如以下代码：

```
<a href="#myPanel" class="ui-btn ui-btn-inline">最喜欢的诗句</a>
```

【例 12.8】(示例文件 ch12\12.8.html)

使用面板：

```
<!DOCTYPE html>
<html>
<head>
<meta name="viewport" content="width=device-width, initial-scale=1">
<link rel="stylesheet"
href="http://code.jquery.com/mobile/1.4.5/jquery.mobile-1.4.5.min.css">
<script src="http://code.jquery.com/jquery-1.11.3.min.js"></script>
 <script src="http://code.jquery.com/mobile/1.4.5/jquery.mobile-
1.4.5.min.js"></script>
 </head>
<body>
<div data-role="first">
  <div data-role="panel" id="myPanel">
    <h2>长恨歌</h2>
    <p>天生丽质难自弃，一朝选在君王侧。回眸一笑百媚生，六宫粉黛无颜色。</p>
  </div>
  <div data-role="header">
  <h1>使用面板</h1>
  </div>
  <div data-role="content" class="content">
     <a href="#myPanel" class="ui-btn ui-btn-inline">最喜欢的诗句</a>
   </div>
</div>
</body>
</html>
```

在模拟器中的预览效果如图 12-36 所示。单击【最喜欢的诗句】链接，即可打开面板，结果如图 12-37 所示。

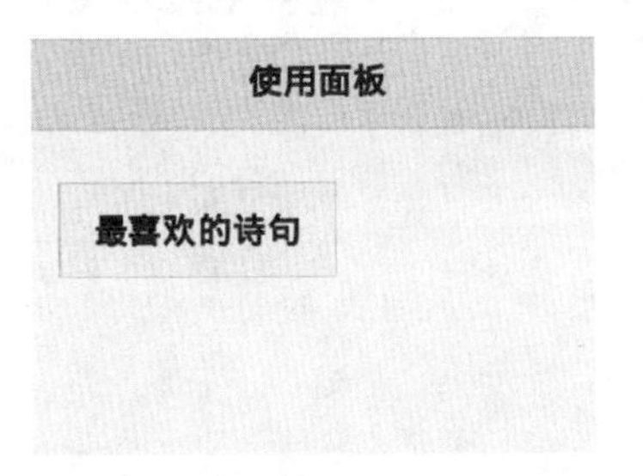

图 12-36　程序预览效果

长恨歌

天生丽质难自弃，一朝选在君王侧。回眸一笑百媚生，六宫粉黛无颜色。

图 12-37　打开面板

面板的展示方式由属性 data-display 来控制，具体分为以下 3 种。

(1) data-display="reveal"：面板的展示方式为从左到右划出，这是面板展示方式的默认值。

(2) data-display="overlay"：在内容上显示面板。

(3) data-display="push"：同时“推动”面板和页面。

这 3 种面板展示方式的代码如下：

```
<div data-role="panel" id="overlayPanel" data-display="overlay">
<div data-role="panel" id="revealPanel" data-display="reveal">
<div data-role="panel" id="pushPanel" data-display="push">
```

在默认情况下，面板会显示在屏幕的左侧。如果想让面板出现在屏幕的右侧，可以指定 data-position="right"属性。代码如下：

```
<div data-role="panel" id="myPanel" data-position="right">
```

在默认情况下，面板是随着页面一起滚动的。如果需要实现面板内容固定不随页面滚动而滚动，可以在面板中添加 the data-position-fixed="true"属性。代码如下：

```
<div data-role="panel" id="myPanel" data-position-fixed="true">
```

12.3.2　可折叠块

通过可折叠块，用户可以隐藏或显示指定的内容，这对于存储部分信息很有用。

创建可折叠块的方法比较简单，只需要在<div>标签中添加 data-role="collapsible"属性即可，添加标题标签为 H1～H6，后面即可添加隐藏的信息。

```
<div data-role="collapsible">
 <h1>折叠块的标题</h1>
 <p>可折叠的具体内容。</p>
 </div>
```

【例 12.9】(示例文件 ch12\12.9.html)

```
<!DOCTYPE html>
<html>
<head>
```

```
<meta name="viewport" content="width=device-width, initial-scale=1">
<link rel="stylesheet"
href="http://code.jquery.com/mobile/1.4.5/jquery.mobile-1.4.5.min.css">
<script src="http://code.jquery.com/jquery-1.11.3.min.js"></script>
 <script src="http://code.jquery.com/mobile/1.4.5/jquery.mobile-
1.4.5.min.js"></script>
 </head>
<body>
<div data-role="first">
  <div data-role="header">
  <h1>可折叠块</h1>
  </div>
  <div data-role="content" class="content">
    <div data-role="collapsible">
      <h1>最喜欢的水果</h1>
      <p>香蕉、橘子、苹果</p>
    </div>
  </div>
</div>
</body>
</html>
```

在模拟器中的预览效果如图 12-38 所示。单击【最喜欢的水果】按钮，即可打开可折叠块，结果如图 12-39 所示。

图 12-38　程序预览效果

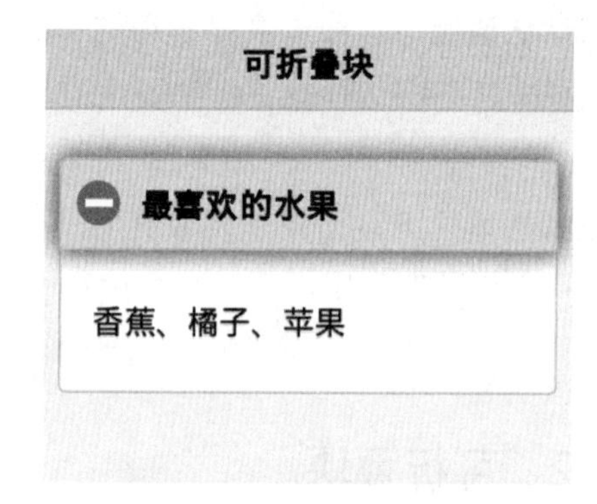

图 12-39　打开可折叠块

在默认情况下，内容是被折叠起来的。如果需要在页面加载时展开内容，添加 data-collapsed="false"属性即可，代码如下：

```
<div data-role="collapsible" data-collapsed="false">
 <h1>折叠块的标题</h1>
 <p>这里显示的内容是展开的</p>
 </div>
```

可折叠块是可以嵌套的，例如以下代码：

```
<div data-role="collapsible">
 <h1>全部智能商品</h1>
```

```
<div data-role="collapsible">
 <h1>智能家居</h1>
 <p>智能办公、智能厨电和智能网络</p>
 </div>
</div>
```

在模拟器中的预览效果如图 12-40 所示。

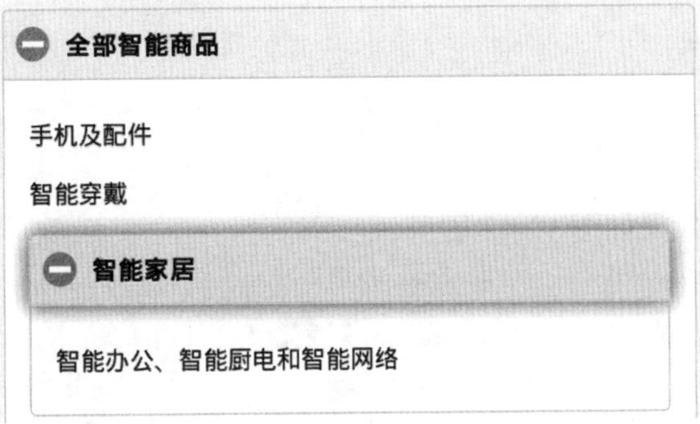

全部智能商品
手机及配件
智能穿戴
智能家居
智能办公、智能厨电和智能网络

图 12-40　程序预览效果

12.4 导 航 条

导航条通常位于页面的头部或尾部，主要作用是便于用户快速访问需要的页面。下面重点学习导航条的使用方法和技巧。

在 jQuery Mobile 中，使用 data-role="navbar" 属性来定义导航栏。需要特别注意的是，导航栏中的链接将自动变成按钮，不需要使用 data-role="button"属性。

例如以下代码：

```
<div data-role="header">
  <h1>鸿鹄网购平台</h1>
  <div data-role="navbar">
    <ul>
      <li><a href="#">主页</a></li>
      <li><a href="#">团购</a></li>
      <li><a href="#">搜索商品</a></li>
    </ul>
  </div>
</div>
```

在模拟器中的预览效果如图 12-41 所示。

鸿鹄网购平台

主页	团购	搜索商品

图 12-41　程序预览效果

通过前面章节的学习，用户还可以为导航添加按钮图标，例如以上代码修改如下：

```
<div data-role="header">
  <h1>鸿鹄网购平台</h1>
  <div data-role="navbar">
    <ul>
      <li><a href="#" data-icon="home">主页</a></li>
      <li><a href="#" data-icon="arrow-d">团购</a></li>
      <li><a href="#" data-icon="search">搜索商品</a></li>
    </ul>
  </div>
</div>
```

在模拟器中的预览效果如图 12-42 所示。

图 12-42　程序预览效果

细心的读者会发现，导航按钮的图标默认位置是位于文字的上方，这跟普通的按钮图片是不一样的。如果需要修改导航按钮图标的位置，可以通过设置 data-iconpos 属性来指定位置，包括 left(左侧)、right(右侧)和 bottom(底部)。

例如下面修改导航按钮图标的位置为文本的左侧，代码如下：

```
<div data-role="header">
  <h1>鸿鹄网购平台</h1>
  <div data-role="navbar" data-iconpos="left">
    <ul>
      <li><a href="#" data-icon="home" >主页</a></li>
      <li><a href="#" data-icon="arrow-d" >团购</a></li>
      <li><a href="#" data-icon="search">搜索商品</a></li>
    </ul>
  </div>
</div>
```

在模拟器中的预览效果如图 12-43 所示。

图 12-43　程序预览效果

注意　和设置普通按钮图标位置不同的是，这里 data-iconpos="left"属性只能添加到<div>标签中，而不能添加到<li>标签中，否则是无效的，读者可以自行检测。

在默认情况下，当单击导航按钮时，按钮的样式会发生变换。例如这里单击【搜索商品】导航按钮，发现按钮的底纹颜色变成了蓝色，如图 12-44 所示。

图 12-44 导航按钮的样式变化

如果用户想取消上面的样式变化，可以添加 class="ui-btn-active"属性即可。例如以下代码：

```
<li><a href="#anylink" class="ui-btn-active">首页</a></li>
```

修改完成后，再次单击【主页】导航按钮时，样式不会发生变化。

对于多个页面的情况，往往用户希望显示哪个页面，对应导航按钮处于被选中状态，下面通过一个案例来讲解。

【例 12.10】(示例文件 ch12\12.10.html)

```
<!DOCTYPE html>
<html>
<head>
<meta name="viewport" content="width=device-width, initial-scale=1">
<link rel="stylesheet"
href="http://code.jquery.com/mobile/1.4.5/jquery.mobile-1.4.5.min.css">
<script src="http://code.jquery.com/jquery-1.11.3.min.js"></script>
 <script src="http://code.jquery.com/mobile/1.4.5/jquery.mobile-
1.4.5.min.js"></script>
 </head>
<body>
<div data-role="page" id="first">
  <div data-role="header">
  <h1>鸿鹄购物平台</h1>
    <div data-role="navbar">
      <ul>
        <li><a href="#" class="ui-btn-active ui-state-persist">主页</a></li>
        <li><a href="#second">团购</a></li>
        <li><a href="#">搜索商品</a></li>
      </ul>
    </div>
  </div>
  <div data-role="content" class="content">
<p>这里是首页显示的内容</p>
 </div>
 <div data-role="footer">
    <h1>首页</h1>
  </div>
</div>

<div data-role="page" id="second">
```

```
  <div data-role="header">
  <h1>鸿鹄购物平台</h1>
    <div data-role="navbar">
      <ul>
        <li><a href="#first">主页</a></li>
        <li><a href="#" class="ui-btn-active ui-state-persist">团购</a></li>
        <li><a href="#">搜索商品</a></li>
      </ul>
  </div>
  </div>
  <div data-role="content" class="content">
<p>这里是团购显示的内容</p>
 </div>
 <div data-role="footer">
    <h1>团购页面</h1>
  </div>
</div>
</body>
</html>
```

在模拟器中的预览效果如图 12-45 所示。此时默认显示首页的内容，【主页】导航按钮处于选中状态。切换到团购页面后，【团购】导航按钮处于选中状态，如图 12-46 所示。

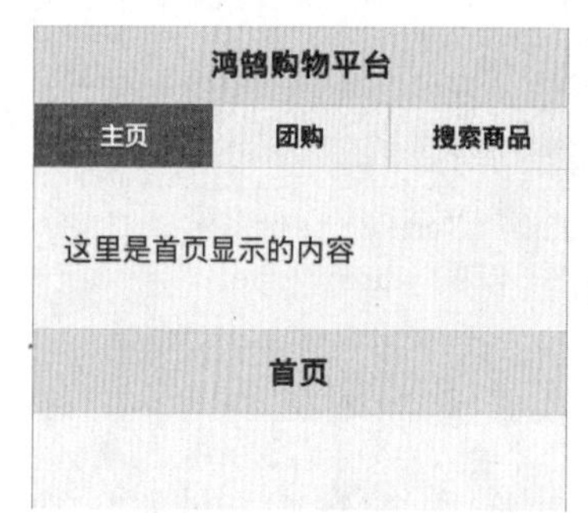

图 12-45 程序预览效果

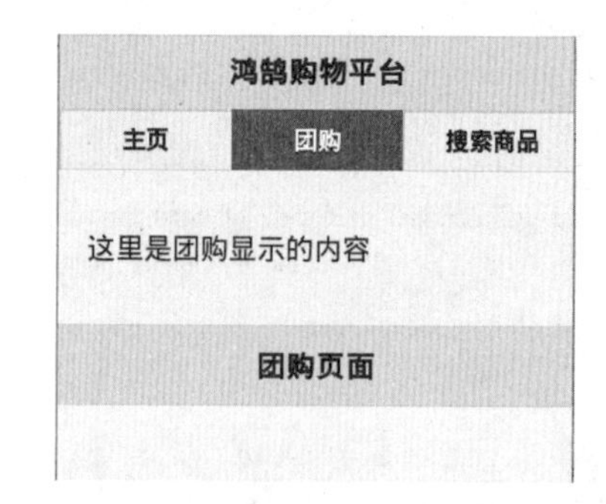

图 12-46 【团购】导航按钮处于选中状态

12.5 实战演练——使用 jQuery Mobile 主题

用户在设计移动网站时，往往需要配置背景颜色、导航颜色、布局颜色等，这些工作是非常耗费时间的。为此，jQuery Mobile 提供了 2 种不同的主题样式，每种主题颜色的按钮、导航、内容等颜色都是配置好的，效果也不相同。

这两种主题分别为 a 和 b，通过设置 data-theme 属性来引用主题 a 或主题 b，代码如下：

```
<div data-role="page" id="first" data-theme="a">
<div data-role="page" id="first" data-theme="b">
```

1. 主题 a

页面为灰色背景、黑色文字；头部与底部均为灰色背景、黑色文字；按钮为灰色背景、黑色文字；激活的按钮和链接为白色文本、蓝色背景；input 输入框中 placeholder 属性值为浅灰色，value 值为黑色。

下面通过一个案例来讲解主题 a 的样式效果。

【例 12.11】(示例文件 ch12\12.11.html)

```
<!DOCTYPE html>
<html>
<head>
<meta name="viewport" content="width=device-width, initial-scale=1">
<link rel="stylesheet"
href="http://code.jquery.com/mobile/1.4.5/jquery.mobile-1.4.5.min.css">
<script src="http://code.jquery.com/jquery-1.11.3.min.js"></script>
 <script src="http://code.jquery.com/mobile/1.4.5/jquery.mobile-
1.4.5.min.js"></script>
 </head>
<body>
<div data-role="page" id="first" data-theme="a">
  <div data-role="header">
    <h1>古诗鉴赏</h1>
  </div>

  <div data-role="content " class="content">
    <p>秋风起兮白云飞，草木黄落兮雁南归。兰有秀兮菊有芳，怀佳人兮不能忘。泛楼船兮济汾河，
横中流兮扬素波。</p>
    <a href="#">秋风辞</a>
    <a href="#" class="ui-btn">更多古诗</a>
    <p>唐诗:</p>
    <ul data-role="listview" data-autodividers="true" data-inset="true">
      <li><a href="#">将进酒</a></li>
      <li><a href="#">春望</a></li>
    </ul>
    <label for="fullname">请输入喜欢诗的名字:</label>
      <input type="text" name="fullname" id="fullname" placeholder="诗词名称..">
    <label for="switch">切换开关:</label>
      <select name="switch" id="switch" data-role="slider">
        <option value="on">On</option>
        <option value="off" selected>Off</option>
      </select>
  </div>

  <div data-role="footer">
    <h1>经典诗歌</h1>
  </div>
</div>
</body>
</html>
```

主题 a 的样式效果如图 12-47 所示。

2. 主题 b

页面为黑色背景、白色文字；头部与底部均为黑色背景、白色文字；按钮为白色文字、木炭背景；激活的按钮和链接为白色文本、蓝色背景；input 输入框中 placeholder 属性值为浅灰色、value 值为白色。

为了对比主题 a 的样式效果，请将上面案例中的代码：

```
<div data-role="page" id="first" data-theme="a">
```

修改如下：

```
<div data-role="page" id="first" data-theme="b">
```

主题 b 的样式效果如图 12-48 所示。

图 12-47　主题 a 的样式效果

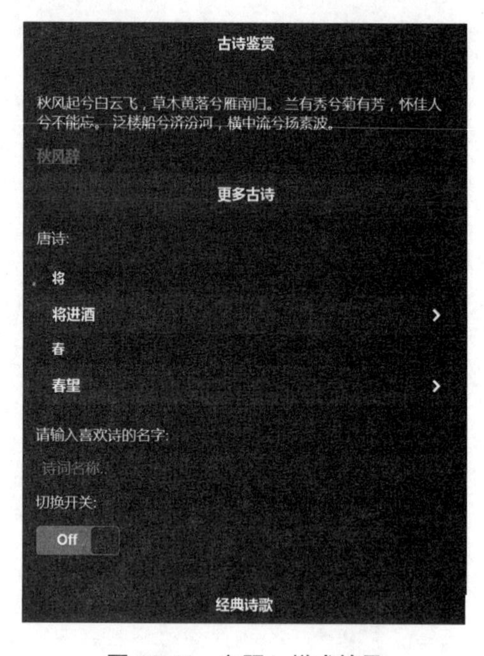

图 12-48　主题 b 样式效果

主题样式 a 和主题样式 b 不仅仅可以应用到页面，也可以单独地应用到页面的头部、内容、底部、导航条、按钮、面板、列表、表单等元素上。

例如，将主题样式 b 添加到页面的头部和底部，代码如下：

```
<div data-role="header" data-theme="b"></div>
<div data-role="footer" data-theme="b"></div>
```

将主题样式 b 添加到对话框的头部和底部，代码如下：

```
<div data-role="page" data-dialog="true" id="second">
  <div data-role="header" data-theme="b"></div>
  <div data-role="footer" data-theme="b"></div>
</div>
```

将主题样式 b 添加到按钮上时，需要使用 class="ui-btn- a/b "来设置按钮颜色为灰色或黑色。例如，将样式b的样式应用到按钮上，代码如下：

```
<a href="#" class="ui-btn">灰色按钮(默认)</a>
<a href="#" class="ui-btn ui-btn-b">黑色按钮</a>
```

预览效果如图12-49所示。

图 12-49　按钮添加主题后的效果

在弹窗上应用主题样式的代码如下：

```
<div data-role="popup" id="myPopup" data-theme="b">
```

在头部和底部的按钮上也可以添加主题样式，例如以下代码：

```
<div data-role="header">
  <a href="#" class="ui-btn ui-btn-b">主页</a>
  <h1>古诗欣赏</h1>
  <a href="#" class="ui-btn">搜索</a>
</div>

<div data-role="footer">
   <a href="#" class="ui-btn ui-btn-b">上传古诗图文</a>
  <a href="#" class="ui-btn">名句欣赏鉴别</a>
  <a href="#" class="ui-btn ui-btn-b">联系我们</a>
</div>
```

预览效果如图12-50所示。

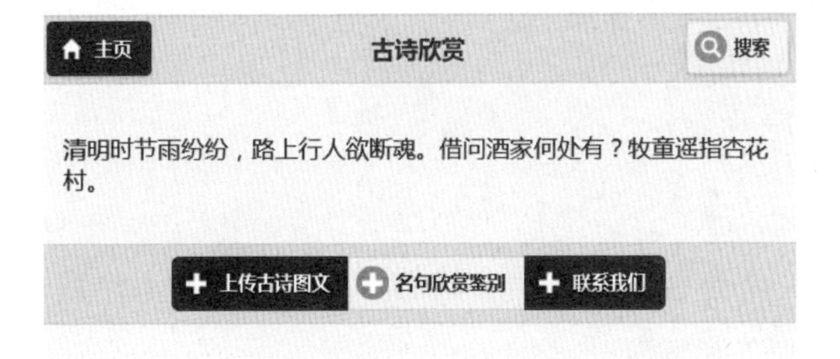

图 12-50　头部和底部的按钮添加主题后的效果

12.6 疑难解惑

疑问1：如何制作一个后退按钮？

答：如果需要创建后退按钮，请使用 data-rel="back" 属性(这会忽略锚的 href 值)：

```
<a href="#" data-role="button" data-rel="back">返回</a>
```

疑问 2：如何在面板上添加主题样式 b？

答：在主题上添加主题样式的方法比较简单，代码如下：

```
<div data-role="panel" id="myPanel" data-theme="b">
```

面板添加主题样式 b 后的效果如图 12-51 所示。

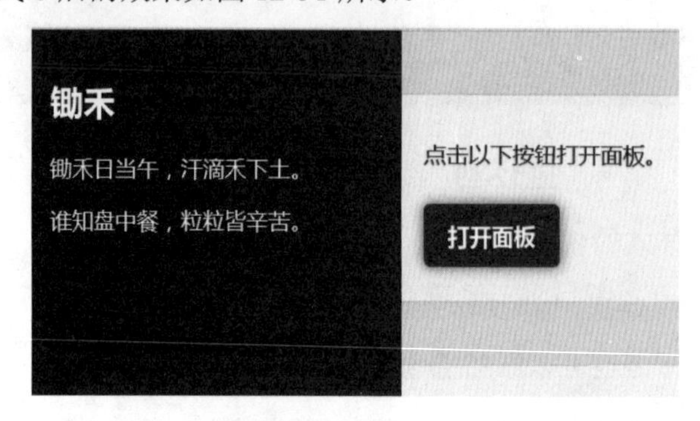

图 12-51　面板添加主题样式 b 后的效果

第 13 章 jQuery Mobile 事件

页面有了事件就有了“灵魂”，可见事件对于页面是多么重要。这是因为事件使页面具有了动态性和响应性，如果没有事件将很难完成页面与用户之间的交互。jQuery Mobile 针对移动端提供了各种浏览器事件，包括页面事件、触摸事件、滚屏事件、定位事件等。本章主要介绍如何使用 jQuery Mobile 事件。

13.1 页 面 事 件

jQuery Mobile 针对各个页面生命周期的事件可以分为以下几种。

(1) 初始化事件：分别在页面初始化之前，页面创建时和页面初始化之后触发的事件。

(2) 外部页面加载事件：外部页面加载时触发事件。

(3) 页面过渡事件：页面过渡时触发事件。

使用 jQuery Mobile 事件的方法比较简单，只需要使用 on()方法指定要触发的时间并设定事件处理函数即可，语法格式如下：

```
$(document).on(事件名称,选择器,事件处理函数)
```

其中选择器为可选参数，如果省略该参数，表示事件应用于整个页面而不限定哪一个组件。

13.1.1 初始化事件

初始化事件发生的时间包括页面初始化之前、页面创建时和页面创建后。下面详细介绍初始化事件。

1. mobileinit 事件

当 jQuery Mobile 开始执行时，首先会触发 mobileinit 事件。如果想更改 jQuery Mobile 的默认值时，就可以将函数绑定到 mobileinit 事件。其语法格式如下：

```
$(document).on("mobileinit",function(){
  // jQuery 事件
});
```

例如，jQuery Mobile 开始执行任何操作时都会使用 Ajax 的方式，如果不想使用 Ajax，可以在 mobileinit 事件中将$.mobile.ajaxEnabled 更改为 false，代码如下：

```
$(document).on("mobileinit",function(){
 $.mobile.ajaxEnabled=false;
});
```

这里需要注意的是，上述代码要放在引用 jquery.mobile.js 之前。

2. jQuery Mobile Initialization 事件

jQuery Mobile Initialization 事件主要包括 pagebeforecreate 事件、pagecreate 事件和 pageinit 事件，它们的区别如下。

(1) pagebeforecreate 事件：发生在页面 DOM 加载后，正在初始化时，语法格式如下：

```
$(document).on("pagebeforecreate",function(){
  // 程序语句
});
```

(2) pagecreate 事件：发生在页面 DOM 加载完成，初始化也完成时，语法格式如下：

```
$(document).on("pagecreate",function(){
  // 程序语句
});
```

(3) pageinit 事件：发生在页面初始化完成以后，语法格式如下：

```
$(document).on("pageinit",function(){
  // 程序语句
});
```

下面通过一个综合案例来学习上面 3 个事件触发的时机。

【例 13.1】(示例文件 ch13\13.1.html)

```
<!DOCTYPE html>
<html>
<head>
<meta name="viewport" content="width=device-width, initial-scale=1">
<link rel="stylesheet"
href="http://code.jquery.com/mobile/1.4.5/jquery.mobile-1.4.5.min.css">
<script src="http://code.jquery.com/jquery-1.11.3.min.js"></script>
 <script src="http://code.jquery.com/mobile/1.4.5/jquery.mobile-
1.4.5.min.js"></script>
<script>
$(document).on("pagebeforecreate",function(){
   alert("注意: pagebeforecreate 事件开始触发");
});
$(document).on("pagecreate",function(){
  alert("注意: pagecreate 事件触发开始触发");
});
$(document).on("pageinit",function(){
   alert("注意: pageinit 事件开始触发");
});
</script>
</head>
<body>
<div data-role="page" id="first">
  <div data-role="header">
    <h1>古诗欣赏</h1>
  </div>
  <div data-role="main" class="ui-content">
    <p>几回花下坐吹箫，银汉红墙入望遥。</p>
    <a href="#second">下一页</a>
  </div>
  <div data-role="footer">
    <h1>清代诗人</h1>
  </div>
</div>
<div data-role="page" id="second">
  <div data-role="header">
```

```
    <h1>古诗欣赏</h1>
  </div>
  <div data-role="main" class="ui-content">
    <p>似此星辰非昨夜，为谁风露立中宵。</p>
    <a href="#first">上一页</a>
  </div>
  <div data-role="footer">
    <h1>经典诗词</h1>
  </div>
</div>
</body>
</html>
```

在模拟器中预览程序的效果，各个事件的执行顺序如图 13-1 所示。3 次单击【确认】按钮后，结果如图 13-2 所示。

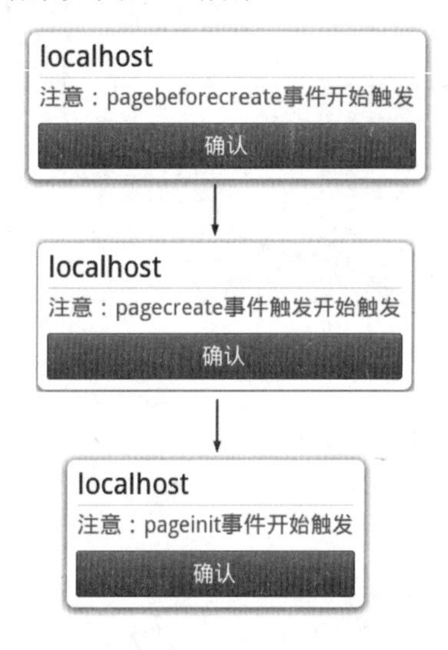

图 13-1　初始化事件

图 13-2　页面最终效果

13.1.2　外部页面加载事件

外部页面加载时，最常见的加载事件如下。

1. pagebeforeload 事件

pagebeforeload 事件在外部页面加载前触发。其语法格式如下：

```
<script>
$(document).on("pagebeforeload",function(){
  alert("有外部文件将要被加载);
});
</script>
```

2. pageload 事件

当页面加载成功时，触发 pageload 事件。其语法格式如下：

```
<script>
$(document).on("pageload",function(event,data){
  alert("pageload 事件触发!\nURL: " + data.url);
});
</script>
```

pageload 事件的函数的参数含义如下。

(1) event：任何 jQuery 的事件属性，如 event.type、event.pageX、target 等。

(2) data：包含以下属性。

(3) url：页面的 url 地址，是字符串类型。

(4) absUrl：绝对地址，是字符串类型。

(5) dataUrl：地址栏 URL，是字符串类型。

(6) options: $.mobile.loadPage()指定的选项，是对象类型。

(7) xhr:XMLHttpRequest 对象，是对象类型。

(8) textStatus：对象状态或空值，返回状态。

3. pageloadfailed 事件

如果页面载入失败，触发 pageloadfailed 事件。默认地，将显示 Error Loading Page 消息。其语法格式如下：

```
$(document).on("pageloadfailed",function(event,data){
  alert("抱歉，被请求页面不存在。");
});
</script>
```

下面通过一个例子来理解上述事件触发时机。

【例 13.2】(示例文件 ch13\13.2.html)

```
<!DOCTYPE html>
<html>
<head>
<meta name="viewport" content="width=device-width, initial-scale=1">
<link rel="stylesheet"
href="http://code.jquery.com/mobile/1.4.5/jquery.mobile-1.4.5.min.css">
<script src="http://code.jquery.com/jquery-1.11.3.min.js"></script>
 <script src="http://code.jquery.com/mobile/1.4.5/jquery.mobile-
1.4.5.min.js"></script>
<script>
$(document).on("pageload",function(event,data){
alert("pageload 事件触发!\nURL: " + data.url);
});
$(document).on("pageloadfailed",function(){
  alert("抱歉，被请求页面不存在。");
});
```

```
</script>
</head>
<body>
<div data-role="page" id="first">
  <div data-role="header">
    <h1>古诗欣赏</h1>
  </div>
  <div data-role="content" class="content">
    <p>众鸟高飞尽，孤云独去闲。相看两不厌，只有敬亭山。</p>
    <a href="123.html">下一页</a>
  </div>
  <div data-role="footer">
    <h1>经典诗词</h1>
  </div>
</div>
</body>
</html>
```

在模拟器中的预览效果如图 13-3 所示。单击【下一页】超链接，结果如图 13-4 所示。

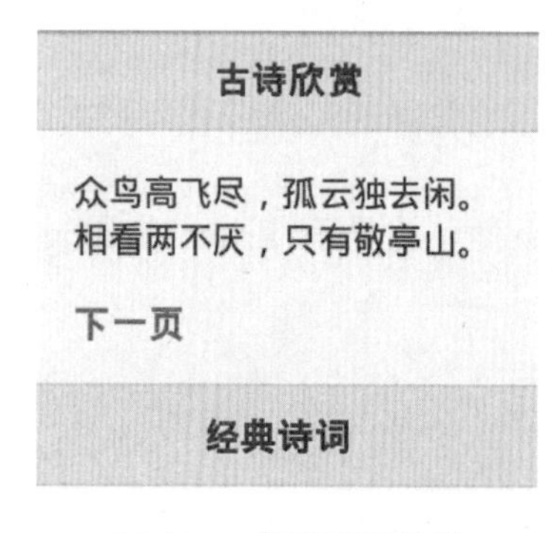

图 13-3　程序预览效果

图 13-4　触发 pageloadfailed 事件

13.1.3　页面过渡事件

在 jQuery Mobile 中，在当前页面过渡到下一页时，会触发以下几个事件。

(1)　pagebeforeshow 事件：在当前页面触发，在过渡动画开始前。

(2)　pageshow 事件：在当前页面触发，在过渡动画完成后。

(3)　pagebeforehide 事件：在下一页触发，在过渡动画开始前。

(4)　pagehide 事件：在下一页触发，在过渡动画完成后。

下面通过一个案例来学习页面过渡事件的触发时机。

【例 13.3】(示例文件 ch13\13.3.html)

```
<!DOCTYPE html>
<html>
<head>
<meta name="viewport" content="width=device-width, initial-scale=1">
<link rel="stylesheet"
```

```
href="http://code.jquery.com/mobile/1.4.5/jquery.mobile-1.4.5.min.css">
<script src="http://code.jquery.com/jquery-1.11.3.min.js"></script>
 <script src="http://code.jquery.com/mobile/1.4.5/jquery.mobile-
1.4.5.min.js"></script>
<script>
$(document).on("pagebeforeshow","#second",function(){
  alert("触发 pagebeforeshow 事件，下一页即将显示");
});
$(document).on("pageshow","#second",function(){
  alert("触发 pageshow 事件，现在显示下一页");
});
$(document).on("pagebeforehide","#second",function(){
  alert("触发 pagebeforehide 事件，下一页即将隐藏");
});
$(document).on("pagehide","#second",function(){
  alert("触发 pagehide 事件，现在隐藏下一页");
});</script>
</head>
<body>
<div data-role="page" id="first">
  <div data-role="header">
    <h1>古诗欣赏</h1>
  </div>
  <div data-role="content" class="content">
    <p>众鸟高飞尽，孤云独去闲。相看两不厌，只有敬亭山。</p>
    <a href="#second">下一页</a>
  </div>
  <div data-role="footer">
    <h1>经典诗词</h1>
  </div>
  </div>

<div data-role="page" id="second">
  <div data-role="header">
    <h1>古诗欣赏</h1>
  </div>
  <div data-role="content" class="content">
    <p>众鸟高飞尽，孤云独去闲。相看两不厌，只有敬亭山。</p>
    <a href="#first">上一页</a>
  </div>
  <div data-role="footer">
    <h1>经典诗词</h1>
  </div>
</div>
</body>
</html>
```

在模拟器中的预览效果如图 13-5 所示。单击【下一页】超链接，事件触发顺序如图 13-6 所示。

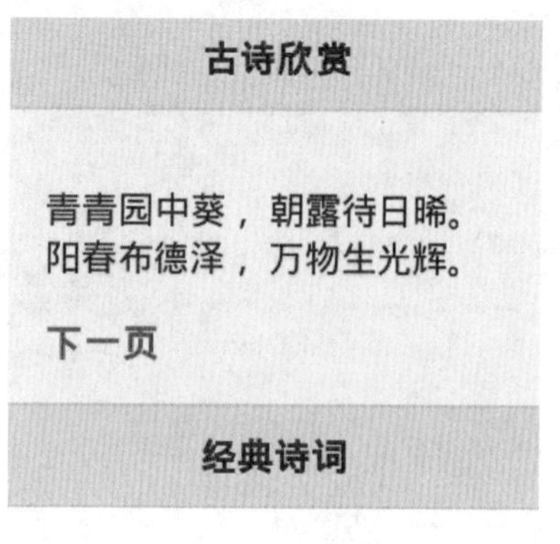

图 13-5　程序预览效果

图 13-6　当前页面触发事件顺序

单击【确认】按钮，进入下一页中，如图 13-7 所示。单击【上一页】超链接，事件触发顺序如图 13-8 所示。

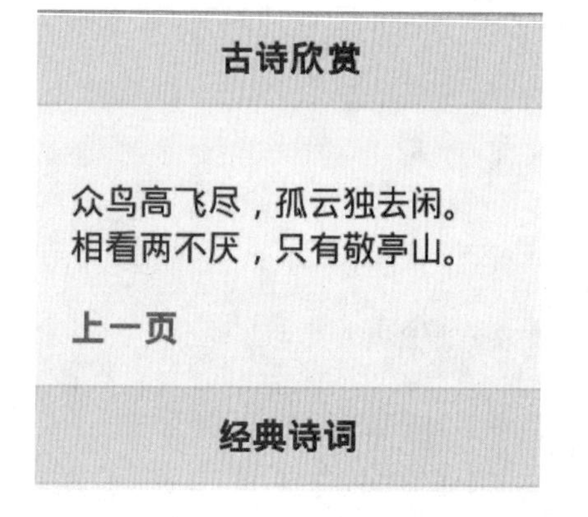

图 13-7　下一页预览效果

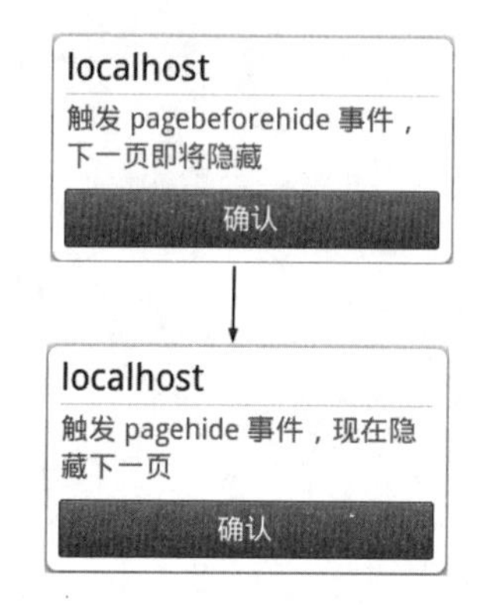

图 13-8　下一页触发事件顺序

13.2 触 摸 事 件

针对移动端浏览器提供了触摸事件，表示当用户触摸屏幕时触发的事件，包括点击事件和滑动事件。

13.2.1 点击事件

点击事件包括 tap 事件和 taphold 事件。下面详细介绍它们的用法和区别。

1. tap 事件

当用户点击页面上的元素时，会触发点击(tap)事件。其语法格式如下：

```
$("p").on("tap",function(){
  $(this).hide();
});
```

上述代码的作用是点击 p 组件后，将会把该组件隐藏。

下面通过一个案例来讲解点击事件的使用方法。

【例 13.4】(示例文件 ch13\13.4.html)

```
<!DOCTYPE html>
<html>
<head>
<meta name="viewport" content="width=device-width, initial-scale=1">
<link rel="stylesheet"
href="http://code.jquery.com/mobile/1.4.5/jquery.mobile-1.4.5.min.css">
<script src="http://code.jquery.com/jquery-1.11.3.min.js"></script>
 <script src="http://code.jquery.com/mobile/1.4.5/jquery.mobile-
1.4.5.min.js"></script>
<script>
$("div").on("tap",function(){
  $(this).css("color","green");
});
</script>
</head>
<body>
<div data-role="page" id="first">
  <div data-role="header">
    <h1>古诗欣赏</h1>
  </div>
  <div data-role="content" class="content">
        <p>黄师塔前江水东，春光懒困倚微风。桃花一簇开无主，可爱深红爱浅红。</p>
  </div>
  <div data-role="footer">
    <h1>经典诗词</h1>
  </div>
</div>
</body>
</html>
```

在模拟器中的预览效果如图 13-9 所示。在页面中的诗词上面点击，即可发现 div 块内文字的颜色变成了绿色，如图 13-10 所示。

古诗欣赏

黄师塔前江水东，春光懒困倚微风。
桃花一簇开无主，可爱深红爱浅红。

经典诗词

图 13-9　程序预览效果

古诗欣赏

黄师塔前江水东，春光懒困倚微风。
桃花一簇开无主，可爱深红爱浅红。

经典诗词

图 13-10　触发 tap 事件

2. taphold 事件

如果点击页面并按住不放，则会触发 taphold 事件。其语法格式如下：

```
$("p").on("taphold",function(){
  $(this).hide();
});
```

在默认情况下，按住不放 750ms 之后触发 taphold 事件。用户也可以修改这个时间的长短。代码如下：

```
$(document).on("mobileinit",function(){
  $.event.special.tap.tapholdThreshold=5000;
});
```

修改后需要按住 5 秒以后才会触发 taphold 事件。

【例 13.5】(示例文件 ch13\13.5.html)

```
<!DOCTYPE html>
<html>
<head>
<meta name="viewport" content="width=device-width, initial-scale=1">
<link rel="stylesheet"
href="http://code.jquery.com/mobile/1.4.5/jquery.mobile-1.4.5.min.css">
<script src="http://code.jquery.com/jquery-1.11.3.min.js"></script>
 <script src="http://code.jquery.com/mobile/1.4.5/jquery.mobile-
1.4.5.min.js"></script>
<script>
$(document).on("mobileinit",function(){
  $.event.special.tap.tapholdThreshold=1000
});
$(function(){
  $("img").on("taphold",function(){
   $(this).hide();
});
});
</script>
</head>
<body>
<div data-role="page" id="first">
  <div data-role="header">
    <h1>可爱宠物</h1>
  </div>
  <div data-role="content" class="content">
<img src=13.1.jpg > <br>
         <p>按住图片 1 秒后隐藏图片哦！</p>
  </div>
  <div data-role="footer">
    <h1>动物天地</h1>
  </div>
</div>
```

```
</body>
</html>
```

在模拟器中的预览效果如图 13-11 所示。按住图片 1 秒后，即可发现图片被隐藏了，如图 13-12 所示。

图 13-11　程序预览效果

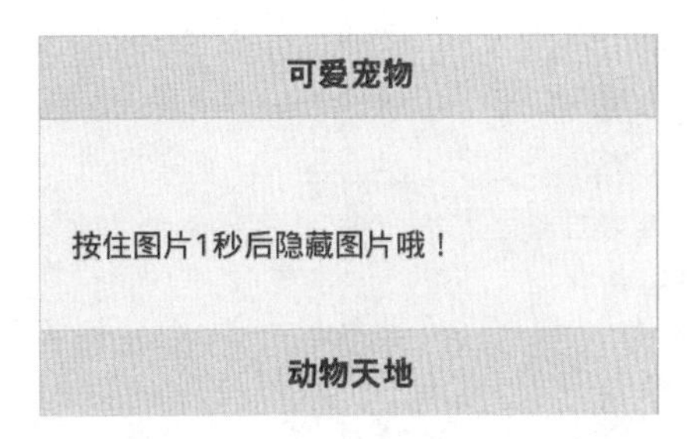

图 13-12　触发 taphold 事件

13.2.2　滑动事件

滑动事件是在用户 1 秒内水平拖曳大于 30px，或者纵向拖曳小于 20px 的事件发生时触发的事件。滑动事件使用 swipe 语法来捕捉，语法格式如下：

```
$("p").on("swipe",function(){
  $("span").text("滑动检测!");
});
```

上述语法是捕捉 p 组件的滑动事件，并将消息显示在 span 组件中。

向左滑动事件在用户向左拖动元素大于 30px 时触发，使用 swipeleft 语法来捕捉，语法格式如下：

```
$("p").on("swipeleft",function(){
  $("span").text("向左滑动检测!");
});
```

向右滑动事件在用户向右拖动元素大于 30px 时触发，使用 swiperight 语法来捕捉，语法格式如下：

```
$("p").on("swiperight,function(){
  $("span").text("向右滑动检测!");
});
```

下面以向右滑动事件为例进行讲解。

【例 13.6】(示例文件 ch13\13.6.html)

```
<!DOCTYPE html>
<html>
```

```
<head>
<meta name="viewport" content="width=device-width, initial-scale=1">
<link rel="stylesheet"
href="http://code.jquery.com/mobile/1.4.5/jquery.mobile-1.4.5.min.css">
<script src="http://code.jquery.com/jquery-1.11.3.min.js"></script>
 <script src="http://code.jquery.com/mobile/1.4.5/jquery.mobile-
1.4.5.min.js"></script>
<script>
$(document).on("pagecreate","#first",function(){
   $("img").on("swiperight",function(){
    alert("干嘛向右滑动我!!");
   });
});
</script>
</head>
<body>
<div data-role="page" id="first">
  <div data-role="header">
    <h1>可爱宠物</h1>
  </div>
  <div data-role="content" class="content">
<img src=13.2.jpg > <br>
        <p>向右滑动图片查看效果</p>
  </div>
  <div data-role="footer">
    <h1>动物天地</h1>
  </div>
</div>
</body>
</html>
```

在模拟器中的预览效果如图 13-13 所示。向右滑动图片，效果如图 13-14 所示。

图 13-13　程序预览效果

图 13-14　触发向右滑动事件

13.3 滚屏事件

jQuery Mobile 提供了两种滚屏事件，分别是滚动开始时触发 Scrollstart 事件和滚动结束时触发 Scrollstop 事件。

1. Scrollstart 事件

Scrollstart 事件是在用户开始滚动页面时触发。其语法格式如下：

```
$(document).on("scrollstart",function(){
  alert("屏幕开始滚动了！");
});
```

下面通过一个案例来理解 Scrollstart 事件。

【例 13.7】(示例文件 ch13\13.7.html)

```
<!DOCTYPE html>
<html>
<head>
<meta name="viewport" content="width=device-width, initial-scale=1">
<link rel="stylesheet"
href="http://code.jquery.com/mobile/1.4.5/jquery.mobile-1.4.5.min.css">
<script src="http://code.jquery.com/jquery-1.11.3.min.js"></script>
 <script src="http://code.jquery.com/mobile/1.4.5/jquery.mobile-
1.4.5.min.js"></script>
<script>
$(document).on("pagecreate","#first",function(){
  $(document).on("scrollstart",function(){
    alert("屏幕开始滚动了！");
  });
});
</script>
</head>
<body>
<div data-role="page" id="first">
  <div data-role="header">
    <h1>古诗欣赏</h1>
  </div>
  <div data-role="content" class="content">
    <p>西施越溪女，出自苎萝山。</p>
    <p>秀色掩今古，荷花羞玉颜。</p>
    <p>浣纱弄碧水，自与清波闲。</p>
    <p>皓齿信难开，沉吟碧云间。</p>
    <p>勾践徵绝艳，扬蛾入吴关。</p>
    <p>提携馆娃宫，杳渺讵可攀。</p>
    <p>一破夫差国，千秋竟不还。</p>
    <p>西施越溪女，出自苎萝山。</p>
    <p>秀色掩今古，荷花羞玉颜。</p>
```

```
    <p>浣纱弄碧水，自与清波闲。</p>
    <p>皓齿信难开，沉吟碧云间。</p>
    <p>勾践徵绝艳，扬蛾入吴关。</p>
    <p>提携馆娃宫，杳渺讵可攀。</p>
    <p>一破夫差国，千秋竟不还。</p>
  </div>
  <div data-role="footer">
    <h1>经典诗词</h1>
  </div>
</div>
</body>
</html>
```

在模拟器中的预览效果如图 13-15 所示。向上滚动屏幕，效果如图 13-16 所示。

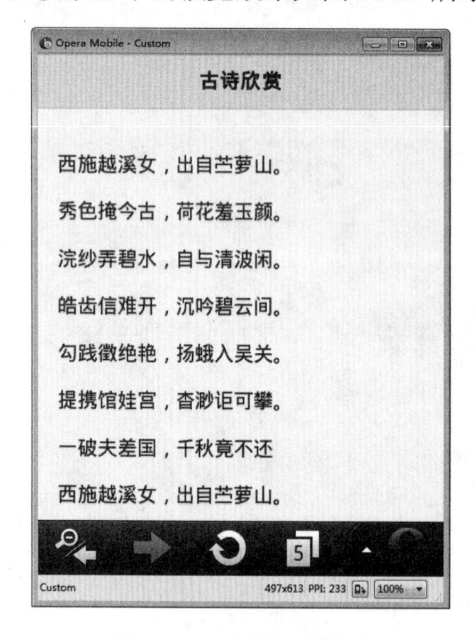

图 13-15　程序预览效果

图 13-16　触发滚屏事件

2. Scrollstop 事件

Scrollstop 事件是在用户停止滚动页面时触发。其语法格式如下：

```
$(document).on("scrollstop",function(){
 alert("停止滚动!");
});
```

下面通过一个案例来理解 Scrollstop 事件。

【例 13.8】(示例文件 ch13\13.8.html)

```
<!DOCTYPE html>
<html>
<head>
<meta name="viewport" content="width=device-width, initial-scale=1">
<link rel="stylesheet"
```

```
href="http://code.jquery.com/mobile/1.4.5/jquery.mobile-1.4.5.min.css">
<script src="http://code.jquery.com/jquery-1.11.3.min.js"></script>
 <script src="http://code.jquery.com/mobile/1.4.5/jquery.mobile-
1.4.5.min.js"></script>
<script>
$(document).on("pagecreate","#first",function(){
  $(document).on("scrollstop",function(){
    alert("屏幕已经停止滚动了!");
  });
});
</script>
</head>
<body>
<div data-role="page" id="first">
  <div data-role="header">
    <h1>古诗欣赏</h1>
  </div>
  <div data-role="content" class="content">
    <p>噫吁嚱，危乎高哉！</p>
    <p>蜀道之难，难于上青天！</p>
    <p>蚕丛及鱼凫，开国何茫然！</p>
    <p>尔来四万八千岁，不与秦塞通人烟。</p>
    <p>西当太白有鸟道，可以横绝峨嵋巅。</p>
    <p>地崩山摧壮士死，然后天梯石栈方钩连。</p>
    <p>上有六龙回日之高标，下有冲波逆折之回川。</p>
    <p>黄鹤之飞尚不得过，猿猱欲度愁攀援。</p>
    <p>青泥何盘盘，百步九折萦岩峦。</p>
    <p>扪参历井仰胁息，以手抚膺坐长叹。</p>
    <p>问君西游何时还？畏途巉岩不可攀。</p>
    <p>但见悲鸟号古木，雄飞从雌绕林间。</p>
    <p>又闻子规啼夜月，愁空山。</p>
    <p>蜀道之难，难于上青天，使人听此凋朱颜。</p>
    <p>连峰去天不盈尺，枯松倒挂倚绝壁。</p>
    <p>飞湍瀑流争喧豗，砯崖转石万壑雷。</p>
    <p>其险也若此，嗟尔远道之人，胡为乎来哉。</p>
    <p>剑阁峥嵘而崔嵬，一夫当关，万夫莫开。</p>
    <p>所守或匪亲，化为狼与豺。</p>
    <p>朝避猛虎，夕避长蛇，磨牙吮血，杀人如麻。</p>
    <p>锦城虽云乐，不如早还家。</p>
    <p>蜀道之难，难于上青天，侧身西望长咨嗟。</p>
  </div>
  <div data-role="footer">
    <h1>经典诗词</h1>
  </div>
</div>
</body>
</html>
```

在模拟器中的预览效果如图 13-17 所示。向上滚动屏幕，停止后的效果如图 13-18 所示。

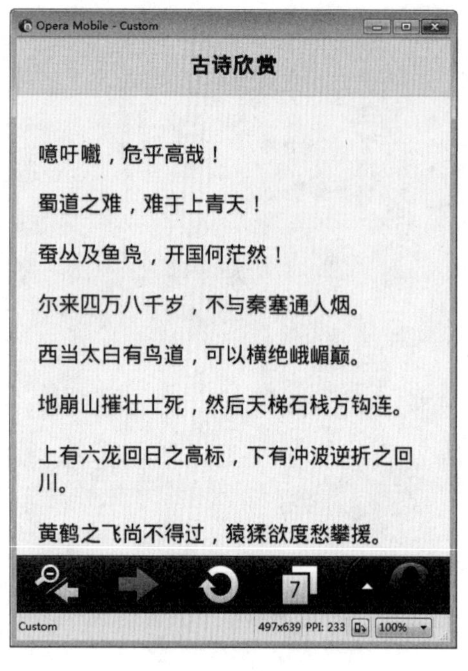

图 13-17　程序预览效果

图 13-18　触发滚屏停止事件

13.4　定 位 事 件

当移动设备水平或垂直翻转时触发定位事件，也就是常说的方向改变(orientationchange)事件。

在使用定位事件时，请将 orientationchange 事件绑定到 window 对象上，语法格式如下：

```
$(window).on("orientationchange",function(event){
alert("设备的方向改变为"+ event.orientation);
});
```

这里的 event 对象用来接收 orientation 属性值。用 event.orientation 返回的是设备是水平还是垂直，类型为字符串，如果是横向，返回值为 landscape；如果是纵向，返回值为 portrait。

下面通过一个案例来理解 orientationchange 事件。

【例 13.9】(示例文件 ch13\13.9.html)

```
<!DOCTYPE html>
<html>
<head>
<meta name="viewport" content="width=device-width, initial-scale=1">
<link rel="stylesheet"
href="http://code.jquery.com/mobile/1.4.5/jquery.mobile-1.4.5.min.css">
<script src="http://code.jquery.com/jquery-1.11.3.min.js"></script>
<script src="http://code.jquery.com/mobile/1.4.5/jquery.mobile-
1.4.5.min.js"></script>
```

```
<script type="text/javascript">
    $(document).on("pageinit",function(event){
        $( window ).on( "orientationchange", function( event ) {
         if(event.orientation == "landscape")
           $( "#orientation" ).text( "现在是水平模式!" ).css({"background-
color":"yellow","font-size":"300%"});
         if(event.orientation == "portrait")
            $( "#orientation" ).text( "现在是垂直模式!" ).css({"background-
color":"green","font-size":"200%"});
         });
        })
</script>
</head>
<body>
<div data-role="page" id="first">
  <div data-role="header">
    <h1>古诗欣赏</h1>
  </div>
  <div data-role="content" class="content">
  <span id="orientation"></span><br>
    <p>燕草如碧丝，秦桑低绿枝。当君怀归日，是妾断肠时。春风不相识，何事入罗帏</p>
  </div>
  <div data-role="footer">
    <h1>经典诗词</h1>
  </div>
</div>
</body>
</html>
```

在模拟器中的预览效果如图 13-19 所示。单击模拟器上的方向改变按钮，此时方向改变为水平方向，效果如图 13-20 所示。

图 13-19 程序预览效果

图 13-20 设备水平方向

再次单击模拟器上的方向改变按钮，此时方向改变为垂直方向，效果如图 13-21 所示。

图 13-21　设备垂直方向

13.5　疑 难 解 惑

疑问 1：绑定事件的方法 on()和 one()的区别？

答：绑定事件的方法 on()和 one()的作用相似，它们唯一的区别在于 one()只能执行一次。例如，当在按钮上绑定单击鼠标事件时，on()方法的程序如下：

```
<script>
$(document).on('click',function(){
   alert("这是使用 on()方法绑定的事件")
});
</script>
```

疑问 2：如何在设备方向改变时获取移动设备的高度和宽度？

答：如果设备方向改变时要获取移动设备的长度和宽度，可以绑定 resize 事件。该事件在页面大小改变时将触发，语法格式如下：

```
$(window).on("resize",function(){
  var win= $(this);    //this 指的是 window
  alert("宽度为"+win.width()+"高度为"+ win.height());
});
```

第4篇

项目实战

- 第14章　项目演练1——开发时钟特效系统
- 第15章　项目演练2——开发动态字符演示系统
- 第16章　项目演练3——开发图片堆叠系统
- 第17章　项目演练4——开发商品信息展示系统
- 第18章　项目演练5——开发连锁酒店移动网站

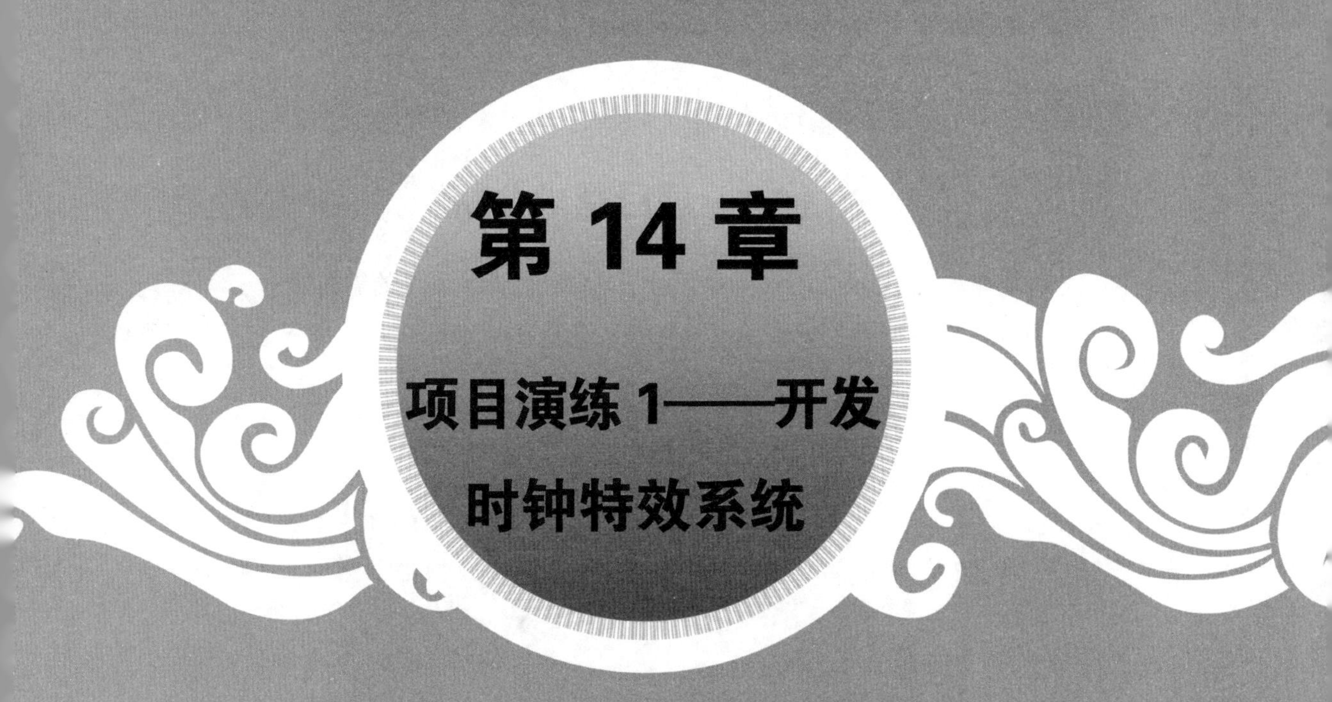

第 14 章 项目演练 1——开发时钟特效系统

该项目主要是实现时钟特效演示效果。用户每次单击不同的特效链接，将显示不同的时钟转化效果。特效主要包括翻转特效、弹跳特效、抖动特效和默认特效。通过本项目的练习，读者可以学习了解 jQuery 如何和 ES6 结合使用，并深入学习和了解 ES6 的类如何声明和使用，通过 ES6 让 JavaScript 具有面向对象编程的能力。

14.1 项目需求分析

需求分析是开发项目的必要环节。下面分析时钟特效演示系统的需求。

(1) 该项目利用 jQuery 构造了一个时钟特效演示系统，在 Firefox 53.0 中查看效果，如图 14-1 所示。

图 14-1 时钟特效演示系统默认效果

(2) 从图 14-1 可以看出，时钟特效演示系统中提供了 4 种特效，包括翻转、弹跳、抖动和默认。单击【翻转】按钮，即可查看翻转效果，如图 14-2 所示。

图 14-2 翻转效果

(3) 单击【抖动】按钮，即可查看抖动效果，如图 14-3 所示。同理，用户可以自行查看弹跳效果。

图 14-3 抖动效果

14.2 项目技术分析

该项目利用 javascript 的 Date 对象获取当前时间，并用 setInterval()每秒递增 1 秒钟，每 60 秒触发更新分钟数事件，每 60 分钟触发更新小时数事件。分别用 DigitalView.js 实现了数字时钟的视图，CircleView.js 实现了指针时钟的视图。多种时钟视图可以同时添加到同一时钟实例中。对于数字时钟，我们利用 animate.css 实现多种不同动画效果。分别点击下面的翻转、弹跳、抖动、默认(fadeIn/fadeOut)可以看到数字变化的不同效果。

该项目使用的 jQuery 方法相对较多，有利于初学者对 jQuery 的学习。同时该案例使用了 ES6 的语法规则。ES6 又被称为 ECMAScript 2015，顾名思义，它是 ECMAScript 在 2015 年发布的新标准。它为 JavaScript 语法带来了重大变革。ES6 包含了许多新的语言特性，使 JavaScript 变得更加强大。比如其中一个主要变革是 ES6 中添加了对类的支持，引入了 class 关键字。从而让类的声明和继承更加直观。

随着 JavaScript 的快速发展，JavaScript 和 jQuery 配合使用，加上 ES6 新语法，JavaScript 可以以更简单和更高效的方式开发复杂的应用。本案例中使用了类，这里读者需要多学习和理解。

14.3 系统的代码实现

下面分析时钟特效演示系统的代码是如何实现的。

14.3.1 设计首页

首页中显示了时钟各种特效的效果。代码如下：

```
<!DOCTYPE html>
<html lang="en">
<head>
    <meta charset="UTF-8">
    <meta name="viewport" content="width=device-width, initial-scale=1.0,
maximum-scale=1.0, user-scalable=no">
    <title>基于 jquery 的多种时钟演示系统</title>
    <link href="css/animate.css" rel="stylesheet">
    <link href="css/circle.css" rel="stylesheet">
    <link href="css/jsclock.css" rel="stylesheet">

    <script src="js/jquery-3.1.1.min.js"></script>
    <script src="js/dist/clock.js"></script>
    <!--<script src="js/jquery.clock.js"></script>-->
</head>
<body>
    <style id="clock-animations"></style>

    <h2 style="text-align: center;color:grey;">基于 jquery 的多种时钟演示系统
</h2>

    <div id="circleClock" style="height:300px"></div>

    <div class="nclock">
        <div class="dclock"></div>
        <div class="day"></div>
        <div class="effect">
            <a data-effectin="rotateIn" data-effectout="rotateOut"
href="#"><span>翻转</span></a>
            <a data-effectin="bounceIn" data-effectout="bounceOut"
href="#"><span>弹跳</span></a>
            <a data-effectin="tada" data-effectout="lightSpeedOut"
href="#"><span>抖动</span></a>
            <a data-effectin="fadeIn" data-effectout="fadeOut"
href="#"><span>默认</span></a>
        </div>
    </div>




    <script>
        $(document).ready(function(){
            var circleview = new CircleView({container:"#circleClock"});
```

```
        var digitalview = new DigitalView({
            container:$('.dclock'),
            effectIn:"fadeIn",
            effectOut:"fadeOut"});
        var clock = new Clock([circleview, digitalview]);

        $('.effect a').on('click tap', function(e){
            var effectIn = $(this).data('effectin');
            var effectOut = $(this).data('effectout');
            digitalview.effectIn = effectIn;
            digitalview.effectOut = effectOut;
        });

        var days = ['星期日','星期一','星期二','星期三','星期四','星期五','星期六'];
        var date = new Date();
        var i = date.getDay();
        $('.day').text(days[i]);
    });
  </script>
</body>
</html>
```

14.3.2 定义时钟类

Clock.js 文件位于项目的 js 文件夹下，使用 ES6 语法定义了 Clock 类。

该类主要实现如下方法。

(1) Constructor()：初始化时钟。

(2) updateSecond()：每秒执行一次，并调用所有视图的 updateSecond()方法。

(3) updateMinute()：每分钟执行一次，并调用所有视图的 updateMinute()方法。

(4) updateHour()：每小时执行一次，并调用所有视图的 updateHour()方法。

Clock.js 文件的代码如下：

```
/**
 * Created by francis on 17-6-12.
 * 该类实现时钟的逻辑
 * 该类采用单例模式,因为同一时区内所有的时间都应该是一样的，我们只考虑同一时区内的时间
 */
let ClockInstance = null;
class Clock{
    /**
     * 构造器
     * @param array views 视图的数组,可以传入多个视图,每个视图可以渲染一种类型的时钟,
如数字式时钟、指针式时钟
     * @returns {*} ClockInstance 单例对象
     */
    constructor(views){
        if(!ClockInstance){
```

```
        var self = this;
        var now = new Date();
        self.hour   = now.getHours(); //当前时间的小时数
        self.minute  = now.getMinutes(); //当前时间的分钟数
        self.second = now.getSeconds(); //当前时间的秒数
        self.secondMaxEvent = new CustomEvent('maxSecond');
          //秒数是 60 秒时触发该事件
        self.minuteMaxEvent = new CustomEvent('maxMinute');
          //分钟数是 60 分钟时触发该事件
        self.views = []; //存储视图的数组,可以有多个视图
        for(var view of views){ //存储多个视图
            view.init(self.hour, self.minute, self.second);
            self.views.push(view);
        }
        ClockInstance = self; //把当前对象赋值给变量 ClockInstance,形成单例模式
        //添加事件句柄,当秒数达到 60 时,分钟数加 1
        addEventListener('maxSecond',function(event){
            self._updateMinute();
        });
        //添加事件句柄,当分钟数达到 60 时,小时数加 1
        addEventListener('maxMinute',function(event){
            self._updateHour();
        });
        //设置每秒执行一次 self._updateSecond()
        setInterval(function(){
            self._updateSecond();
        }, 1000);
    }
    return ClockInstance;
}

/**
 * 每秒执行一次所有视图的 updateSecond()方法
 * 每秒执行一次,当前秒数是 59, 下一秒重新从 0 开始计数
 * @private
 */
_updateSecond(){
    var self = this;
    if(self.second == 59){
        dispatchEvent(self.secondMaxEvent);
        self.second = 0;
    }else{
        self.second += 1;
    }
    for(var view of self.views){
        view.updateSecond(self.second);
    }
}
/**
 * 每分钟执行一次所有视图的 updateMinute()方法
```

```
     * @private
     */
    _updateMinute(){
       var self = this;
       if(self.minute == 59){
          dispatchEvent(self.minuteMaxEvent);
          self.minute = 0;
       }else{
          self.minute++;
       }
       for(var view of self.views){
          view.updateMinute(self.minute);
       }
    }
    /**
     * 每小时执行一次所有视图的 updateHour()方法
     * @private
     */
    _updateHour(){
       var self = this;
       if(self.hour == 23){
          self.hour = 0;
       }else{
          self.hour++;
       }
       for(var view of self.views){
          view.updateHour(self.hour);
       }
    }
}
```

14.3.3 定义数字时钟的视图类

DigitalView.js 文件位于项目的 js 文件夹下，使用 ES6 语法定义了数字时钟的视图类。该类主要实现如下方法。

(1) Constructor()：初始化时钟。

(2) updateSecond()：每秒执行一次，并调用所有视图的 updateSecond()方法。

(3) updateMinute()：每分钟执行一次，并调用所有视图的 updateMinute()方法。

(4) updateHour()：每小时执行一次，并调用所有视图的 updateHour()方法。

(5) Constructor()：初始化数字时钟的视图类。

(6) Init()：初始化时钟，分解每一个字符。

(7) animateUpdate()：动画更新，当“时:分:秒”变化时，使用动画对其更新。

(8) updateMinute()：更新分钟数。

(9) updateHour()：更新小时数。

(10) set effectIn()：从外部设置数字进入的效果。

(11) set effectOut()：从外部设置数字退出的效果。

DigitalView.js 的代码如下：

```
/**
 * Created by francis on 17-6-13
 * 数字时钟的视图类
 */
class DigitalView{
    /**
     * 构造器,可接收 options 对象
     * @param Object options 视图的设置选项
     */
    constructor(options){
        var self = this;
        var default_options = { //视图默认设置选项
            container:".dclock", //视图默认容器
            selector:{h0:"#h0",h1:"#h1",m0:"#m0",m1:"#m1",s0:"#s0",s1:"#s1"},
              //设置时:分:秒的选择符
            effectIn: 'fadeIn', //数字进入效果
            effectOut: 'fadeOut'//数字退出效果
        };
        self.options = Object.assign(default_options, options||{});
              //用户自定义选项可覆盖默认选项
        //时钟的 HTML 模板
        self.template = `<div class="clock">
            <div class="digitalfont">
                <p id="h0"><span class="placeHolder">8</span></p>
                <p id="h1"><span class="placeHolder">8</span></p>
                <p class="separator">:</p>
                <p id="m0"><span class="placeHolder">8</span></p>
                <p id="m1"><span class="placeHolder">8</span></p>
                <p class="separator">:</p>
                <p id="s0"><span class="placeHolder">8</span></p>
                <p id="s1"><span class="placeHolder">8</span></p>
            </div>
        </div>`;
        $(self.options.container).append(self.template);//添加模板到容器中
    }

    /**
     * 初始化时钟,分解每一个字符
     * @param hour
     * @param minute
     * @param second
     */
    init(hour, minute, second){
        var self = this;
        var hourChar = self.valTochars(hour);
        var minuteChar = self.valTochars(minute);
        var secondChar = self.valTochars(second);
```

```
        self.animateUpdate(self.options.selector.h0, hourChar[0]);
        self.animateUpdate(self.options.selector.h1, hourChar[1]);
        self.animateUpdate(self.options.selector.m0, minuteChar[0]);
        self.animateUpdate(self.options.selector.m1, minuteChar[1]);
        self.animateUpdate(self.options.selector.s0, secondChar[0]);
        self.animateUpdate(self.options.selector.s1, secondChar[1]);
    }

    /**
     * 把给定的数值分解成字符
     * @param val 传入数值
     * @returns {Array|*} 字符数字
     */
    valTochars(val){
        var chars = val.toString().split('');
        if(chars.length < 2){
            chars.unshift('0');
        }
        return chars;
    }

    /**
     * 获取选择符内的数值
     * @param selector  时:分:秒选择符
     * @returns {*|jQuery}
     */
    getVal(selector){
        return $(selector).find('span.real:eq(0)').text();
    }

    /**
     * 动画更新,当时:分:秒变化时, 使用动画对其更新
     * @param selector 时:分:秒选择符
     * @param newVal 需要更新到的时间数值
     */
    animateUpdate(selector, newVal){
        var self = this;
        var $container = $(selector);
        var $old = $(selector).find('span.real').addClass(`animated
${self.options.effectOut}`);
        $('<span>').text(newVal).addClass(`real animated
${self.options.effectIn}`).css({position:'absolute'}).appendTo($container);
        $old.one('webkitAnimationEnd mozAnimationEnd MSAnimationEnd
oanimationend animationend', function () {
            $old.remove();
        });
    }

    /**
     * 更新秒数
```

```
 * @param second
 */
updateSecond(second){
    var self = this;
    var currentChars = self.currentSecond();
    var targetChars = self.valTochars(second);
    for(var i = 0; i<2; i++){
        if(currentChars[i] != targetChars[i]){
            self.animateUpdate(self.options.selector['s'+i], targetChars[i]);
        }
    }
}
/**
 * 更新分钟数
 * @param second
 */
updateMinute(minute){
    var self = this;
    var currentChars = self.currentMinute();
    var targetChars = self.valTochars(minute);
    for(var i = 0; i<2; i++){
        if(currentChars[i] != targetChars[i]){
            self.animateUpdate(self.options.selector['m'+i], targetChars[i]);
        }
    }
}
/**
 * 更新小时数
 * @param second
 */
updateHour(hour){
    var self = this;
    var currentChars = self.currentHour();
    var targetChars = self.valTochars(hour);
    for(var i = 0; i<2; i++){
        if(currentChars[i] != targetChars[i]){
            self.animateUpdate(self.options.selector['h'+i], targetChars[i]);
        }
    }
}

/**
 * 获取当前已经显示的秒数
 * @returns {Array} 秒数的字符数组
 */
currentSecond(){
    var self = this;
    var chars = [];
    chars.push(self.getVal(self.options.selector.s0));
    chars.push(self.getVal(self.options.selector.s1));
```

```
        return chars;
    }
    /**
     * 获取当前已经显示的分钟数
     * @returns {Array} 分钟数的字符数组
     */
    currentMinute(){
        var self = this;
        var chars = [];
        chars.push(self.getVal(self.options.selector.m0));
        chars.push(self.getVal(self.options.selector.m1));
        return chars;
    }
    /**
     * 获取当前已经显示的小时数
     * @returns {Array} 小时数的字符数组
     */
    currentHour(){
        var self = this;
        var chars = [];
        chars.push(self.getVal(self.options.selector.h0));
        chars.push(self.getVal(self.options.selector.h1));
        return chars;
    }

    /**
     * 从外部设置数字进入的效果
     * @param effect
     */
    set effectIn(effect){
        this.options.effectIn = effect;
    }
    /**
     * 从外部设置数字退出的效果
     * @param effect
     */
    set effectOut(effect){
        this.options.effectOut = effect;
    }
}
```

14.3.4 定义圆形指针时钟的视图类

CircleView.js 文件位于项目的 js 文件夹下，使用 ES6 语法定义了圆形指针时钟的视图类。该文件定义 Init()方法，根据当前时间计算出时针、分针、秒针应该处于多少角度。并用 CSS 定义时针、分针、秒针的动画效果。

CircleView.js 文件的代码如下：

```
/**
 * Created by francis on 17-6-13
 */
class CircleView{
    constructor(options){
        var self = this;
        var default_options = {
            container:"#circleClock",
        };
        self.options = Object.assign(default_options, options||{});
        self.template =
        `<div class="clock-wrapper">
            <div class="clock-base">
                <div class="click-indicator">
                    <div></div>
                    <div></div>
                    <div></div>
                    <div></div>
                    <div></div>
                    <div></div>
                    <div></div>
                    <div></div>
                    <div></div>
                    <div></div>
                    <div></div>
                    <div></div>
                </div>
                <div class="clock-hour"></div>
                <div class="clock-minute"></div>
                <div class="clock-second"></div>
                <div class="clock-center"></div>
            </div>
        </div>`;
        $(self.options.container).append(self.template);
    }


    init(hour, minute, second){
        var hourDeg   = hour / 12 * 360 + minute / 60 * 30;
        var minuteDeg = minute / 60 * 360 + second / 60 * 6;
        var secondDeg = second / 60 * 360;
        var stylesDeg = [
            "@-webkit-keyframes rotate-hour{from{transform:rotate(" + hourDeg
+ "deg);}to{transform:rotate(" + (hourDeg + 360) + "deg);}}",
            "@-webkit-keyframes rotate-minute{from{transform:rotate(" +
minuteDeg + "deg);}to{transform:rotate(" + (minuteDeg + 360) + "deg);}}",
            "@-webkit-keyframes rotate-second{from{transform:rotate(" +
secondDeg + "deg);}to{transform:rotate(" + (secondDeg + 360) + "deg);}}",
            "@-moz-keyframes rotate-hour{from{transform:rotate(" + hourDeg +
"deg);}to{transform:rotate(" + (hourDeg + 360) + "deg);}}",
```

```
        "@-moz-keyframes rotate-minute{from{transform:rotate(" +
minuteDeg + "deg);}to{transform:rotate(" + (minuteDeg + 360) + "deg);}}",
        "@-moz-keyframes rotate-second{from{transform:rotate(" +
secondDeg + "deg);}to{transform:rotate(" + (secondDeg + 360) + "deg);}}"
      ].join("");
      document.getElementById("clock-animations").innerHTML = stylesDeg;
   }

   updateSecond(second){
   }

   updateMinute(minute){
   }

   updateHour(hour){
   }
}
```

14.3.5 合并多个 js 文件

gulpfile.js 文件位于项目根目录下，主要功能是把 ES6 文件转换成 ES5 文件，并把多个 js 文件合并成一个 js 文件，以及把 less 文件编译成 CSS 文件。

gulpfile.js 文件的代码如下：

```
/**
 * Created by francis on 17-5-11.
 */
const gulp = require('gulp');
const strip = require('gulp-strip-comments');
const babel = require("gulp-babel");
const concat = require('gulp-concat');
//css
const less = require('gulp-less');
const autoprefixer = require('gulp-autoprefixer');

/*******************************
 Tasks
 *******************************/
var src_less = './public/css/*.less';
var css_dist = './public/css';

gulp.task('css', function () {
   return gulp.src(src_less)
      .pipe(less())
      // .pipe(minifyCSS())
      .pipe(autoprefixer({
         browsers: ['last 2 versions', 'ie >= 9']
      }))
      .pipe(gulp.dest(css_dist));
```

```
});

var jsfiles = [
    'public/js/Clock.js',
    'public/js/DigitalView.js',
    'public/js/CircleView.js',
];
gulp.task('js',function(){
    gulp.src(jsfiles)
        .pipe(babel({presets: ['es2015']}))
        // .pipe(stripDebug())
        .pipe(concat('clock.js'))
        // .pipe(strip())
        // .pipe(uglify({'mangle':false}))
        .pipe(gulp.dest('public/js/dist/'));
});
```

14.3.6 合并 Clock.js、DigitalView.js 和 CircleView.js 文件

clock.js 文件位于 dist 文件夹下，主要功能是把 Clock.js、DigitalView.js 和 CircleView.js 合并到 clock.js 文件里。

clock.js 文件的代码如下：

```
'use strict';

var _createClass = function () { function defineProperties(target, props)
{ for (var i = 0; i < props.length; i++) { var descriptor = props[i];
descriptor.enumerable = descriptor.enumerable || false;
descriptor.configurable = true; if ("value" in descriptor)
descriptor.writable = true; Object.defineProperty(target, descriptor.key,
descriptor); } } return function (Constructor, protoProps, staticProps)
{ if (protoProps) defineProperties(Constructor.prototype, protoProps); if
(staticProps) defineProperties(Constructor, staticProps); return
Constructor; }; }();

function _classCallCheck(instance, Constructor) { if (!(instance instanceof
Constructor)) { throw new TypeError("Cannot call a class as a
function"); } }

/**
 * Created by francis on 17-6-12
 * 该类实现时钟的逻辑
 * 该类采用单例模式,因为同一时区内所有的时间都应该是一样的，我们只考虑同一时区内的时间
 */
var ClockInstance = null;

var Clock = function () {
    /**
     * 构造器
     * @param array views 视图的数组,可以传入多个视图,每个视图可以渲染一种类型的时钟,
```

```
如数字式时钟、指针式时钟
     * @returns {*} ClockInstance 单例对象
     */
    function Clock(views) {
        _classCallCheck(this, Clock);

        if (!ClockInstance) {
            var self = this;
            var now = new Date();
            self.hour = now.getHours(); //当前时间的小时数
            self.minute = now.getMinutes(); //当前时间的分钟数
            self.second = now.getSeconds(); //当前时间的秒数
            self.secondMaxEvent = new CustomEvent('maxSecond');
              //秒数是60秒时触发该事件
            self.minuteMaxEvent = new CustomEvent('maxMinute');
              //分钟数是60分钟时触发该事件
            self.views = []; //存储视图的数组,可以有多个视图
            var _iteratorNormalCompletion = true;
            var _didIteratorError = false;
            var _iteratorError = undefined;

            try {
                for (var _iterator = views[Symbol.iterator](), _step; !
(_iteratorNormalCompletion = (_step = _iterator.next()).done);
_iteratorNormalCompletion = true) {
                    var view = _step.value;
                    //存储多个视图
                    view.init(self.hour, self.minute, self.second);
                    self.views.push(view);
                }
            } catch (err) {
                _didIteratorError = true;
                _iteratorError = err;
            } finally {
                try {
                    if (!_iteratorNormalCompletion && _iterator.return) {
                        _iterator.return();
                    }
                } finally {
                    if (_didIteratorError) {
                        throw _iteratorError;
                    }
                }
            }

            ClockInstance = self; //把当前对象赋值给变量ClockInstance,形成单例模式
            //添加事件句柄,当秒数达到60时,分钟数加1
            addEventListener('maxSecond', function (event) {
                self._updateMinute();
            });
            //添加事件句柄,当分钟数达到60时,小时数加1
            addEventListener('maxMinute', function (event) {
```

```
            self._updateHour();
        });
        //设置每秒执行一次 self._updateSecond()
        setInterval(function () {
            self._updateSecond();
        }, 1000);
    }
    return ClockInstance;
}

/**
 * 每秒执行一次所有视图的 updateSecond()方法
 * 每秒执行一次,当前秒数是 59, 下一秒重新从 0 开始计数
 * @private
 */


_createClass(Clock, [{
    key: '_updateSecond',
    value: function _updateSecond() {
        var self = this;
        if (self.second == 59) {
            dispatchEvent(self.secondMaxEvent);
            self.second = 0;
        } else {
            self.second += 1;
        }
        var _iteratorNormalCompletion2 = true;
        var _didIteratorError2 = false;
        var _iteratorError2 = undefined;

        try {
            for (var _iterator2 = self.views[Symbol.iterator](), _step2; !
(_iteratorNormalCompletion2 = (_step2 = _iterator2.next()).done);
_iteratorNormalCompletion2 = true) {
                var view = _step2.value;

                view.updateSecond(self.second);
            }
        } catch (err) {
            _didIteratorError2 = true;
            _iteratorError2 = err;
        } finally {
            try {
                if (!_iteratorNormalCompletion2 && _iterator2.return) {
                    _iterator2.return();
                }
            } finally {
                if (_didIteratorError2) {
                    throw _iteratorError2;
                }
            }
```

```
        }
    }
    /**
     * 每分钟执行一次所有视图的updateMinute()方法
     * @private
     */

}, {
    key: '_updateMinute',
    value: function _updateMinute() {
        var self = this;
        if (self.minute == 59) {
            dispatchEvent(self.minuteMaxEvent);
            self.minute = 0;
        } else {
            self.minute++;
        }
        var _iteratorNormalCompletion3 = true;
        var _didIteratorError3 = false;
        var _iteratorError3 = undefined;

        try {
            for (var _iterator3 = self.views[Symbol.iterator](),
_step3; !(_iteratorNormalCompletion3 = (_step3 = _iterator3.next()).done);
_iteratorNormalCompletion3 = true) {
                var view = _step3.value;

                view.updateMinute(self.minute);
            }
        } catch (err) {
            _didIteratorError3 = true;
            _iteratorError3 = err;
        } finally {
            try {
                if (!_iteratorNormalCompletion3 && _iterator3.return) {
                    _iterator3.return();
                }
            } finally {
                if (_didIteratorError3) {
                    throw _iteratorError3;
                }
            }
        }
    }
    /**
     * 每小时执行一次所有视图的updateHour()方法
     * @private
     */

}, {
    key: '_updateHour',
    value: function _updateHour() {
```

```
            var self = this;
            if (self.hour == 23) {
                self.hour = 0;
            } else {
                self.hour++;
            }
            var _iteratorNormalCompletion4 = true;
            var _didIteratorError4 = false;
            var _iteratorError4 = undefined;

            try {
                for (var _iterator4 = self.views[Symbol.iterator](), _step4; !
(_iteratorNormalCompletion4 = (_step4 = _iterator4.next()).done);
_iteratorNormalCompletion4 = true) {
                    var view = _step4.value;

                    view.updateHour(self.hour);
                }
            } catch (err) {
                _didIteratorError4 = true;
                _iteratorError4 = err;
            } finally {
                try {
                    if (!_iteratorNormalCompletion4 && _iterator4.return) {
                        _iterator4.return();
                    }
                } finally {
                    if (_didIteratorError4) {
                        throw _iteratorError4;
                    }
                }
            }
        }
    }]);

    return Clock;
}();
"use strict";

var _createClass = function () { function defineProperties(target, props)
{ for (var i = 0; i < props.length; i++) { var descriptor = props[i];
descriptor.enumerable = descriptor.enumerable || false;
descriptor.configurable = true; if ("value" in descriptor)
descriptor.writable = true; Object.defineProperty(target, descriptor.key,
descriptor); } } return function (Constructor, protoProps, staticProps)
{ if (protoProps) defineProperties(Constructor.prototype, protoProps); if
(staticProps) defineProperties(Constructor, staticProps); return
Constructor; }; }();

function _classCallCheck(instance, Constructor) { if (!(instance instanceof
Constructor)) { throw new TypeError("Cannot call a class as a
function"); } }
```

```
/**
 * Created by francis on 17-6-13
 * 数字时钟的视图类
 */
var DigitalView = function () {
    /**
     * 构造器,可接收 options 对象
     * @param Object options 视图的设置选项
     */
    function DigitalView(options) {
        _classCallCheck(this, DigitalView);

        var self = this;
        var default_options = {      //视图默认设置选项
            container: ".dclock",     //视图默认容器
            selector: { h0: "#h0", h1: "#h1", m0: "#m0", m1: "#m1", s0: "#s0",
s1: "#s1" }, //设置时:分:秒的选择符
            effectIn: 'fadeIn', //数字进入效果
            effectOut: 'fadeOut' //数字退出效果
        };
        self.options = Object.assign(default_options, options || {});
         //用户自定义选项可覆盖默认选项
         //时钟的 HTML 模板
        self.template = "<div class=\"clock\">\n          <div
class=\"digitalfont\">\n             <p id=\"h0\"><span
class=\"placeHolder\">8</span></p>\n              <p id=\"h1\"><span
class=\"placeHolder\">8</span></p>\n              <p
class=\"separator\">:</p>\n             <p id=\"m0\"><span
class=\"placeHolder\">8</span></p>\n              <p id=\"m1\"><span
class=\"placeHolder\">8</span></p>\n              <p
class=\"separator\">:</p>\n             <p id=\"s0\"><span
class=\"placeHolder\">8</span></p>\n              <p id=\"s1\"><span
class=\"placeHolder\">8</span></p>\n          </div>\n      </div>";
        $(self.options.container).append(self.template); //添加模板到容器中
    }

    /**
     * 初始化时钟,分解每一个字符
     * @param hour
     * @param minute
     * @param second
     */


    _createClass(DigitalView, [{
        key: "init",
        value: function init(hour, minute, second) {
            var self = this;
            var hourChar = self.valTochars(hour);
            var minuteChar = self.valTochars(minute);
            var secondChar = self.valTochars(second);
```

```
            self.animateUpdate(self.options.selector.h0, hourChar[0]);
            self.animateUpdate(self.options.selector.h1, hourChar[1]);
            self.animateUpdate(self.options.selector.m0, minuteChar[0]);
            self.animateUpdate(self.options.selector.m1, minuteChar[1]);
            self.animateUpdate(self.options.selector.s0, secondChar[0]);
            self.animateUpdate(self.options.selector.s1, secondChar[1]);
        }

        /**
         * 把给定的数值分解成字符
         * @param val 传入数值
         * @returns {Array|*} 字符数字
         */

    }, {
        key: "valTochars",
        value: function valTochars(val) {
            var chars = val.toString().split('');
            if (chars.length < 2) {
                chars.unshift('0');
            }
            return chars;
        }

        /**
         * 获取选择符内的数值
         * @param selector  时:分:秒选择符
         * @returns {*|jQuery}
         */

    }, {
        key: "getVal",
        value: function getVal(selector) {
            return $(selector).find('span.real:eq(0)').text();
        }

        /**
         * 动画更新,当时:分:秒变化时, 使用动画对其更新
         * @param selector 时:分:秒选择符
         * @param newVal 需要更新到的时间数值
         */

    }, {
        key: "animateUpdate",
        value: function animateUpdate(selector, newVal) {
            var self = this;
            var $container = $(selector);
            var $old = $(selector).find('span.real').addClass("animated " +
self.options.effectOut);
            $('<span>').text(newVal).addClass("real animated " +
self.options.effectIn).css({ position: 'absolute' }).appendTo($container);
            $old.one('webkitAnimationEnd mozAnimationEnd MSAnimationEnd
```

```
oanimationend animationend', function () {
                $old.remove();
            });
        }

        /**
         * 更新秒数
         * @param second
         */

    }, {
        key: "updateSecond",
        value: function updateSecond(second) {
            var self = this;
            var currentChars = self.currentSecond();
            var targetChars = self.valTochars(second);
            for (var i = 0; i < 2; i++) {
                if (currentChars[i] != targetChars[i]) {
                    self.animateUpdate(self.options.selector['s' + i], targetChars[i]);
                }
            }
        }
        /**
         * 更新分钟数
         * @param second
         */

    }, {
        key: "updateMinute",
        value: function updateMinute(minute) {
            var self = this;
            var currentChars = self.currentMinute();
            var targetChars = self.valTochars(minute);
            for (var i = 0; i < 2; i++) {
                if (currentChars[i] != targetChars[i]) {
                    self.animateUpdate(self.options.selector['m' + i], targetChars[i]);
                }
            }
        }
        /**
         * 更新小时数
         * @param second
         */

    }, {
        key: "updateHour",
        value: function updateHour(hour) {
            var self = this;
            var currentChars = self.currentHour();
            var targetChars = self.valTochars(hour);
            for (var i = 0; i < 2; i++) {
                if (currentChars[i] != targetChars[i]) {
```

```
                self.animateUpdate(self.options.selector['h' + i], targetChars[i]);
            }
        }
    }

    /**
     * 获取当前已经显示的秒数
     * @returns {Array} 秒数的字符数组
     */

}, {
    key: "currentSecond",
    value: function currentSecond() {
        var self = this;
        var chars = [];
        chars.push(self.getVal(self.options.selector.s0));
        chars.push(self.getVal(self.options.selector.s1));
        return chars;
    }
    /**
     * 获取当前已经显示的分钟数
     * @returns {Array} 分钟数的字符数组
     */

}, {
    key: "currentMinute",
    value: function currentMinute() {
        var self = this;
        var chars = [];
        chars.push(self.getVal(self.options.selector.m0));
        chars.push(self.getVal(self.options.selector.m1));
        return chars;
    }
    /**
     * 获取当前已经显示的小时数
     * @returns {Array} 小时数的字符数组
     */

}, {
    key: "currentHour",
    value: function currentHour() {
        var self = this;
        var chars = [];
        chars.push(self.getVal(self.options.selector.h0));
        chars.push(self.getVal(self.options.selector.h1));
        return chars;
    }

    /**
     * 从外部设置数字进入的效果
     * @param effect
     */
```

```
    }, {
        key: "effectIn",
        set: function set(effect) {
            this.options.effectIn = effect;
        }
        /**
         * 从外部设置数字退出的效果
         * @param effect
         */

    }, {
        key: "effectOut",
        set: function set(effect) {
            this.options.effectOut = effect;
        }
    }]);

    return DigitalView;
}();
"use strict";

var _createClass = function () { function defineProperties(target, props)
{ for (var i = 0; i < props.length; i++) { var descriptor = props[i];
descriptor.enumerable = descriptor.enumerable || false;
descriptor.configurable = true; if ("value" in descriptor)
descriptor.writable = true; Object.defineProperty(target, descriptor.key,
descriptor); } } return function (Constructor, protoProps, staticProps)
{ if (protoProps) defineProperties(Constructor.prototype, protoProps); if
(staticProps) defineProperties(Constructor, staticProps); return
Constructor; }; }();

function _classCallCheck(instance, Constructor) { if (!(instance instanceof
Constructor)) { throw new TypeError("Cannot call a class as a function"); } }

/**
 * Created by francis on 17-6-13.
 */
var CircleView = function () {
    function CircleView(options) {
        _classCallCheck(this, CircleView);

        var self = this;
        var default_options = {
            container: "#circleClock"
        };
        self.options = Object.assign(default_options, options || {});
        self.template = "<div class=\"clock-wrapper\">\n          <div
class=\"clock-base\">\n              <div class=\"click-indicator\">\n
<div></div>\n                   <div></div>\n                   <div></div>\n
<div></div>\n                   <div></div>\n                   <div></div>\n
<div></div>\n                   <div></div>\n                   <div></div>\n
```

```
<div></div>\n                    <div></div>\n                    <div></div>\n
</div>\n            <div class=\"clock-hour\"></div>\n              <div
class=\"clock-minute\"></div>\n             <div class=\"clock-
second\"></div>\n            <div class=\"clock-center\"></div>\n
</div>\n      </div>";
        $(self.options.container).append(self.template);
    }

    _createClass(CircleView, [{
        key: "init",
        value: function init(hour, minute, second) {
            var hourDeg = hour / 12 * 360 + minute / 60 * 30;
            var minuteDeg = minute / 60 * 360 + second / 60 * 6;
            var secondDeg = second / 60 * 360;
            var stylesDeg = ["@-webkit-keyframes rotate-
hour{from{transform:rotate(" + hourDeg + "deg);}to{transform:rotate(" +
(hourDeg + 360) + "deg);}}", "@-webkit-keyframes rotate-
minute{from{transform:rotate(" + minuteDeg + "deg);}to{transform:rotate(" +
(minuteDeg + 360) + "deg);}}", "@-webkit-keyframes rotate-
second{from{transform:rotate(" + secondDeg + "deg);}to{transform:rotate(" +
(secondDeg + 360) + "deg);}}", "@-moz-keyframes rotate-
hour{from{transform:rotate(" + hourDeg + "deg);}to{transform:rotate(" +
(hourDeg + 360) + "deg);}}", "@-moz-keyframes rotate-
minute{from{transform:rotate(" + minuteDeg + "deg);}to{transform:rotate(" +
(minuteDeg + 360) + "deg);}}", "@-moz-keyframes rotate-
second{from{transform:rotate(" + secondDeg + "deg);}to{transform:rotate(" +
(secondDeg + 360) + "deg);}}"].join("");
            document.getElementById("clock-animations").innerHTML = stylesDeg;
        }
    }, {
        key: "updateSecond",
        value: function updateSecond(second) {}
    }, {
        key: "updateMinute",
        value: function updateMinute(minute) {}
    }, {
        key: "updateHour",
        value: function updateHour(hour) {}
    }]);

    return CircleView;
}();
```

第 15 章 项目演练 2——开发动态字符演示系统

该项目主要是实现动态字符演示效果。通过 jQuery 构建出炫酷的文字动画效果。通过本项目的练习，读者可以了解使用 ES6 语法实现面向对象语言类的概念的方法，通过该案例最终封装成了一个 jQuery 插件。另外，读者还可以掌握如何使用 CSS 实现网页上的动画效果。

15.1 项目需求分析

需求分析是开发项目的必要环节。下面分析动态字符演示系统的需求。

(1) 动态效果一：一段英文文字按单词分割，进行动画循环展示。该案例，在绿线上方展示了两个动态效果。

(2) 动态效果二：四段中英文混合的文字按字符分割，循环进行动画展示。在 Firefox 53.0 中查看效果，如图 15-1 所示。

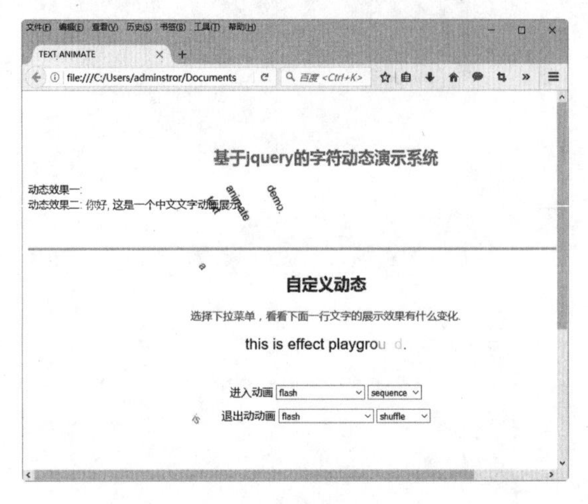

图 15-1　动态字符演示系统主页

(3) 在绿线的下方是一个小测试区，用户可以选择测试 animate.css 提供的各种动画效果。其中“进入动画”，是指文字显示时的动画效果；“退出动画”，是指文字隐藏时的动画效果。其中文字显示动画的顺序包括 sequence(顺序)、reverse(倒序)、sync(同步显示)和 shuffle(乱序显示)，如图 15-2 所示。

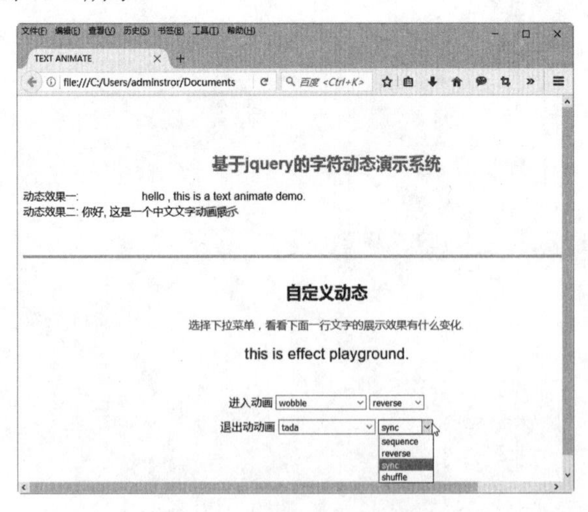

图 15-2　自定义动画动态效果

15.2 项目技术分析

该项目利用 jQuery 并结合 animate.css 构建出炫酷的文字动画效果。同时，该案例使用了 ES6 的语法规则。ES6 包含了许多新的语言特性，使 JavaScript 变得更加强大。比如其中一个主要变革是 ES6 中添加了对类的支持，引入了 class 关键字，从而使类的声明和继承更加直观。

随着 CSS 3 的出现，用户可以不再依赖 JavaScript，通过使用 CSS 样式表，也可以制作出网页动画，而且还可以实现跨平台显示动画效果。本案例将使用 CSS 文件来实现动画效果。

本项目将教会读者如何开发 jQuery 插件，并且学习使用 animate.css 实现各种动画效果。

15.3 系统的代码实现

下面来分析动态文字演示系统的代码是如何实现的。

15.3.1 设计首页

首页中显示了动态文字的动画效果。代码如下：

```
<!DOCTYPE html>
<html >
<head>
    <meta charset="UTF-8">
    <meta name="viewport" content="width=device-width, initial-scale=1.0,
maximum-scale=1.0, user-scalable=no">
    <title>TEXT ANIMATE</title>
    <link href="css/animate.css" rel="stylesheet">
    <script src="js/jquery-3.1.1.min.js"></script>
    <script src="js/jquery.lettering.js"></script>
    <script src="js/dist/textanimate.js"></script>
    <style>
        .container{width:900px;margin:5em auto;}
        .header{}
        .examples{border-bottom:solid 4px #99cb84;height:6em;}
        .example2>div{display:inline-block;}
        .playground{margin:2em 0;text-align: center;}
        .gray{color:#66512c;}
        .playground form div{padding:0.4em 0;}
        .playgroundTxt{height:3em;font-size:1.4em;}

    </style>
</head>
<body>
    <div class="container">
```

```
        <h2 style="text-align: center;color:grey;">基于jquery的字符动态演示系统</h2>

        <div class="examples">
            <div class="header">
                动态效果一:
                <span class="txtam">
                    hello , this is a text animate demo.
                </span>
            </div>

            <div class="example2">
                动态效果二:
                <div class="tlt">
                    <div class="texts">
                        <h2 data-in-effect="rollIn" data-out-effect="fadeOut">
你好，这是一个中文文字动画展示</h2>
                        <h3 data-in-effect="fadeInLeftBig" data-out-
effect="fadeOut">这个动画采用animate.css的fadeInLeftBig和fadeOut的效果</h3>
                        <h4 data-in-effect="rollIn" data-out-
effect="hinge">this is english characters animation demostration.</h4>
                        <p data-in-effect="rollIn" data-out-effect="fadeOut">这
是最后一条展示,后面的循环进行</p>
                    </div>
                </div>
            </div>



        </div>



        <div class="playground">
            <h2>自定义动态</h2>
            <p class="gray">选择下拉菜单，看看下面一行文字的展示效果有什么变化.</p>
            <div class="playgroundTxt">
                <ul class="texts" style="display: none">
                    <li data-id="wizard">这儿是效果游乐场</li>
                    <li data-id="fox">this is effect playground.</li>
                </ul>
            </div>

            <div>
                <form>
                    <div>
                        <label>进入动画</label>
                        <select data-key="effect" data-type="in"></select>
                        <select data-key="type" data-type="in">
                            <option value="">sequence</option>
                            <option value="reverse">reverse</option>
                            <option value="sync">sync</option>
                            <option value="shuffle">shuffle</option>
```

```
                    </select>
                </div>
                <div class="control col-1-2">
                    <label>退出动画</label>
                    <select data-key="effect" data-type="out"></select>
                    <select data-key="type" data-type="out">
                        <option value="">sequence</option>
                        <option value="reverse">reverse</option>
                        <option value="sync">sync</option>
                        <option selected="selected" value="shuffle">shuffle
</option>
                    </select>
                </div>
            </form>
        </div>

    </div>
</div>


<script>
    $(document).ready(function(){
        $('.header .txtam').textanimate({type:"word",in:
{sync:false,reverse:true,effect: 'rollIn'},loop:true});
        $('.tlt').textanimate({loop:true});


        var log = function (msg) {
            return function () {
                if (console) console.log(msg);
            }
        }
        var animateClasses = 'flash bounce shake tada swing wobble pulse
flip flipInX flipOutX flipInY flipOutY fadeIn fadeInUp fadeInDown
fadeInLeft fadeInRight fadeInUpBig fadeInDownBig fadeInLeftBig
fadeInRightBig fadeOut fadeOutUp fadeOutDown fadeOutLeft fadeOutRight
fadeOutUpBig fadeOutDownBig fadeOutLeftBig fadeOutRightBig bounceIn
bounceInDown bounceInUp bounceInLeft bounceInRight bounceOut bounceOutDown
bounceOutUp bounceOutLeft bounceOutRight rotateIn rotateInDownLeft
rotateInDownRight rotateInUpLeft rotateInUpRight rotateOut
rotateOutDownLeft rotateOutDownRight rotateOutUpLeft rotateOutUpRight hinge
rollIn rollOut';
        var $form = $('.playground form');
        var getFormData = function () {
            var data = {
                loop: true,
                in: { callback: log('in callback called.') },
                out: { callback: log('out callback called.') }
            };
            $form.find('[data-key="effect"]').each(function () {
```

```
                var $this = $(this)
                    , key = $this.data('key')
                    , type = $this.data('type');
                data[type][key] = $this.val();
            });
            $form.find('[data-key="type"]').each(function () {
                var $this = $(this)
                    , key = $this.data('key')
                    , type = $this.data('type')
                    , val = $this.val();
                data[type].shuffle = (val === 'shuffle');
                data[type].reverse = (val === 'reverse');
                data[type].sync = (val === 'sync');
            });
            return data;
        };

        $.each(animateClasses.split(' '), function (i, value) {
            var type = '[data-type]'
                , option = '<option value="' + value + '">' + value +
'</option>';
            if (/Out/.test(value) || value === 'hinge') {
                type = '[data-type="out"]';
            } else if (/In/.test(value)) {
                type = '[data-type="in"]';
            }
            if (type) {
                $form.find('[data-key="effect"]' + type).append(option);
            }
        });

//          $('.playgoundTxt').textanimate({loop:true});
        $form.on('change', function () {
            var obj = getFormData();
            console.log('getFormData:', obj);
            $('.playgroundTxt').textanimate(obj);
        }).trigger('change');

    });
  </script>
</body>
</html>
```

15.3.2 定义动画的类和执行动画的类

class.js 文件位于项目根目录下，主要定义动画的类。代码如下：

```
/**
 * Created by francis on 17-6-9.
```

```
 */
var Textillate = function (element, options) {
   var base = this
      , $element = $(element);

   console.log('$element:',$element.html());

   base.init = function () {
      base.$texts = $element.find(options.selector);

      if (!base.$texts.length) {
         base.$texts = $('<ul class="texts"><li>' + $element.html() +
'</li></ul>');
         $element.html(base.$texts);
      }

      console.log(':first-child:',base.$texts.find(':first-child').html());

      base.$texts.hide();

      base.$current = $('<span>')
         .html(base.$texts.find(':first-child').html())
         .prependTo($element);

      if (isInEffect(options.in.effect)) {
         base.$current.css('visibility', 'hidden');
      } else if (isOutEffect(options.out.effect)) {
         base.$current.css('visibility', 'visible');
      }

      base.setOptions(options);

      base.timeoutRun = null;

      setTimeout(function () {
         base.options.autoStart && base.start();
      }, base.options.initialDelay)
   };

   base.setOptions = function (options) {
      base.options = options;
   };

   base.triggerEvent = function (name) {
      var e = $.Event(name + '.tlt');
      $element.trigger(e, base);
      return e;
   };

   base.in = function (index, cb) {
```

```
        index = index || 0;

        var $elem = base.$texts.find(':nth-child(' + ((index||0) + 1) + ')')
            , options = $.extend(true, {}, base.options, $elem.length ?
                            getData($elem[0]) : {})
            , $tokens;

        $elem.addClass('current');

        base.triggerEvent('inAnimationBegin');
        $element.attr('data-active', $elem.data('id'));

        base.$current
            .html($elem.html())
            .lettering('words');

        // split words to individual characters if token type is set to 'char'
        if (base.options.type == "char") {
            base.$current.find('[class^="word"]')
                .css({
                    'display': 'inline-block',
                    // fix for poor ios performance
                    '-webkit-transform': 'translate3d(0,0,0)',
                    '-moz-transform': 'translate3d(0,0,0)',
                    '-o-transform': 'translate3d(0,0,0)',
                    'transform': 'translate3d(0,0,0)'
                })
                .each(function () { $(this).lettering() });
        }

        $tokens = base.$current
            .find('[class^="' + base.options.type + '"]')
            .css('display', 'inline-block');

        if (isInEffect(options.in.effect)) {
            $tokens.css('visibility', 'hidden');
        } else if (isOutEffect(options.in.effect)) {
            $tokens.css('visibility', 'visible');
        }

        base.currentIndex = index;

        animateTokens($tokens, options.in, function () {
            base.triggerEvent('inAnimationEnd');
            if (options.in.callback) options.in.callback();
            if (cb) cb(base);
        });
    };

    base.out = function (cb) {
```

```
        var $elem = base.$texts.find(':nth-child(' + ((base.currentIndex||0)
+ 1) + ')')
            , $tokens = base.$current.find('[class^="' + base.options.type + '"]')
            , options = $.extend(true, {}, base.options, $elem.length ?
getData($elem[0]) : {})

        base.triggerEvent('outAnimationBegin');

        animateTokens($tokens, options.out, function () {
            $elem.removeClass('current');
            base.triggerEvent('outAnimationEnd');
            $element.removeAttr('data-active');
            if (options.out.callback) options.out.callback();
            if (cb) cb(base);
        });
    };

    base.start = function (index) {
        setTimeout(function () {
            base.triggerEvent('start');

            (function run (index) {
                base.in(index, function () {
                    var length = base.$texts.children().length;

                    index += 1;

                    if (!base.options.loop && index >= length) {
                        if (base.options.callback) base.options.callback();
                        base.triggerEvent('end');
                    } else {
                        index = index % length;

                        base.timeoutRun = setTimeout(function () {
                            base.out(function () {
                                run(index)
                            });
                        }, base.options.minDisplayTime);
                    }
                });
            }(index || 0));
        }, base.options.initialDelay);
    };

    base.stop = function () {
        if (base.timeoutRun) {
            clearInterval(base.timeoutRun);
            base.timeoutRun = null;
        }
    };
```

```
    base.init();
}
```

TextAnimate.js 文件位于项目的 js 文件夹下，使用 ES6 语法定义执行动画的类。该类主要实现方法的功能如下。

(1) Constructor()：初始化界面。
(2) In ()：执行显示文字动画。
(3) Out ()：执行文字消失动画。
(4) Start()：开始执行动画。
(5) _isInEffect()：判断是否是显示文字效果。
(6) _isOutEffect()：判断是否是文字消失效果。
(7) _shuffle()：把给定的数组打乱顺序。
(8) getData()：解析 data-*属性，把属性值解析成对象。
(9) Animate()：对单个碎片执行动画。
(10) animateFragments()：对碎片数组执行动画。

TextAnimate.js 文件的代码如下：

```
/**
 * Created by francis on 17-6-10
 */
class TextAnimate{
    /**
     * 初始化
     * @param element 执行动画的元素
     * @param options 选项对象
     */
    constructor(element, options){
        var self = this;
        self.$element = $(element);
        self.options = options;
        self.$texts = self.$element.find(self.options.selector); // content
of search engine?
        if (!self.$texts.length) {
            self.$texts = $('<ul class="texts"><li>' + self.$element.html() +
'</li></ul>'); // create content for search engine?
            self.$element.html(self.$texts);
        }
        self.$texts.hide(); // hide content.

        // original text, include html tags.
        self.$current = $('<span>')
            .html(self.$texts.find(':first-child').html())
            .prependTo(self.$element);

        if (self._isInEffect(self.options.in.effect)) {
            self.$current.css('visibility', 'hidden');
```

```
        } else if (self._isOutEffect(self.options.out.effect)) {
            self.$current.css('visibility', 'visible');
        }

        self.timeoutRun = null;
        setTimeout(function () {
            self.options.autoStart && self.start();
        }, self.options.initialDelay)
    }

    /**
     * 重新设置选项
     * @param options
     */
    setOptions(options) {
        this.options = options;
    };

    triggerEvent(name) {
        var self = this;
        var event = $.Event(name + '.tlt');
        self.$element.trigger(event, self);
        return event;
    };

    /**
     * 执行显示文字动画
     * @param integer index 执行动画的字符索引
     * @param cb 回调函数
     */
    in(index, cb) {
        var self = this;
        index = index || 0;
        var $elem = self.$texts.find(':nth-child(' + ((index||0) + 1) + ')');
        var options = $.extend(true, {}, self.options, $elem.length ?
self.constructor.getData($elem[0]) : {});
        var $tokens;

        $elem.addClass('current');

        self.triggerEvent('inAnimationBegin');
        self.$element.attr('data-active', $elem.data('id'));

        // split to words
        self.$current
            .html($elem.html())
            .lettering('words');

        // split words to individual characters if token type is set to 'char'
        if (options.type == "char") {
```

```
            self.$current.find('[class^="word"]')
                .css({
                    'display': 'inline-block',
                    // fix for poor ios performance
                    '-webkit-transform': 'translate3d(0,0,0)',
                    '-moz-transform': 'translate3d(0,0,0)',
                    '-o-transform': 'translate3d(0,0,0)',
                    'transform': 'translate3d(0,0,0)'
                })
                .each(function () { $(this).lettering() });
        }

        // span.char elements
        $tokens = self.$current
            .find('[class^="' + options.type + '"]')
            .css('display', 'inline-block');

        if (self._isInEffect(options.in.effect)) {
            $tokens.css('visibility', 'hidden');
        } else if (self._isOutEffect(options.in.effect)) {
            $tokens.css('visibility', 'visible');
        }

        self.currentIndex = index;

        self.animateFragments($tokens, options.in, function () {
            self.triggerEvent('inAnimationEnd');
            if (options.in.callback) options.in.callback();
            if (cb) cb(self);
        });
    }

    /**
     * 执行文字消失动画
     * @param cb
     */
    out(cb){
        var self = this;
        var $elem = self.$texts.find(':nth-child(' + ((self.currentIndex||0)
+ 1) + ')')
            , $tokens = self.$current.find('[class^="' + self.options.type + '"]')
            , options = $.extend(true, {}, self.options, $elem.length ?
self.constructor.getData($elem[0]) : {})

        self.triggerEvent('outAnimationBegin');

        self.animateFragments($tokens, options.out, function () {
            $elem.removeClass('current');
            self.triggerEvent('outAnimationEnd');
```

```
        self.$element.removeAttr('data-active');
        if (options.out.callback) options.out.callback();
        if (cb) cb(self);
    });
};

/**
 * 开始执行动画
 * @param integer index 执行动画的文字索引
 */
start(index) {
    var self = this;
    setTimeout(function () {
        self.triggerEvent('start');

        (function run (index) {
            self.in(index, function () {
                var length = self.$texts.children().length;

                index += 1;

                if (!self.options.loop && index >= length) {
                    if (self.options.callback) self.options.callback();
                    self.triggerEvent('end');
                } else {
                    index = index % length;

                    self.timeoutRun = setTimeout(function () {
                        self.out(function () {
                            run(index)
                        });
                    }, self.options.minDisplayTime);
                }
            });
        }(index || 0));
    }, self.options.initialDelay);
};

stop() {
    var self = this;
    if (self.timeoutRun) {
        clearInterval(self.timeoutRun);
        self.timeoutRun = null;
    }
};

/**
 * 是否是显示文字效果
 * @param effect 效果名称
 * @returns {boolean}
```

```
     * @private
     */
    _isInEffect (effect) {
        var self = this;
        return /In/.test(effect) || self.options.inEffects.indexOf(effect) >= 0;
    };

    /**
     * 是否是文字消失效果
     * @param effect 效果名称
     * @returns {boolean}
     * @private
     */
    _isOutEffect (effect) {
        var self = this;
        return /Out/.test(effect) || self.options.outEffects.indexOf(effect) >= 0;
    };

    /**
     * 把给定的数组打乱顺序
     * Shuffle order of gaven array
     * @param array An array
     * @returns {*}
     */
    _shuffle(arr) {
        for (let i = arr.length; i; i--) {
            let j = Math.floor(Math.random() * i);
            [arr[i - 1], arr[j]] = [arr[j], arr[i - 1]];
        }
        return arr;
    }

    static stringToBoolean(str) {
        if (str !== "true" && str !== "false") return str;
        return (str === "true");
    };
    /**
     * 解析data-*属性,把属性值解析成对象
     * parse data-* attribute to an object. for example: <h1 class="tlt"
data-in-effect="rollIn">Title</h1>
     * @param node htmlNode
     * @returns {{}}
     */
    static getData (node) {
        var attrs = node.attributes || [];
        var data = {};
        if (!attrs.length) return data;
        $.each(attrs, function (i, attr) {
            var nodeName = attr.nodeName.replace(/delayscale/, 'delayScale');
```

```
            if (/^data-in-*/.test(nodeName)) {
                data.in = data.in || {};
                data.in[nodeName.replace(/data-in-/, '')] =
TextAnimate.stringToBoolean(attr.nodeValue);
            } else if (/^data-out-*/.test(nodeName)) {
                data.out = data.out || {};
                data.out[nodeName.replace(/data-out-/, '')] =
TextAnimate.stringToBoolean(attr.nodeValue);
            } else if (/^data-*/.test(nodeName)) {
                data[nodeName.replace(/data-/, '')] =
TextAnimate.stringToBoolean(attr.nodeValue);
            }
        })
        return data;
    }

    /**
     * 对碎片执行动画
     * @param $fragement 碎片可以是一个字符或一个英文单词.
     * @param effect 动画效果
     * @param cb
     */
    animate ($fragement, effect, cb) {
        $fragement.addClass('animated ' + effect)
            .css('visibility', 'visible')
            .show();
        $fragement.one('webkitAnimationEnd mozAnimationEnd MSAnimationEnd
oanimationend animationend', function () {
            $fragement.removeClass('animated ' + effect);
            cb && cb();
        });
    }

    /**
     * 对碎片数组执行动画
     * animate all fragments. Lettering.js separate sentence into
fragments(characters or words).
     * @param array $fragements 需要执行动画的碎片数组，碎片可以是一个字符或一个英文单词
     * @param options 选项
     * @param cb 回调函数
     */
    animateFragments ($fragements, options, cb) {
        var self = this;
        var count = $fragements.length;
        if (!count) {
            cb && cb();
            return;
        }
        if (options.shuffle) $fragements = self._shuffle($fragements);
        if (options.reverse) $fragements = $fragements.toArray().reverse();
```

```
        $.each($fragements, function (i, fragment) {
            var $fragement = $(fragment);
            // animation on singal charactor completed.
            function complete () {
                if (self._isInEffect(options.effect)) {
                    $fragement.css('visibility', 'visible');
                } else if (self._isOutEffect(options.effect)) {
                    $fragement.css('visibility', 'hidden');
                }
                count -= 1;
                if (!count && cb) cb();
            }
            var delay = options.sync ? options.delay : options.delay * i *
options.delayScale;
            $fragement.text() ?
                setTimeout(function () { self.animate($fragement,
options.effect, complete) }, delay) :
                complete();
        });
    };
}
```

15.3.3 封装 jQuery 插件

jquery.textanimate.js 文件位于项目的 js 文件夹下，通过调用 TextAnimate 类，封装成 jquery 插件。

jquery.textanimate.js 的代码如下：

```
/**
 * Created by francis on 17-6-10
 */
(function($){
    $.fn.textanimate = function (settings) {
        return this.each(function () {
            var $this = $(this);
            var data = $this.data('textanimate');
            var options = $.extend(true, {}, $.fn.textanimate.defaults,
TextAnimate.getData(this), typeof settings == 'object' && settings);

            if (!data) {
                $this.data('textanimate', (data = new TextAnimate(this, options)));
            } else if (typeof settings == 'string') {
                data[settings].apply(data, [].concat(args));
            } else {
                data.setOptions.call(data, options);
            }
        })
    };
    /**
```

```
     * 默认参数
     */
    $.fn.textanimate.defaults = {
        selector: '.texts',
        loop: false,
        minDisplayTime: 2000,
        initialDelay: 0,
        in: {
            effect: 'fadeInLeftBig',
            delayScale: 1.5,
            delay: 50,
            sync: false,
            reverse: false,
            shuffle: false,
            callback: function () {}
        },
        out: {
            effect: 'hinge',
            delayScale: 1.5,
            delay: 50,
            sync: false,
            reverse: false,
            shuffle: false,
            callback: function () {}
        },
        autoStart: true,
        inEffects: [],
        outEffects: [ 'hinge' ],
        callback: function () {},
        type: 'char'
    };
}(jQuery));
```

15.3.4 合并 js 文件和编译 CSS 文件

gulpfile.js 文件位于项目根目录下，主要定义了如何把 ES6 文件转换成 ES5 文件，并把多个 js 文件合并成一个 js 文件，以及把 less 文件编译成 CSS 文件。

gulpfile.js 文件的代码如下：

```
/**
 * Created by francis on 17-5-11
 */
const gulp = require('gulp');
const strip = require('gulp-strip-comments');
const babel = require("gulp-babel");
const concat = require('gulp-concat');

/*******************************
 Tasks
```

```
*******************************/

var jsfiles = [
    'js/TextAnimate.js',
    'js/jquery.textanimate.js',
];
gulp.task('js',function(){
    gulp.src(jsfiles)
        .pipe(babel({presets: ['es2015']}))
        // .pipe(stripDebug())
        .pipe(concat('textanimate.js'))
        .pipe(strip())
        // .pipe(uglify({'mangle':false}))
        .pipe(gulp.dest('js/dist/'));
});
```

15.3.5 合并 TextAnimate.js 和 jquery.textanimate.js 文件

Textanimate.js 文件位于 dist 文件夹下，主要功能是把 TextAnimate.js 和 jquery.textanimate.js 合并到这一个文件里。

Textanimate.js 文件的代码如下：

```
'use strict';

var _createClass = function () { function defineProperties(target, props)
{ for (var i = 0; i < props.length; i++) { var descriptor = props[i];
descriptor.enumerable = descriptor.enumerable || false;
descriptor.configurable = true; if ("value" in descriptor)
descriptor.writable = true; Object.defineProperty(target, descriptor.key,
descriptor); } } return function (Constructor, protoProps, staticProps)
{ if (protoProps) defineProperties(Constructor.prototype, protoProps); if
(staticProps) defineProperties(Constructor, staticProps); return
Constructor; }; }();

function _classCallCheck(instance, Constructor) { if (!(instance instanceof
Constructor)) { throw new TypeError("Cannot call a class as a function"); } }

var TextAnimate = function () {
    function TextAnimate(element, options) {
        _classCallCheck(this, TextAnimate);

        var self = this;
        self.$element = $(element);
        self.options = options;
        self.$texts = self.$element.find(self.options.selector);
        if (!self.$texts.length) {
            self.$texts = $('<ul class="texts"><li>' + self.$element.html() +
'</li></ul>');
```

```
            self.$element.html(self.$texts);
        }
        self.$texts.hide();

        self.$current = $('<span>').html(self.$texts.find(':first-
child').html()).prependTo(self.$element);

        if (self._isInEffect(self.options.in.effect)) {
            self.$current.css('visibility', 'hidden');
        } else if (self._isOutEffect(self.options.out.effect)) {
            self.$current.css('visibility', 'visible');
        }

        self.timeoutRun = null;
        setTimeout(function () {
            self.options.autoStart && self.start();
        }, self.options.initialDelay);
    }

    _createClass(TextAnimate, [{
        key: 'setOptions',
        value: function setOptions(options) {
            this.options = options;
        }
    }, {
        key: 'triggerEvent',
        value: function triggerEvent(name) {
            var self = this;
            var event = $.Event(name + '.tlt');
            self.$element.trigger(event, self);
            return event;
        }
    }, {
        key: 'in',
        value: function _in(index, cb) {
            var self = this;
            index = index || 0;
            var $elem = self.$texts.find(':nth-child(' + ((index || 0) + 1) + ')');
            var options = $.extend(true, {}, self.options, $elem.length ?
self.constructor.getData($elem[0]) : {});
            var $tokens;

            $elem.addClass('current');

            self.triggerEvent('inAnimationBegin');
            self.$element.attr('data-active', $elem.data('id'));

            self.$current.html($elem.html()).lettering('words');

            if (options.type == "char") {
```

```
                self.$current.find('[class^="word"]').css({
                    'display': 'inline-block',
                    '-webkit-transform': 'translate3d(0,0,0)',
                    '-moz-transform': 'translate3d(0,0,0)',
                    '-o-transform': 'translate3d(0,0,0)',
                    'transform': 'translate3d(0,0,0)'
                }).each(function () {
                    $(this).lettering();
                });
            }

            $tokens = self.$current.find('[class^="' + options.type +
'"]').css('display', 'inline-block');

            if (self._isInEffect(options.in.effect)) {
                $tokens.css('visibility', 'hidden');
            } else if (self._isOutEffect(options.in.effect)) {
                $tokens.css('visibility', 'visible');
            }

            self.currentIndex = index;

            self.animateFragments($tokens, options.in, function () {
                self.triggerEvent('inAnimationEnd');
                if (options.in.callback) options.in.callback();
                if (cb) cb(self);
            });
        }
    }, {
        key: 'out',
        value: function out(cb) {
            var self = this;
            var $elem = self.$texts.find(':nth-child(' + ((self.currentIndex
|| 0) + 1) + ')'),
                $tokens = self.$current.find('[class^="' + self.options.type + '"]'),
                options = $.extend(true, {}, self.options, $elem.length ?
self.constructor.getData($elem[0]) : {});

            self.triggerEvent('outAnimationBegin');

            self.animateFragments($tokens, options.out, function () {
                $elem.removeClass('current');
                self.triggerEvent('outAnimationEnd');
                self.$element.removeAttr('data-active');
                if (options.out.callback) options.out.callback();
                if (cb) cb(self);
            });
        }
    }, {
        key: 'start',
```

```
    value: function start(index) {
        var self = this;
        setTimeout(function () {
            self.triggerEvent('start');

            (function run(index) {
                self.in(index, function () {
                    var length = self.$texts.children().length;

                    index += 1;

                    if (!self.options.loop && index >= length) {
                        if (self.options.callback) self.options.callback();
                        self.triggerEvent('end');
                    } else {
                        index = index % length;

                        self.timeoutRun = setTimeout(function () {
                            self.out(function () {
                                run(index);
                            });
                        }, self.options.minDisplayTime);
                    }
                });
            })(index || 0);
        }, self.options.initialDelay);
    }
}, {
    key: 'stop',
    value: function stop() {
        var self = this;
        if (self.timeoutRun) {
            clearInterval(self.timeoutRun);
            self.timeoutRun = null;
        }
    }
}, {
    key: '_isInEffect',
    value: function _isInEffect(effect) {
        var self = this;
        return (/In/.test(effect) || self.options.inEffects.indexOf(effect) >= 0
        );
    }
}, {
    key: '_isOutEffect',
    value: function _isOutEffect(effect) {
        var self = this;
        return (/Out/.test(effect) || self.options.outEffects.indexOf(effect) >= 0
        );
    }
```

```
    }, {
        key: '_shuffle',


        value: function _shuffle(arr) {
            for (var i = arr.length; i; i--) {
                var j = Math.floor(Math.random() * i);
                var _ref = [arr[j], arr[i - 1]];
                arr[i - 1] = _ref[0];
                arr[j] = _ref[1];
            }
            return arr;
        }
    }, {
        key: 'animate',
        value: function animate($fragement, effect, cb) {
            $fragement.addClass('animated ' + effect).css('visibility',
'visible').show();
            $fragement.one('webkitAnimationEnd mozAnimationEnd MSAnimationEnd
oanimationend animationend', function () {
                $fragement.removeClass('animated ' + effect);
                cb && cb();
            });
        }


    }, {
        key: 'animateFragments',
        value: function animateFragments($fragements, options, cb) {
            var self = this;
            var count = $fragements.length;
            if (!count) {
                cb && cb();
                return;
            }
            if (options.shuffle) $fragements = self._shuffle($fragements);
            if (options.reverse) $fragements = $fragements.toArray().reverse();
            $.each($fragements, function (i, fragment) {
                var $fragement = $(fragment);
                function complete() {
                    if (self._isInEffect(options.effect)) {
                        $fragement.css('visibility', 'visible');
                    } else if (self._isOutEffect(options.effect)) {
                        $fragement.css('visibility', 'hidden');
                    }
                    count -= 1;
                    if (!count && cb) cb();
                }
                var delay = options.sync ? options.delay : options.delay * i
* options.delayScale;
```

```
                $fragement.text() ? setTimeout(function () {
                    self.animate($fragement, options.effect, complete);
                }, delay) : complete();
            });
        }
    }], [{
        key: 'stringToBoolean',
        value: function stringToBoolean(str) {
            if (str !== "true" && str !== "false") return str;
            return str === "true";
        }
    }, {
        key: 'getData',

        value: function getData(node) {
            var attrs = node.attributes || [];
            var data = {};
            if (!attrs.length) return data;
            $.each(attrs, function (i, attr) {
                var nodeName = attr.nodeName.replace(/delayscale/, 'delayScale');
                if (/^data-in-*/.test(nodeName)) {
                    data.in = data.in || {};
                    data.in[nodeName.replace(/data-in-/, '')] =
TextAnimate.stringToBoolean(attr.nodeValue);
                } else if (/^data-out-*/.test(nodeName)) {
                    data.out = data.out || {};
                    data.out[nodeName.replace(/data-out-/, '')] =
TextAnimate.stringToBoolean(attr.nodeValue);
                } else if (/^data-*/.test(nodeName)) {
                    data[nodeName.replace(/data-/, '')] =
TextAnimate.stringToBoolean(attr.nodeValue);
                }
            });
            return data;
        }
    }]);

    return TextAnimate;
}();
'use strict';

var _typeof = typeof Symbol === "function" && typeof Symbol.iterator ===
"symbol" ? function (obj) { return typeof obj; } : function (obj) { return
obj && typeof Symbol === "function" && obj.constructor === Symbol &&
obj !== Symbol.prototype ? "symbol" : typeof obj; };

(function ($) {
    $.fn.textanimate = function (settings) {
        return this.each(function () {
            var $this = $(this);
```

```
            var data = $this.data('textanimate');
            var options = $.extend(true, {}, $.fn.textanimate.defaults,
TextAnimate.getData(this), (typeof settings === 'undefined' ? 'undefined' :
_typeof(settings)) == 'object' && settings);

            console.log('data:', data);

            if (!data) {
                $this.data('textanimate', data = new TextAnimate(this, options));
            } else if (typeof settings == 'string') {
                data[settings].apply(data, [].concat(args));
            } else {
                data.setOptions.call(data, options);
            }
        });
    };

    $.fn.textanimate.defaults = {
        selector: '.texts',
        loop: false,
        minDisplayTime: 2000,
        initialDelay: 0,
        in: {
            effect: 'fadeInLeftBig',
            delayScale: 1.5,
            delay: 50,
            sync: false,
            reverse: false,
            shuffle: false,
            callback: function callback() {}
        },
        out: {
            effect: 'hinge',
            delayScale: 1.5,
            delay: 50,
            sync: false,
            reverse: false,
            shuffle: false,
            callback: function callback() {}
        },
        autoStart: true,
        inEffects: [],
        outEffects: ['hinge'],
        callback: function callback() {},
        type: 'char'
    };
})(jQuery);
```

第 16 章 项目演练 3——开发图片堆叠系统

该项目主要是实现图片堆叠效果，同时用户可以移动任意一张图片，还可以实现图片的放大和缩小效果。通过本项目的练习，读者可以学习了解 jQuery 插件是如何开发的。该项目运用了较多的 jQuery 方法，有利于初学者对 jQuery 的学习，同时进一步掌握 JavaScript 和 jQuery 配合使用的技能。

16.1 项目需求分析

需求分析是开发项目的必要环节。下面分析图片堆叠系统的需求。

(1) 该项目利用 jQuery 构造了一个图片堆叠插件，该插件让图片散乱地显示在屏幕上。在 Firefox 53.0 中查看效果，如图 16-1 所示。

图 16-1 图片堆叠系统主页

(2) 在主页中，用户可以用鼠标单击图片后按住不放，然后任意移动图片的位置，如图 16-2 所示。

图 16-2 任意移动图片的位置

(3) 用户可以单击任意一张图片，图片会放大显示，模拟用户拿起照片的效果，如图 16-3 所示。用户在屏幕任意地方再次单击，图片会缩小到原来的位置，模拟用户放下照片的效果。

图 16-3 图片放大后的效果

从演示效果可以看出，JavaScript 和 jQuery 配合使用，可以制作出更加复杂的网页效果。

16.2 系统的代码实现

下面来分析图片堆叠系统的代码是如何实现的。

16.2.1 设计首页

首页中显示了图片堆叠的效果。代码如下：

```
<!DOCTYPE html>
<html>
<head>
   <meta charset="UTF-8">
   <meta name="viewport" content="width=device-width, initial-scale=1.0,
maximum-scale=1.0, user-scalable=no">
   <title>基于 jquery 的图片堆叠演示系统</title>
   <link href="css/animate.css" rel="stylesheet">
   <link href="css/imgpile.css" rel="stylesheet">

   <script src="js/jquery-3.1.1.min.js"></script>
   <script src="js/jquery.ui.touch-punch.min.js"></script>
```

```
    <script src="js/jquery-ui.min.js"></script>
    <script src="js/dist/imagepile.js"></script>
</head>
<body>
    <div class="container" style="width:900px;margin:0 auto;">
        <h2 style="text-align: center;color:grey;">基于 jquery 的图片堆叠系统</h2>

        <!-- Photopile Demo Gallery Markup -->
        <ul id="imgpile">
            <li draggable="true">
                <a href="images/fullsize/01.jpg">
                    <img src="images/thumbs/01.jpg" alt="Barton Dam, Ann Arbor,
Michigan" width="133" height="100" />
                </a>
            </li>
            <li>
                <a href="images/fullsize/02.jpg">
                    <img src="images/thumbs/02.jpg" alt="Building Atlanta,
Georgia" width="133" height="100" />
                </a>
            </li>
            <li>
                <a href="images/fullsize/03.jpg">
                    <img src="images/thumbs/03.jpg" alt="Nice day for a swim"
width="133" height="100" />
                </a>
            </li>
            <li>
                <a href="images/fullsize/04.jpg">
                    <img src="images/thumbs/04.jpg" alt="The plants that never
die" width="100" height="133" />
                </a>
            </li>
            <li>
                <a href="images/fullsize/05.jpg">
                    <img src="images/thumbs/05.jpg" alt="Downtown Atlanta,
Georgia" width="100" height="134" />
                </a>
            </li>
            <li>
                <a href="images/fullsize/06.jpg">
                    <img src="images/thumbs/06.jpg" alt="Atlanta traffic"
width="100" height="137" />
                </a>
            </li>
            <li>
                <a href="images/fullsize/07.jpg">
                    <img src="images/thumbs/07.jpg" alt="A pathetic dog"
width="120" height="120" />
                </a>
            </li>
            <li>
```

```
            <a href="images/fullsize/08.jpg">
                <img src="images/thumbs/08.jpg" alt="Two happy dogs"
width="140" height="100" />
            </a>
        </li>
        <li>
            <a href="images/fullsize/09.jpg">
                <img src="images/thumbs/09.jpg" alt="Antigua, Guatemala"
width="100" height="133" />
            </a>
        </li>
        <li>
            <a href="images/fullsize/10.jpg">
                <img src="images/thumbs/10.jpg" alt="Iximche, Guatemala"
width="135" height="128" />
            </a>
        </li>
        <li>
            <a href="images/fullsize/11.jpg">
                <img src="images/thumbs/11.jpg" alt="The bat cave"
width="119" height="133" />
            </a>
        </li>
        <li>
            <a href="images/fullsize/12.jpg">
                <img src="images/thumbs/12.jpg" alt="All Saints Day Kite
Festival" width="133" height="100" />
            </a>
        </li>
        <li>
            <a href="images/fullsize/13.jpg">
                <img src="images/thumbs/13.jpg" alt="Tikal, Guatemala"
width="130" height="130" />
            </a>
        </li>
        <li>
            <a href="images/fullsize/14.jpg">
                <img src="images/thumbs/14.jpg" alt="Mackinac Island,
Michigan" width="111" height="130" />
            </a>
        </li>
        <li>
            <a href="images/fullsize/15.jpg">
                <img src="images/thumbs/15.jpg" alt="Summer flowers"
width="133" height="100" />
            </a>
        </li>
        <li>
            <a href="images/fullsize/16.jpg">
                <img src="images/thumbs/16.jpg" alt="Full of hot air"
width="133" height="100" />
            </a>
```

```
            </li>
            <li>
                <a href="images/fullsize/17.jpg">
                    <img src="images/thumbs/17.jpg" alt="On the rise"
width="133" height="100" />
                </a>
            </li>
            <li>
                <a href="images/fullsize/18.jpg">
                    <img src="images/thumbs/18.jpg" alt="Amador Causeway,
Panama City" width="133" height="100" />
                </a>
            </li>
            <li>
                <a href="images/fullsize/19.jpg">
                    <img src="images/thumbs/19.jpg" alt="The Panama Canal"
width="140" height="79" />
                </a>
            </li>
            <li>
                <a href="images/fullsize/20.jpg">
                    <img src="images/thumbs/20.jpg" alt="Flags over Brussels"
width="133" height="99" />
                </a>
            </li>
            <li>
                <a href="images/fullsize/21.jpg">
                    <img src="images/thumbs/21.jpg" alt="Eiffel Tower"
width="100" height="134" />
                </a>
            </li>
            <li>
                <a href="images/fullsize/22.jpg">
                    <img src="images/thumbs/22.jpg" alt="Never-ending stairs"
width="99" height="133" />
                </a>
            </li>
            <li>
                <a href="images/fullsize/23.jpg">
                    <img src="images/thumbs/23.jpg" alt="Paris, France"
width="140" height="93" />
                </a>
            </li>
            <li>
                <a href="images/fullsize/24.jpg">
                    <img src="images/thumbs/24.jpg" alt="Rainbow over Belgium"
width="136" height="98" />
                </a>
            </li>
            <li>
                <a href="images/fullsize/25.jpg">
                    <img src="images/thumbs/25.jpg" alt="Playa del Carmen"
```

```
width="136" height="96" />
            </a>
        </li>
        <li>
            <a href="images/fullsize/26.jpg">
                <img src="images/thumbs/26.jpg" alt="Panama City, Panama"
width="133" height="96" />
            </a>
        </li>
        <li>
            <a href="images/fullsize/27.jpg">
                <img src="images/thumbs/27.jpg" alt="Red frog" width="133"
height="100" />
            </a>
        </li>
        <li>
            <a href="images/fullsize/28.jpg">
                <img src="images/thumbs/28.jpg" alt="Dogs on a couch"
width="133" height="99" />
            </a>
        </li>
        <li>
            <a href="images/fullsize/29.jpg">
                <img src="images/thumbs/29.jpg" alt="Tarantula"
width="133" height="97" />
            </a>
        </li>
        <li>
            <a href="images/fullsize/30.jpg">
                <img src="images/thumbs/30.jpg" alt="Ceviche Loco"
width="140" height="93" />
            </a>
        </li>
        <li>
            <a href="images/fullsize/31.jpg">
                <img src="images/thumbs/31.jpg" alt="Playa Blanca, Panama"
width="133" height="100" />
            </a>
        </li>
        <li>
            <a href="images/fullsize/32.jpg">
                <img src="images/thumbs/32.jpg" alt="Pickles" width="133"
height="99" />
            </a>
        </li>
        <li>
            <a href="images/fullsize/33.jpg">
                <img src="images/thumbs/33.jpg" alt="Parrot" width="133"
height="100" />
            </a>
        </li>
        <li>
```

```
                <a href="images/fullsize/34.jpg">
                    <img src="images/thumbs/34.jpg" alt="Panama rains"
width="133" height="100" />
                </a>
            </li>
            <li>
                <a href="images/fullsize/35.jpg">
                    <img src="images/thumbs/35.jpg" alt="Cinta Costera, Panama
City" width="120" height="120" />
                </a>
            </li>
            <li>
                <a href="images/fullsize/36.jpg">
                    <img src="images/thumbs/36.jpg" alt="Afternoon stroll"
width="133" height="100" />
                </a>
            </li>
            <li>
                <a href="images/fullsize/37.jpg">
                    <img src="images/thumbs/37.jpg" alt="Overlooking the
Panama Canal" width="138" height="94" />
                </a>
            </li>
        </ul>

    </div>

    <script>
        $(document).ready(function(){
            $('#imgpile').imgpile();
        });
    </script>
</body>
</html>
```

16.2.2 图片堆叠核心功能

ImgPile.js 文件位于项目的 js 文件夹下，使用 ES6 语法定义了 LargeImg 类和 Thumb 类。

1. LargeImg 类

该类使用单例模式，主要实现方法的功能如下。

(1) Constructor()：初始化界面。

(2) Pickup()：模拟用户拿起一张照片的效果。

(3) putDown()：模拟放下一张照片的效果。

(4) loadImage()：加载照片。

(5) startPosition()：放大照片后的初始化位置。

(6) Enlarge()：显示放大后的照片。浏览器将根据设备和窗口大小等情况自动选择调用

enlargeToFullSize()、enlargeToWindowWidth()或 enlargeToWindowHeight()方法。

(7) enlargeToFullSize()：全尺寸显示图片。

(8) enlargeToWindowWidth()：适应屏幕宽度显示图片。

(9) enlargeToWindowHeight()：适应屏幕高度显示图片。

单例模式是一种常用的软件设计模式。在它的核心结构中只包含一个被称为单例的特殊类。通过单例模式可以保证系统中一个类只有一个实例。即一个类只有一个对象实例。

2. Thumb 类

该类实现索引小图的方法。主要实现方法的功能如下。

(1) Constructor()：初始化索引图。

(2) setZ()：设置索引图的 z-index 值。

(3) bringToTop()：设置索引图的 z-index 值为最大值。

(4) moveDownOne()：降低索引图的 z-index 值。

(5) _setRotation()：设置图片的随机旋转角度。

(6) getRotation()：获取索引图旋转角度。

(7) _setOverlap()：设置图片叠加部分的值。

(8) _bindUIActions()：绑定 mouseover、mouseleave 等事件。

ImgPile.js 文件的代码如下：

```
/**
 * 大图类,该类用于显示控制大图,因大图只有一个显示窗口,所以采用单例模式
 */
let LargeImgInstance = null;
class LargeImg{
    constructor(options){
        if(!LargeImgInstance){
            var self = this;
            self.$container = $( '<div id="imgpile-active-image-container"/>' );
            self.$info = $( '<div id="imgpile-active-image-info"/>');
            self.$image = $( '<img id="imgpile-active-image" />');
            self.isPickedUp = false;
            self.options = options;
            self.fullSizeWidth = null;
            self.fullSizeHeight = null;

            $('body').append( this.$container );
            self.$container.css({
                'display'    : 'none',
                'position'   : 'absolute',
                'padding'    : self.options.thumbBorderWidth,
                'z-index'    : self.options.photoZIndex,
                'background' : self.options.photoBorderColor,
                'background-image'    : 'url(' + self.options.loading + ')',
                'background-repeat'   : 'no-repeat',
```

```
                'background-position' : '50%, 50%'
            });

            self.$container.append( self.$image );
            self.$image.css('display', 'block');

            if ( self.options.showInfo ) {
                self.$container.append( this.info );
                self.$info.append('<p></p>');
                self.$info.css('opacity', '0');
            };

            LargeImgInstance = this;
        }

        return LargeImgInstance;
    }

    /**
     * 模拟拿起一张照片
     * @param thumb 小照片对象
     */
    pickup(thumb){
        var self = this;
        if ( self.isPickedUp ) {
            // photo already picked up. put it down and then pickup the
clicked thumbnail
            self.putDown(thumb, function() { self.pickup( thumb ); });
        } else {
            self.isPickedUp = true;
            self.loadImage(thumb, function() {
                self.$image.fadeTo(self.options.fadeDuration, '1');
                self.enlarge();
                $('body').bind('click', function() { self.putDown(thumb); });
// bind putdown event to body
            });
        }
    }; // pickup
    /**
     * 模拟放下一张照片
     * @param thumb  小照片对象
     * @param callback
     */
    putDown(thumb, callback){
        self = this;
        $('body').off();
        // self.hideInfo();
        thumb.setZ( self.options.numLayers );
        self.$container.stop().animate({
            'top'     : thumb.offset.top + thumb.getShift(),
            'left'    : thumb.offset.left + thumb.getShift(),
            'width'   : thumb.width + 'px',
```

```
        'height'  : thumb.height + 'px',
        'padding' : self.options.thumbBorderWidth + 'px'
    }, self.options.pickupDuration, function() {
        self.isPickedUp = false;
        self.$container.fadeOut( self.options.fadeDuration, function() {
            if (callback) callback();
        });
    });
}

/**
 * 加载照片
 * @param thumb
 * @param callback
 */
loadImage( thumb, callback ) {
    var self = this;
    self.$image.css('opacity', '0');     // Image is not visible until
    self.startPosition(thumb);            // the container is positioned,
    var img = new Image;                  // the source is updated,
    img.src = thumb.imgsrc;               // and the image is loaded.
    img.onload = function() {             // Restore visibility in callback
        self.setImageSource( img.src );
        self.fullSizeWidth = this.width;
        self.fullSizeHeight = this.height;
        console.log('img width:', this.width);
        if (callback) callback();
    }
}

/**
 * 照片初始化位置
 * @param thumb
 */
startPosition(thumb){
    var self = this;
    self.$container.css({
        'top'       : thumb.offset.top + thumb.getShift(),
        'left'      : thumb.offset.left + thumb.getShift(),
        'transform' : 'rotate(' + thumb.getShift() + 'deg)',
        'width'     : thumb.width + 'px',
        'height'    : thumb.height + 'px',
        'padding'   : self.options.thumbBorderWidth
    }).fadeTo(self.options.fadeDuration, '1');
}

setImageSource(src){
    this.$image.attr('src', src).css({
        'width'      : '100%',
        'height'     : '100%',
        'margin-top' : '0'
    });
```

```
    }

    /**
     * 放大照片
     */
    enlarge(){
        var windowHeight = window.innerHeight ? window.innerHeight :
$(window).height(); // mobile safari hack
        var availableWidth = $(window).width() - (2 * this.windowPadding);
        var availableHeight = windowHeight - (2 * this.windowPadding);
        if ((availableWidth < this.fullSizeWidth) && ( availableHeight <
this.fullSizeHeight )) {
            // determine which dimension will allow image to fit completely
within the window
            if ((availableWidth * (this.fullSizeHeight / this.fullSizeWidth)) >
availableHeight) {
                this.enlargeToWindowHeight( availableHeight );
            } else {
                this.enlargeToWindowWidth( availableWidth );
            }
        } else if ( availableWidth < this.fullSizeWidth ) {
            this.enlargeToWindowWidth( availableWidth );
        } else if ( availableHeight < this.fullSizeHeight ) {
            this.enlargeToWindowHeight( availableHeight );
        } else {
            this.enlargeToFullSize();
        }
    } // enlarge

    enlargeToFullSize(){
        self = this;
        self.$container.css('transform', 'rotate(0deg)').animate({
            'top'    : ($(window).scrollTop()) + ($(window).height() / 2) -
(self.fullSizeHeight / 2),
            'left'   : ($(window).scrollLeft()) + ($(window).width() / 2) -
(self.fullSizeWidth / 2),
            'width'  : (self.fullSizeWidth - (2 * self.options.photoBorder)) + 'px',
            'height' : (self.fullSizeHeight - (2 * self.options.photoBorder)) + 'px',
            'padding' : self.options.photoBorder + 'px',
        });
    }

    enlargeToWindowWidth(availableWidth){
        self = this;
        var adjustedHeight = availableWidth * (self.fullSizeHeight /
self.fullSizeWidth);
        self.$container.css('transform', 'rotate(0deg)').animate({
            'top'     : $(window).scrollTop()  + ($(window).height() / 2) -
(adjustedHeight / 2),
            'left'    : $(window).scrollLeft() + ($(window).width() / 2)  -
(availableWidth / 2),
            'width'   : availableWidth + 'px',
```

```
            'height'  : adjustedHeight + 'px',
            'padding' : self.options.photoBorder + 'px'
        });
    }

    enlargeToWindowHeight(availableHeight){
        self = this;
        var adjustedWidth = availableHeight * (self.fullSizeWidth /
self.fullSizeHeight);
        self.container.css('transform', 'rotate(0deg)').animate({
            'top'     : $(window).scrollTop()  + ($(window).height() / 2) -
(availableHeight / 2),
            'left'    : $(window).scrollLeft() + ($(window).width() / 2)  -
(adjustedWidth / 2),
            'width'   : adjustedWidth + 'px',
            'height'  : availableHeight + 'px',
            'padding' : self.options.photoBorder + 'px'
        });
    }
}
/**
 * 定义小照片类
 */
class Thumb{
    /**
     * 构造函数
     * @param element 小照片的容器
     * @param options 初始化参数
     */
    constructor(element, options){
        var self = this;
        self.$element = $(element);
        self.options = options;
        self.activeClass = 'imgpile-active-thumbnail';
        self.$element.children().css('padding', this.options.thumbBorderWidth + 'px');
        self._setRotation();
        self._setOverlap();
        self._bindUIActions();
        self._setRandomZ();

        if(self.options.draggable){
            var x = 0;
            var velocity = 0;
            self.$element.draggable({
                start : function(event, ui) {
                    self.$element.addClass('preventClick');
                    self.$element.css('z-index', self.options.numLayers + 2);
                },
                drag : function( event, ui ) {
                    velocity = (ui.offset.left - x) * 1.2;
                    var ratio = parseInt( velocity * 100 / 360 );
                    self.$element.css('transform','rotateZ('+(ratio)+'deg)');
```

```
                x = ui.offset.left;
            },
            stop: function( event, ui ) {
                self.$element.css('z-index', self.options.numLayers + 1);
            }
        });
    }
}

setZ(layer ){
    this.$element.css( 'z-index', layer );
}

bringToTop(){
    var self = this;
    this.$element.css({
        'z-index'    : self.options.numLayers + 1,
        'background' : self.options.thumbBorderHover,
    });
}

moveDownOne(){
    this.$element.css({
        'z-index'    : this.options.numLayers,
        'background' : this.options.thumbBorderColor
    });
}

get offset(){
    var self = this;
    return self.$element.offset();
}

get height(){
    var self = this;
    return self.$element.height();
}

get width(){
    var self = this;
    return self.$element.width();
}

get imgsrc(){
    var self = this;
    return self.$element.children().first().attr('href');
}

getRotation(){
    var self = this;
    var transform = self.$element.css("transform");
    var values = transform.split('(')[1].split(')')[0].split(',');
```

```
        var angle = Math.round( Math.asin( values[1]) * (180/Math.PI) );
        return angle;
    }

    getShift(){
        return ( this.getRotation() < 0 )
            ? -( this.getRotation(this) * 0.40 )
            : ( this.getRotation(this) * 0.40 );
    }

    _setRotation(){
        var min = -1 * this.options.thumbRotation;
        var max = this.options.thumbRotation;
        var randomRotation = Math.floor( Math.random() * (max - min + 1)) + min;
        this.$element.css({ 'transform' : 'rotate(' + randomRotation + 'deg)' });
    }

    _setOverlap(){
        var self = this;
        self.$element.css( 'margin', ((self.options.thumbOverlap * -1) / 2) + 'px' );
    }

    _setRandomZ(){
        var self = this;
        self.$element.css({ 'z-index' : Math.floor((Math.random() *
self.options.numLayers) + 1) });
    }

    _bindUIActions(){
        var self = this;
        self.$element.on('mouseover', function(event){
            self.bringToTop();
        });
        self.$element.on('mouseleave', function(event){
            self.moveDownOne();
        });

        self.$element.find('a').on('click tap', function(event){
            event.preventDefault();
            if (self.$element.hasClass('preventClick')) {
                self.$element.removeClass('preventClick');
            } else{
                if (self.$element.hasClass(self.activeClass)) return;
                var largeImgContainer = new LargeImg(self.options);
                largeImgContainer.pickup(self);
            }
            return false;
        });

        self.$element.on('mousedown', function(event){
            self.$element.removeClass('preventClick');
        });
```

```
    }
}
```

16.2.3 封装 jQuery 插件

jquery.imgpile.js 文件位于项目的 js 文件夹下，通过调用 LargeImg 类和 Thumb 类，把 TextAnimate 类封装成 jQuery 插件。

jquery.imgpile.js 的代码如下：

```
(function($){
   $.fn.imgpile = function (options) {
       var $self = $(this);
       var settings = $.extend({
              // Thumbnails
              numLayers:          5,
              thumbOverlap:        50,
              thumbRotation:       45,
              thumbBorderWidth:   2,
              thumbBorderColor:   'white',
              thumbBorderHover:   '#EAEAEA',
              draggable:          true,
              // Photo container
              fadeDuration:       200,
              pickupDuration:     500,
              photoZIndex:        100,
              photoBorder:        10,
              photoBorderColor:   'white',
              showInfo:           true,
              // Autoplay
              autoplayGallery:    false,
              autoplaySpeed:      5000,
              // Images
              loading:            'images/loading.gif',
       }, options||{});

       // Initializes Photopile
       function init() {
          var defer = $.Deferred();
          defer.done(function(){
             afterInitialization();
          });

          // display gallery loading image in container div while loading
          function initializeThumbs(){
             $self.addClass('loading')
                 .children()
                 .each( function(index, element) {
                     var thumb = new Thumb(this, settings);
                     $(element).thumb = thumb;
                 });
```

```
            defer.resolve('finished initialization');
        }
        initializeThumbs();

        // after Initialization
        function afterInitialization(){
            $self.removeClass('loading').css({  // style container
                'padding' : settings.thumbOverlap + 'px',
            }).children().css({  // display thumbnails
                'opacity' : '1',
                'display' : 'inline-block'
            });
            if (settings.autoplayGallery) {
                // autoplay();
            }
        }

    }
    init();

    return this;
  };
}(jQuery));
```

16.2.4　合并 js 文件和编译 CSS 文件

gulpfile.js 文件位于项目根目录下，主要定义了如何把 ES6 文件转换成 ES5 文件，并把多个 js 文件合并成一个 js 文件，以及把 less 文件编译成 CSS 文件。

gulpfile.js 文件的代码如下：

```
const gulp = require('gulp');
const strip = require('gulp-strip-comments');
const babel = require("gulp-babel");
const concat = require('gulp-concat');
//css
const less = require('gulp-less');
const autoprefixer = require('gulp-autoprefixer');

var src_less = './public/*.less';
var css_dist = './public/css';

gulp.task('css', function () {
    return gulp.src(src_less)
        .pipe(less())
        // .pipe(minifyCSS())
        .pipe(autoprefixer({
            browsers: ['last 2 versions', 'ie >= 9']
        }))
        .pipe(gulp.dest(css_dist));
```

```
});

var jsfiles = [
    'public/js/ImgPile.js',
    'public/js/jquery.imgpile.js',
];
gulp.task('js',function(){
    gulp.src(jsfiles)
        .pipe(babel({presets: ['es2015']}))
        // .pipe(stripDebug())
        .pipe(concat('imagepile.js'))
        .pipe(strip())
        // .pipe(uglify({'mangle':false}))
        .pipe(gulp.dest('public/js/dist/'));
});
```

16.2.5 合并 ImgPile.js 和 jquery.imgpile.js 文件

imagepile.js 文件位于 dist 文件夹下，主要功能是把 ImgPile.js 和 jquery.imgpile.js 最后合并到这一个文件里。

imagepile.js 文件的代码如下：

```
'use strict';
var _createClass = function () { function defineProperties(target, props)
{ for (var i = 0; i < props.length; i++) { var descriptor = props[i];
descriptor.enumerable = descriptor.enumerable || false;
descriptor.configurable = true; if ("value" in descriptor)
descriptor.writable = true; Object.defineProperty(target, descriptor.key,
descriptor); } } return function (Constructor, protoProps, staticProps)
{ if (protoProps) defineProperties(Constructor.prototype, protoProps); if
(staticProps) defineProperties(Constructor, staticProps); return
Constructor; }; }();

function _classCallCheck(instance, Constructor) { if (!(instance instanceof
Constructor)) { throw new TypeError("Cannot call a class as a function"); } }

var LargeImgInstance = null;

var LargeImg = function () {
    function LargeImg(options) {
        _classCallCheck(this, LargeImg);

        if (!LargeImgInstance) {
            var self = this;
            self.$container = $('<div id="imgpile-active-image-container"/>');
            self.$info = $('<div id="imgpile-active-image-info"/>');
            self.$image = $('<img id="imgpile-active-image" />');
            self.isPickedUp = false;
            self.options = options;
            self.fullSizeWidth = null;
```

```
        self.fullSizeHeight = null;

        $('body').append(this.$container);
        self.$container.css({
            'display': 'none',
            'position': 'absolute',
            'padding': self.options.thumbBorderWidth,
            'z-index': self.options.photoZIndex,
            'background': self.options.photoBorderColor,
            'background-image': 'url(' + self.options.loading + ')',
            'background-repeat': 'no-repeat',
            'background-position': '50%, 50%'
        });

        self.$container.append(self.$image);
        self.$image.css('display', 'block');

        if (self.options.showInfo) {
            self.$container.append(this.info);
            self.$info.append('<p></p>');
            self.$info.css('opacity', '0');
        };

        LargeImgInstance = this;
    }

    return LargeImgInstance;
}

_createClass(LargeImg, [{
    key: 'pickup',
    value: function pickup(thumb) {
        var self = this;
        if (self.isPickedUp) {
            self.putDown(thumb, function () {
                self.pickup(thumb);
            });
        } else {
            self.isPickedUp = true;
            self.loadImage(thumb, function () {
                self.$image.fadeTo(self.options.fadeDuration, '1');
                self.enlarge();
                $('body').bind('click', function () {
                    self.putDown(thumb);
                });
            });
        }
    }
}, {
    key: 'putDown',

    value: function putDown(thumb, callback) {
```

```
            self = this;
            $('body').off();
            thumb.setZ(self.options.numLayers);
            self.$container.stop().animate({
                'top': thumb.offset.top + thumb.getShift(),
                'left': thumb.offset.left + thumb.getShift(),
                'width': thumb.width + 'px',
                'height': thumb.height + 'px',
                'padding': self.options.thumbBorderWidth + 'px'
            }, self.options.pickupDuration, function () {
                self.isPickedUp = false;
                self.$container.fadeOut(self.options.fadeDuration, function () {
                    if (callback) callback();
                });
            });
        }
    }, {
        key: 'loadImage',
        value: function loadImage(thumb, callback) {
            var self = this;
            self.$image.css('opacity', '0');
            self.startPosition(thumb);
            var img = new Image();
            img.src = thumb.imgsrc;
            img.onload = function () {
                self.setImageSource(img.src);
                self.fullSizeWidth = this.width;
                self.fullSizeHeight = this.height;
                console.log('img width:', this.width);
                if (callback) callback();
            };
        }
    }, {
        key: 'startPosition',
        value: function startPosition(thumb) {
            var self = this;
            self.$container.css({
                'top': thumb.offset.top + thumb.getShift(),
                'left': thumb.offset.left + thumb.getShift(),
                'transform': 'rotate(' + thumb.getShift() + 'deg)',
                'width': thumb.width + 'px',
                'height': thumb.height + 'px',
                'padding': self.options.thumbBorderWidth
            }).fadeTo(self.options.fadeDuration, '1');
        }
    }, {
        key: 'setImageSource',
        value: function setImageSource(src) {
            this.$image.attr('src', src).css({
                'width': '100%',
                'height': '100%',
                'margin-top': '0'
```

```
            });
        }
    }, {
        key: 'enlarge',
        value: function enlarge() {
            var windowHeight = window.innerHeight ? window.innerHeight :
$(window).height();
            var availableWidth = $(window).width() - 2 * this.windowPadding;
            var availableHeight = windowHeight - 2 * this.windowPadding;
            if (availableWidth < this.fullSizeWidth && availableHeight <
this.fullSizeHeight) {
                if (availableWidth * (this.fullSizeHeight /
this.fullSizeWidth) > availableHeight) {
                    this.enlargeToWindowHeight(availableHeight);
                } else {
                    this.enlargeToWindowWidth(availableWidth);
                }
            } else if (availableWidth < this.fullSizeWidth) {
                this.enlargeToWindowWidth(availableWidth);
            } else if (availableHeight < this.fullSizeHeight) {
                this.enlargeToWindowHeight(availableHeight);
            } else {
                this.enlargeToFullSize();
            }
        }

    }, {
        key: 'enlargeToFullSize',
        value: function enlargeToFullSize() {
            self = this;
            self.$container.css('transform', 'rotate(0deg)').animate({
                'top': $(window).scrollTop() + $(window).height() / 2 -
self.fullSizeHeight / 2,
                'left': $(window).scrollLeft() + $(window).width() / 2 -
self.fullSizeWidth / 2,
                'width': self.fullSizeWidth - 2 * self.options.photoBorder + 'px',
                'height': self.fullSizeHeight - 2 * self.options.photoBorder + 'px',
                'padding': self.options.photoBorder + 'px'
            });
        }
    }, {
        key: 'enlargeToWindowWidth',
        value: function enlargeToWindowWidth(availableWidth) {
            self = this;
            var adjustedHeight = availableWidth * (self.fullSizeHeight /
self.fullSizeWidth);
            self.$container.css('transform', 'rotate(0deg)').animate({
                'top': $(window).scrollTop() + $(window).height() / 2 -
adjustedHeight / 2,
                'left': $(window).scrollLeft() + $(window).width() / 2 -
availableWidth / 2,
                'width': availableWidth + 'px',
```

```
                'height': adjustedHeight + 'px',
                'padding': self.options.photoBorder + 'px'
            });
        }
    }, {
        key: 'enlargeToWindowHeight',
        value: function enlargeToWindowHeight(availableHeight) {
            self = this;
            var adjustedWidth = availableHeight * (self.fullSizeWidth /
self.fullSizeHeight);
            self.container.css('transform', 'rotate(0deg)').animate({
                'top': $(window).scrollTop() + $(window).height() / 2 -
availableHeight / 2,
                'left': $(window).scrollLeft() + $(window).width() / 2 -
adjustedWidth / 2,
                'width': adjustedWidth + 'px',
                'height': availableHeight + 'px',
                'padding': self.options.photoBorder + 'px'
            });
        }
    }]);

    return LargeImg;
}();

var Thumb = function () {
    function Thumb(element, options) {
        _classCallCheck(this, Thumb);

        var self = this;
        self.$element = $(element);
        self.options = options;
        self.activeClass = 'imgpile-active-thumbnail';
        self.$element.children().css('padding', this.options.thumbBorderWidth + 'px');
        self._setRotation();
        self._setOverlap();
        self._bindUIActions();
        self._setRandomZ();

        if (self.options.draggable) {
            var x = 0;
            var velocity = 0;
            self.$element.draggable({
                start: function start(event, ui) {
                    self.$element.addClass('preventClick');
                    self.$element.css('z-index', self.options.numLayers + 2);
                },
                drag: function drag(event, ui) {
                    velocity = (ui.offset.left - x) * 1.2;
                    var ratio = parseInt(velocity * 100 / 360);
                    self.$element.css('transform', 'rotateZ(' + ratio + 'deg)');
                    x = ui.offset.left;
```

```
            },
            stop: function stop(event, ui) {
                self.$element.css('z-index', self.options.numLayers + 1);
            }
        });
    }
}

_createClass(Thumb, [{
    key: 'setZ',
    value: function setZ(layer) {
        this.$element.css('z-index', layer);
    }
}, {
    key: 'bringToTop',
    value: function bringToTop() {
        var self = this;
        this.$element.css({
            'z-index': self.options.numLayers + 1,
            'background': self.options.thumbBorderHover
        });
    }
}, {
    key: 'moveDownOne',
    value: function moveDownOne() {
        this.$element.css({
            'z-index': this.options.numLayers,
            'background': this.options.thumbBorderColor
        });
    }
}, {
    key: 'getRotation',
    value: function getRotation() {
        var self = this;
        var transform = self.$element.css("transform");
        var values = transform.split('(')[1].split(')')[0].split(',');
        var angle = Math.round(Math.asin(values[1]) * (180 / Math.PI));
        return angle;
    }
}, {
    key: 'getShift',
    value: function getShift() {
        return this.getRotation() < 0 ? -(this.getRotation(this) * 0.40) :
this.getRotation(this) * 0.40;
    }
}, {
    key: '_setRotation',
    value: function _setRotation() {
        var min = -1 * this.options.thumbRotation;
        var max = this.options.thumbRotation;
        var randomRotation = Math.floor(Math.random() * (max - min + 1)) + min;
        this.$element.css({ 'transform': 'rotate(' + randomRotation + 'deg)' });
```

```
            }
        }, {
            key: '_setOverlap',
            value: function _setOverlap() {
                var self = this;
                self.$element.css('margin', self.options.thumbOverlap * -1 / 2 + 'px');
            }
        }, {
            key: '_setRandomZ',
            value: function _setRandomZ() {
                var self = this;
                self.$element.css({ 'z-index': Math.floor(Math.random() *
self.options.numLayers + 1) });
            }
        }, {
            key: '_bindUIActions',
            value: function _bindUIActions() {
                var self = this;
                self.$element.on('mouseover', function (event) {
                    self.bringToTop();
                });
                self.$element.on('mouseleave', function (event) {
                    self.moveDownOne();
                });

                self.$element.find('a').on('click tap', function (event) {
                    event.preventDefault();
                    if (self.$element.hasClass('preventClick')) {
                        self.$element.removeClass('preventClick');
                    } else {
                        if (self.$element.hasClass(self.activeClass)) return;
                        var largeImgContainer = new LargeImg(self.options);
                        largeImgContainer.pickup(self);
                    }
                    return false;
                });

                self.$element.on('mousedown', function (event) {
                    self.$element.removeClass('preventClick');
                });
            }
        }, {
            key: 'offset',
            get: function get() {
                var self = this;
                return self.$element.offset();
            }
        }, {
            key: 'height',
            get: function get() {
                var self = this;
                return self.$element.height();
```

```
        }
    }, {
        key: 'width',
        get: function get() {
            var self = this;
            return self.$element.width();
        }
    }, {
        key: 'imgsrc',
        get: function get() {
            var self = this;
            return self.$element.children().first().attr('href');
        }
    }]);

    return Thumb;
}();
'use strict';

(function ($) {
    $.fn.imgpile = function (options) {
        var $self = $(this);
        var settings = $.extend({
            numLayers: 5,
            thumbOverlap: 50,
            thumbRotation: 45,
            thumbBorderWidth: 2,
            thumbBorderColor: 'white',
            thumbBorderHover: '#EAEAEA',
            draggable: true,
            fadeDuration: 200,
            pickupDuration: 500,
            photoZIndex: 100,
            photoBorder: 10,
            photoBorderColor: 'white',
            showInfo: true,
            autoplayGallery: false,
            autoplaySpeed: 5000,
            loading: 'images/loading.gif'
        }, options || {});

        function init() {
            var defer = $.Deferred();
            defer.done(function () {
                afterInitialization();
            });

            function initializeThumbs() {
                $self.addClass('loading').children().each(function (index, element) {
                    var thumb = new Thumb(this, settings);
                    $(element).thumb = thumb;
                });
```

```
                defer.resolve('finished initialization');
            }
            initializeThumbs();

            function afterInitialization() {
                $self.removeClass('loading').css({
                    'padding': settings.thumbOverlap + 'px'
                }).children().css({
                    'opacity': '1',
                    'display': 'inline-block'
                });
                if (settings.autoplayGallery) {
                }
            }
        }
        init();

        return this;
    };
})(jQuery);
```

第17章 项目演练4——开发商品信息展示系统

该项目利用jQuery并结合ECMAScript的ES6语法构建出MVC结构，在一个单独的HTML页面上实现动态的应用程序。该项目是一个类似于美团网的商品信息展示系统，用户可以在上面浏览美食和电影两种不同的商品。同时，用户还可以切换浏览模式，可以用列表模式、大图模式和地图模式进行浏览。利用该项目，可以学习和了解jQuery如何和ES6结合使用，并深入学习和了解ES6的类如何声明和如何使用，通过ES6让JavaScript具有面向对象编程的能力，并利用MVC结构把复杂的问题简单化。

17.1 项目需求分析

需求分析是开发项目的必要环节。下面分析商品信息展示系统的需求。

(1) 该商品信息展示系统类似于美团的商品展示页面。用户可以在上面浏览美食和电影两种不同的商品。在默认情况下，系统会显示所有的商品，用户可以单击不同的分类按钮来分类浏览商品信息。在 Firefox 53.0 中查看效果，如图 17-1 所示。

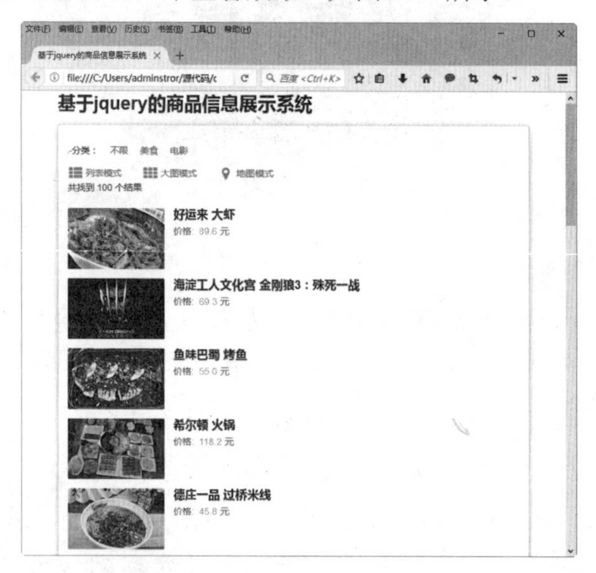

图 17-1 商品信息展示系统主页

(2) 用户可以切换不同的浏览模式，包括默认的列表模式、大图模式和地图模式。如图 17-2 所示为大图模式；如图 17-3 所示为地图模式。

图 17-2 大图模式的效果

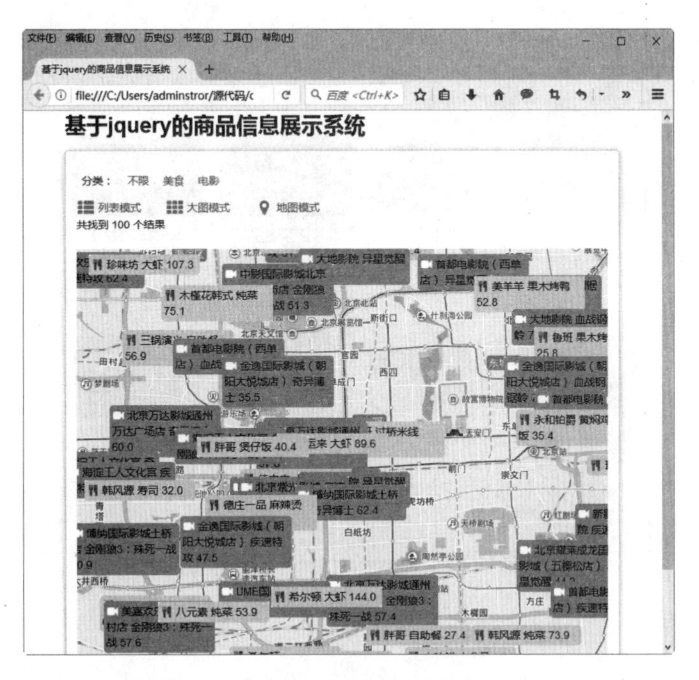

图 17-3 地图模式的效果

(3) 用户单击每个商品，将显示商品放大后的图，并显示商品的详细信息，如图 17-4 所示。

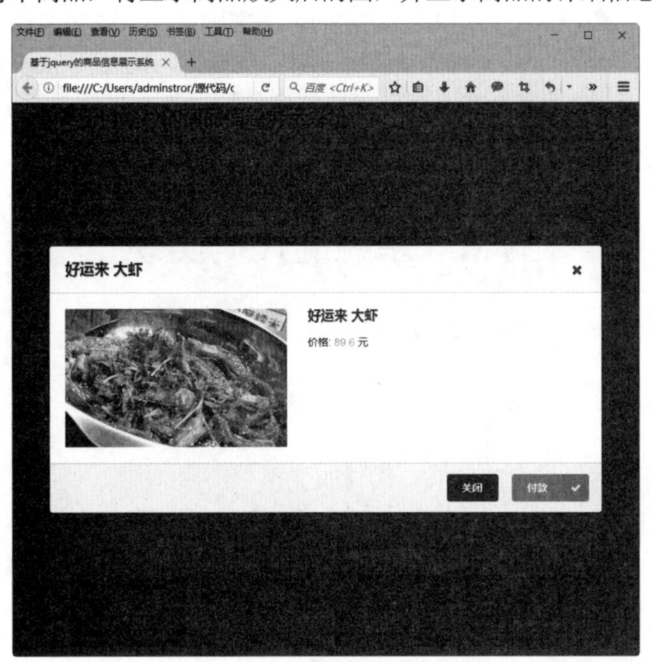

图 17-4 商品的详细信息

(4) 在页面的最下方实现了翻页功能。用户单击不同页码序号，可以实现快速翻页的效果，如图 17-5 所示。

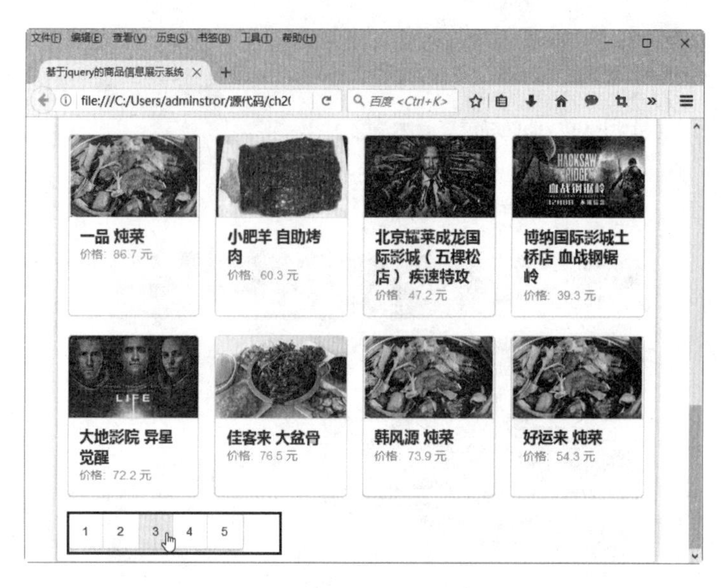

图 17-5　翻页功能

17.2　项目技术分析

和上一章的案例一样，这里也使用了 ES6 的语法规则。

为了便于初学者的学习，该案例没有使用专门的数据库，而是用 nodeJS 生成 data.json 文件，该文件用于提供 json 数据。在没有专门数据库的情况下，仍然实现了动态数据的效果，可见 JavaScript 和 jQuery 搭配使用后的功能是多么强大。

17.3　系统的代码实现

下面来分析商品信息展示系统的代码是如何实现的。

17.3.1　设计首页

首页中显示了商品信息展示的效果。代码如下：

```
<!DOCTYPE html>
<html>
<head>
  <!-- Standard Meta -->
  <meta charset="utf-8" />
  <meta http-equiv="X-UA-Compatible" content="IE=edge,chrome=1" />
  <meta name="viewport" content="width=device-width, initial-scale=1.0,
maximum-scale=1.0">

  <!-- Site Properties -->
  <title>基于 jquery 的商品信息展示系统</title>
```

```
  <link rel="stylesheet" type="text/css" href="css/semantic.min.css">
  <link rel="stylesheet" type="text/css" href="css/style.css">

  <script src="js/jquery-3.1.1.min.js"></script>
  <script type="text/javascript" src="http://api.map.baidu.com/api?v=
2.0&ak=58HRTz2BqR7brG1Ys5qMG6yFdj8A5Gzg"></script>
  <script src="js/RichMarker_min.js"></script>
  <script src="css/semantic.min.js"></script>
  <script src="js/mvc.js"></script>
  <script src="js/app.js"></script>
</head>
<body>

<div class="ui container">
  <h1>基于 jquery 的休闲娱乐信息展示系统</h1>

  <div class="ui raised segment">

    <div class="condition-cont">
      <dl class="condition-area">
        <dt>分类：</dt>
        <dd class="unlimited"><a data-type="">不限</a></dd>
        <dd>
          <a data-type="food">美食</a></dd>
        <dd>
          <a data-type="movie">电影</a></dd>
      </dl>
      <p class="viewtypes">
        <a data-view="list"><i class="large teal list layout icon"></i>列表模
式</a>
        <a data-view="grid"><i class="large teal grid layout icon"></i>大图模
式</a>
        <a data-view="map"><i class="large teal marker icon"></i>地图模式</a>
      </p>
    </div>

    <div class="products">
      <p class="total">共找到 <span class="count"></span> 个结果</p>

      <div class="ui items" id="itemlist"></div>

      <div id="mapContainer"></div>

      <div class="pager">

      </div>
    </div>

    <div id="modalContainer"></div>
```

```
  </div>
</div>
</body>
</html>
```

17.3.2 开发控制器类的文件

Controller.js 文件位于项目的\web\js\es6\目录下，主要内容为控制器类。该类主要实现的方法含义如下。

(1) index()：显示产品列表。

(2) set type()：切换不同类型的产品。

(3) set viewModel()：切换视图模式。

(4) viewDetail()：显示商品详细信息。

(5) Map()：以地图的方式显示商品信息。

Controller.js 文件的代码如下：

```
class Controller{
    constructor(model, view, type){
        this._view = view;
        this.model = model;
        this._type = type;
    }

    /**
     * render index page
     * @param integer page current page number
     * @param string type
     */
    index(page){
        var self = this;
        if(self._view.name == 'map'){
            if(this._type){
                self.model.findType(this._type).then(function(data){
                    self._view.display(data, false);
                });
            }else{
                self.model.findAll().then(function(data){
                    self._view.display(data, false);
                });
            }
        }else{
            self.model.find({page:page,type:self._type}).then(function(data){
                self._view.display(data, false);
            });
        }
    }
```

```
    set type(val){
        this._type = val;
    }

    /**
     * change view model. list or grid or map.
     * @param view
     */
    set viewModel(view){
        var self = this;
        this._view = view;
        if(self._view.name == 'map'){
            self._view.init();
            if(this._type){
                self.model.findType(this._type).then(function(data){
                    self._view.display(data, false);
                });
            }else{
                self.model.findAll().then(function(data){
                    self._view.display(data, false);
                });
            }
        }else{
            this._view.display(this.model.cache.last, false);
        }
    }

    get viewModel(){
        return this._view;
    }

    viewDetail(id){
        var self = this;
        var model = self.model.findById(id);
        if(model){
            this._view.detail(model);
        }
    }

    map(type){
        type = type?type:this._type;
        var map = new MapView();
        if(type){
            map.addMarks(this.model.cache[type]);
        }else{
            map.addMarks(this.model.cache.all);
        }
    }
}
```

17.3.3 开发数据模型类文件

Model.js 位于项目的\web\js\es6\目录下，主要内容为数据模型类。该类主要实现的方法含义如下。

(1) findAll()：返回所有商品的数据信息。

(2) findType()：返回指定类型的商品数据信息。

(3) Find()：根据条件返回数据，并把数据裁剪成 pagination 中定义的 pageSize 的数量。

(4) findById()：根据给定 ID 返回一条数据。

Model.js 的代码如下：

```
class Model{
    constructor(pagination){
        var self = this;
        self.data_file = 'data.json';
        self.pagination = pagination;
        self.data = $.getJSON(self.data_file);
        self.cache = {};
        self.cache.last = null; //last found items, pagination pageSized items.
        self.cache.all = null;
        self.cache.food = null;
        self.cache.movie = null;
    }

    _paginationCut(data, condition){
        var self = this;
        self.pagination.totalCount = data.length;
        self.pagination.page = condition.page;
        console.log('self.pagination.page:', self.pagination.page);
        console.log('self.pagination.offset:', self.pagination.offset);
        var pagination = self.pagination;
        console.log('pagination.offset:', pagination.offset);
        data = data.slice(pagination.offset, pagination.offset+pagination.limit);
        self.cache.last = data;
        return data;
    }

    find(condition){
        var self = this;
        var defer = $.Deferred();
        if(condition.type){
            this.findType(condition.type).then(function(data){
                data = self._paginationCut(data, condition);
                defer.resolve(data);
            });
        }else{
            this.findAll().then(function(data){
                data = self._paginationCut(data, condition);
```

```
            defer.resolve(data);
        });
    }
    return defer;
}

findAll(){
    var self = this;
    var defer = $.Deferred();
    if(self.cache.all){
        self.pagination.totalCount = self.cache.all.length;
        defer.resolve(self.cache.all);
    }else{
        self.data.done(function(data){
            var items = data.items;
            self.cache.all = items;
            self.pagination.totalCount = self.cache.all.length;
            defer.resolve(self.cache.all);
        });
    }
    return defer;
}

findType(type){
    var self = this;
    var defer = $.Deferred();
    if(self.cache[type]){
        self.pagination.totalCount = self.cache[type].length;
        defer.resolve(self.cache[type]);
    }else{
        self.data.done(function(data){
            var items = data.items.filter(function(elem, index, self) {
                return elem.type == type;
            });
            self.cache[type] = items;
            self.pagination.totalCount = items.length;
            defer.resolve(self.cache[type]);
        });
    }
    return defer;
}

findById(id){
    var self = this;
    var result = self.cache.all.filter(function(element, index, array) {
        return element['id'] == id;
    });
    // console.log('findById:', result);
    if(result.length>0)
        return result[0];
```

```
        else
            return false;
    }
}
```

17.3.4 开发视图抽象类的文件

AbstractView.js 位于项目的\web\js\es6\目录下，主要内容为视图抽象类。该类主要实现的方法含义如下。

(1) Detail()：以 modal 窗口显示详细信息。

(2) displayTotalCount()：显示一共找到多少条满足条件。

(3) displayPager()：显示翻页按钮。

(4) replaceProducts()：删除前一页内容，显示当前页内容。

(5) Display()：公共方法。该方法调用方法 displayTotalCount()、appendProducts()和 displayPager()。

(6) _models2HtmlStr()：抽象方法。根据当前页数组对象返回 html 字符串。

AbstractView.js 文件的代码如下：

```
/**
* 该类是视图的抽象类. ListView , GridView 继承该类
 */
class AbstractView{
    constructor(pagination, options){
        this.settings = $.extend({
            listContainer:      '#itemlist',
            mapContainer:      '#mapContainer',
            modalContainer:     '#modalContainer',
            pagerCountainer:    '.pager',
            totalCount:         '.total .count'
        }, options||{});
        this.pagination = pagination;
        this.name = 'abstract';
    }
    static get noSearchData(){
        return `<li class="no-data">当前查询条件下没有信息，去试试其他查询条件吧！</li>`;
    }

    /**
     * 以 modal 窗口显示详细信息
     */
    detail(model){
        var self = this;
        $(self.settings.modalContainer).empty().append(
            `<div class="ui modal">
              <i class="close icon"></i>
              <div class="header">
```

```
            ${model.name}
          </div>
          <div class="image content">
            <div class="ui medium image">
              <img src="${model.img}">
            </div>
            <div class="description">
              <div class="ui header">${model.name}</div>
              <p>
                  <span class="priceLabel">价格: </span>
                  <span class="price">${model.price}</span> 元
              </p>
            </div>
          </div>
          <div class="actions">
            <div class="ui black deny button">
              关闭
            </div>
            <div class="ui positive right labeled icon button">
              付款
              <i class="checkmark icon"></i>
            </div>
          </div>
        </div>`
    );
    $('.ui.modal').modal('show');
    $('.ui.modal').modal({onHidden:function(event){
        $(this).remove();
    }});
}

/**
 * 显示一共找到多少条满足条件
 */
displayTotalCount(){
    console.log('this.pagination.totalCount:',this.pagination.totalCount);
    $(this.settings.totalCount).text(this.pagination.totalCount);
}

/**
 * 显示翻页按钮
 */
displayPager(){
    var $pages = this._pageButtons();
    $(this.settings.pagerCountainer).empty().append($pages);
}

_pageButtons(){
    var $container = $('<div>').addClass('ui pagination menu');
    var pagerange = this.pagination.pageRange;
```

```
    for(var i = pagerange[0]; i <= pagerange[1]; i++){
        var $btn = $('<a>').addClass('item').text(i+1);
        if(i == this.pagination.page){
            $btn.addClass('active');
        }
        $container.append($btn);
    }
    return $container;
}

/**
 * 删除前一页内容，显示当前页内容
 * @param array items 当前页的数组对象
 */
replaceProducts(items){
    var self = this;
    $(self.settings.listContainer).empty().removeAttr('style');
    var htmlString = this._models2HtmlStr(items);
    if(htmlString)
        $(self.settings.listContainer).html(htmlString);
    else{
        $(self.settings.listContainer).html(this.constructor.noSearchData);
    }
    window.scrollTo(0,0);
}
/**
 * 保持前一页内容，在前页内容的后面继续添加当前页的内容
 * @param array items 当前页的数组对象
 */
appendProducts(items){
    var self = this;
    var htmlString = this._models2HtmlStr(items);
    if(htmlString)
        $(self.settings.listContainer).append(htmlString);
}

/**
 * public方法，该方法调用displayTotalCount()、appendProducts()和displayPager()
 * @param items 当前页的数组对象
 * @param append 默认保留前一页内容，在前一页后面继续添加当前页内容
 */
display(items, append=true){
    $(this.settings.mapContainer).empty().removeAttr('style');
    this.displayTotalCount();
    if(append){
        this.appendProducts(items);
    }else{
        this.replaceProducts(items);
    }
    this.displayPager();
```

```
    }

    /**
     * 抽象方法，根据当前页数组对象返回 html 字符串
     * @param array models 当前页的数组对象
     * @private
     */
    _models2HtmlStr(models){

    }
}
```

17.3.5 项目中的其他 js 文件说明

由于篇幅限制，这里不再对每个 js 文件进行详细说明，读者可以参照源文件进行查看。除了上述 js 文件以外，还有以下几个比较重要的 js 文件。

(1) ListView.js：定义列表视图类，继承抽象视图类。

(2) GridView.js：定义大图视图类，继承抽象视图类。

(3) MapView.js：定义地图视图类。

(4) Pagination.js：定义翻页功能类。

(5) generateData.js：用于生成 data.json 数据文件，该文件用 nodejs 执行，注意该文件不能在浏览器上直接运行。

(6) gulpfile.js：定义了如何把 ES6 文件转换成 ES5 文件，并把多个 js 文件合并成一个 js 文件，以及把 less 文件编译成 CSS 文件。

(7) mvc.js：将 Model.js、AbstractView.js、ListView.js、GridView.js、MapView.js 和 Pagination.js 最后合并到 mvc.js 文件里。

第 18 章
项目演练 5——
开发连锁酒店
移动网站

本章介绍一个酒店订购系统的开发，这里将使用前面学习的 local Storage 来处理订单的存储和查询。该系统的主要功能为订购房间、查询连锁分店、查询订单、查看酒店介绍等功能。通过本章的学习，用户可以了解在线订购系统的制作方法、使用 localStorage 模拟在线订购和查询订单的方法和技巧。

18.1 连锁酒店订购的需求分析

需求分析是连锁酒店订购系统开发的必要环节，该系统的需求如下。

(1) 用户可以预订不同的房间级别，定制个性化的房间，而且还可以快速搜索自己需要的房间类型。

(2) 用户可以查看全国连锁酒店的分店情况，并且可以自主联系酒店的分店。

(3) 用户可以查看预订过的订单详情，还可以删除不需要的订单。

(4) 用户可以查看连锁酒店的介绍。

制作完成后的首页效果如图 18-1 所示。

图 18-1 首页效果

18.2 网站的结构

分析完网站的功能后，开始分析整个网站的结构，主要分为以下 5 个页面，如图 18-2 所示。

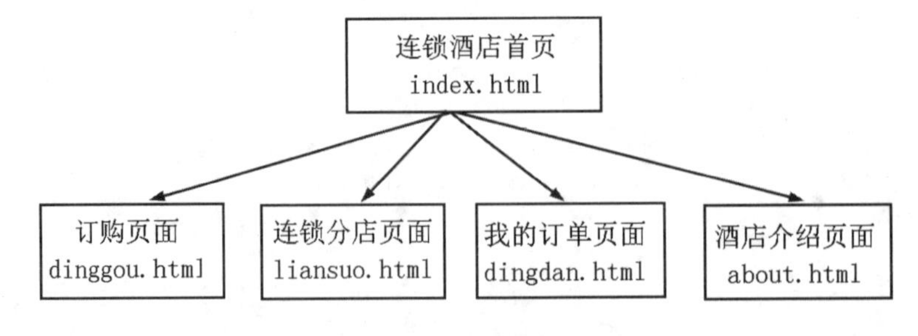

图 18-2 网站的结构

各个页面的主要功能如下。

(1) index.html：该页面是系统的主页面，是网站的入口，通过主页可以链接到订购页面、连锁分店页面、我的订单页面和酒店介绍页面。

(2) dinggou.html：该页面是酒店订购页面，主要包括 3 个 page，第一个 page 是选择房间类型；第二个 page 是选择房间的具体参数；第三个 page 是显示订单完成信息。

(3) liansuo.html：该页面主要显示连锁分店的具体信息。

(4) dingdan.html：该页面主要显示用户已经订购的订单信息。

(5) about.html：该页面主要显示关于连锁酒店的简单介绍。

18.3 连锁酒店系统的代码实现

下面分析连锁酒店系统的代码是如何实现的。

18.3.1 设计首页

首页中主要包括 1 张图片和 4 个按钮，分别链接到订购页面、连锁分店页面、我的订单页面和酒店介绍页面。主要代码如下：

```
<div data-role="page" data-title="Happy" id="first" data-theme="a">
<div data-role="header">
<h1>千谷连锁酒店系统</h1>
</div>
<div data-role="content" id="content" class="firstcontent">
    <img src="images/zhu.png" id="logo"><br/>
    <a href="caigou.html" data-ajax="false" data-role="button" data-icon="home" data-iconpos="top" data-mini="true" data-inline="true"><img src="images/cai.png"><br>立即预订</a>
    <a href="liansuo.html" data-ajax="false" data-role="button" data-icon="search" data-iconpos="top" data-mini="true" data-inline="true"><img src="images/lian.png"><br>连锁分店</a>
    <a href="dingdan.html" data-ajax="false" data-role="button" data-icon="gear" data-iconpos="top" data-mini="true" data-inline="true"><img src="images/ding.png"><br>我的订单</a>
       <a href="about.html" data-ajax="false" data-role="button" data-icon="gear" data-iconpos="top" data-mini="true" data-inline="true"><img src="images/ding.png"><br>关于千谷</a>
</div>
<div data-role="footer" data-position="fixed" style="text-align:center">
  订购专线：12345678
</div>
</div>
```

其中 data-ajax="false"表示停用 Ajax 加载网页；data-role="button"表示该链接的外观以按钮的形式显示；data-icon="home"表示按钮的图标效果；data-iconpos="top"表示小图标在按钮上方显示；data-inline="true"表示以最小宽度显示。效果如图 18-3 所示。

图 18-3　链接的样式效果

其中页脚部分通过设置属性 data-position="fixed"，可以让页脚内容一直显示在页面的最下方。通过设置 style="text-align:center"，可以让页脚内容居中显示，如图 18-4 所示。

图 18-4　页脚的样式效果

18.3.2　订购页面

订购页面主要包含 3 个 page，主要包括选择房间类型 page(id=first)、选择房间的具体参数 page(id=second)和显示订单完成信息 page(id=third)。

1. 选择房间类型 page

其中选择房间类型 page 中包括房间列表、返回到上一页、快速搜索房间等功能。代码如下：

```
<div data-role="page" data-title="房间列表" id="first" data-theme="a">
<div data-role="header">
<a href="index.html" data-icon="arrow-l" data-iconpos="left" data-
ajax="false">Back</a> <h1>房间列表</h1>
</div>
<div data-role="content" id="content">
    <ul data-role="listview" data-inset="true" data-filter="true" data-
filter-placeholder="快速搜索房间">
        <li>
            <a href="#second">
            <img src="images/putong.png" />
            <h3>普通间</h3>
            <p>24 小时有热水</p>
            </a>
            <a href="#second" data-icon="plus"></a>
        </li>
        <li>
            <a href="#second">
             <img src="images/wangluo.png" />
             <h3>网络间</h3>
             <p>有网络和电脑、24 小时热水</p>
            </a>
            <a href="#second" data-icon="plus"></a>
```

```
        </li>
        <li>
            <a href="#second">
              <img src="images/haohua.png" />
              <h3>豪华间</h3>
              <p>免费提供三餐、有网络和电脑、24 小时热水</p>
            </a>
            <a href="#second" data-icon="plus"></a>
        </li>
        <li>
            <a href="#second">
              <img src="images/zongtong.png" />
              <h3>总统间</h3>
              <p>24 小时客服、有网络和电脑、24 小时热水、免费提供三餐</p>
            </a>
            <a href="#second" data-icon="plus"></a>
        </li>
    </ul>
        </div>
<div data-role="footer" data-position="fixed" style="text-align:center">
  订购专线：12345678
</div>
</div>
```

房间列表页面的效果如图 18-5 所示。

页面中有一个 Back 按钮，主要作用是返回到主页上，通过以下代码来控制：

```
<a href="index.html" data-icon="arrow-l" data-iconpos="left" data-
ajax="false">Back</a>
```

房间列表使用 listview 组件，通过设置 data-filter="true"，就会在列表上方显示搜索框；通过设置 data-inset="true"，可以让 listview 组件添加圆角效果，而且不与屏幕同宽；其中 data-filter-placeholder 属性用于设置搜索框内显示的内容，当输入搜索内容，将查询出相关的记账信息，如图 18-6 所示。

图 18-5　房间列表页面效果

图 18-6　快速搜索房间

2. 选择房间的具体参数 page

选择房间的具体参数 page 的 id 为 second，主要让用户选择楼层、是否带窗口、是否需要接送、订购数量和客户联系方式，如图 18-7 所示。

这个页面的 Back 按钮的设置方法和上一个 page 不同，通过设置属性 data-add-back-btn="true"实现返回上一页的功能，代码如下：

```
<div data-role="page" data-title="选择房间"
id="second" data-theme="a" data-add-back-
btn="true">
```

该页面中包含选择菜单(Select menu)、2 个单选按钮组件(Radio button)、范围滑块(Slider)、文本框(text)和按钮组件(button)。

其中添加选择菜单(Select menu)的代码如下：

```
<div data-role="content" id="content">
    选择楼层:
    <select name="selectitem"
id="selectitem">
       <option value="一楼">一楼</option>
       <option value="二楼">二楼</option>
       <option value="三楼">三楼</option>
    </select>
```

图 18-7　选择房间页面

预览效果如图 18-8 所示。

2 个单选按钮组的代码如下：

```
<fieldset data-role="controlgroup">
        <legend>选择是否带窗口: </legend>
            <input type="radio" name="flavoritem" id="radio-choice-1"
value="有窗口" checked />
            <label for="radio-choice-1">有窗户</label>
            <input type="radio" name="flavoritem" id="radio-choice-2"
value="无窗户"  />
            <label for="radio-choice-2">无窗户</label>
<fieldset data-role="controlgroup1">
        <legend>选择是否接送: </legend>
            <input type="radio" name="flavoritem1" id="radio-choice-3"
value="需要接送" checked />
            <label for="radio-choice-3">需要接送</label>
            <input type="radio" name="flavoritem1" id="radio-choice-4"
value="无需接送"  />
            <label for="radio-choice-4">无需接送</label>
```

预览效果如图 18-9 所示。

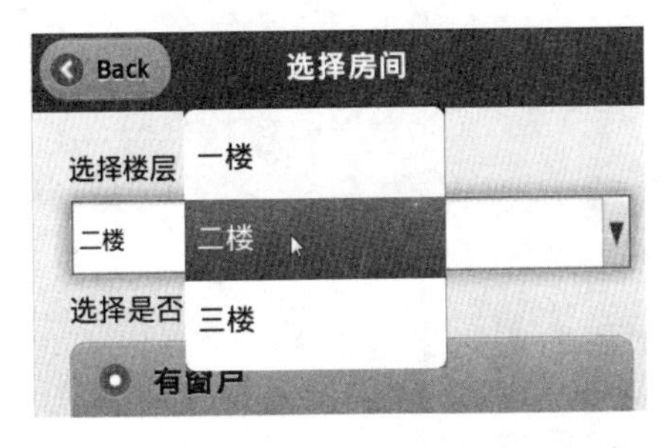

图 18-8　选择菜单效果

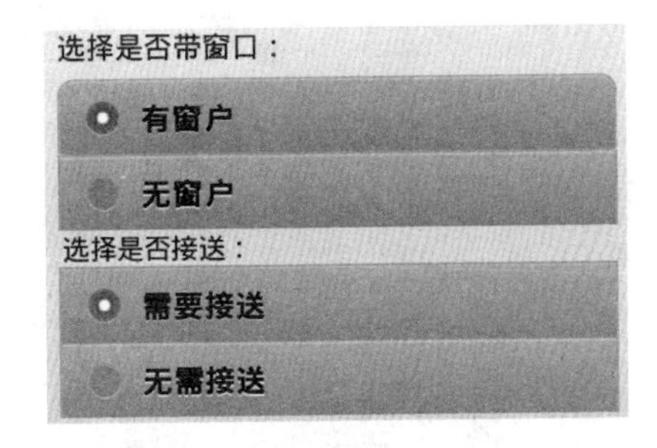

图 18-9　单选按钮组效果

使用<fieldset>标记创建单选按钮组，通过设置属性 data-role="controlgroup"，可以让各个单选按钮外观像一个组合，整体效果比较好。

范围滑块的代码如下：

```
<input type="range" name="num" id="num" value="1" min="0" max="100" data-
highlight="true" />
```

预览效果如图 18-10 所示。

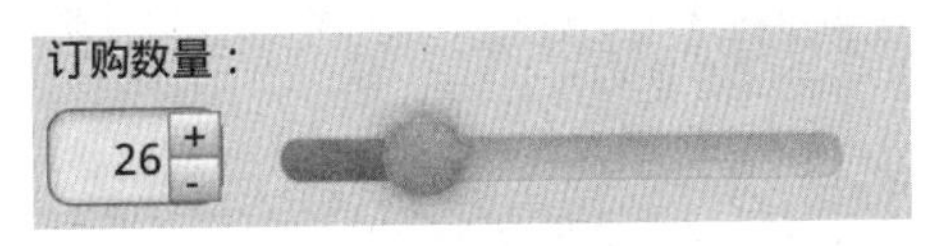

图 18-10　范围滑块效果

文本框的代码如下：

```
<input type="text" name="text1" id="text1" size="10" maxlength="10" />
```

其中 size 属性用于设置文本框的长度；maxlength 属性用于设置输入的最大值。

预览效果如图 18-11 所示。

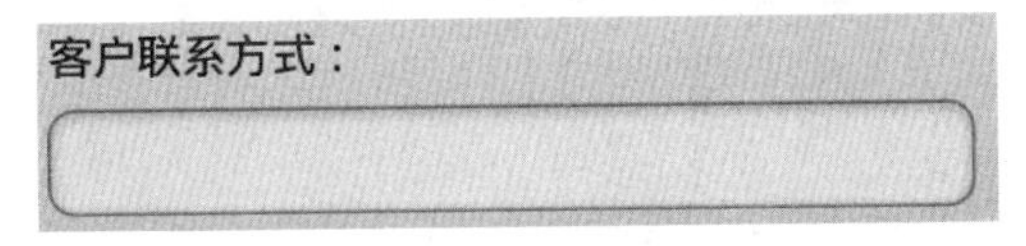

图 18-11　文本框效果

确认按钮的代码如下：

```
<input type="button" id="addToStorage" value="确认订单" />
```

预览效果如图 18-12 所示。

图 18-12　确认按钮效果

3. 显示订单完成信息 page

显示订单完成信息 page 的代码如下：

```
<div data-role="page" id="third">
<div data-role="header">
<a href="index.html" data-icon="arrow-l" data-iconpos="left" data-
ajax="false">回首页</a> <h1>订购完成</h1>
</div>
<div data-role="content" id="content">
<img src="images/ding.png" /><br>
<font style="font-size:20px;">感谢您选择我们酒店<br>
以下为您的订购房间信息：</font>
<p><div id="message" style="font-size:25px;color:#ff0000"></div>
</div>
<div data-role="footer" data-position="fixed" style="text-align:center">
 订购专线：12345678
</div>
</div>
```

预览效果如图 18-13 所示。

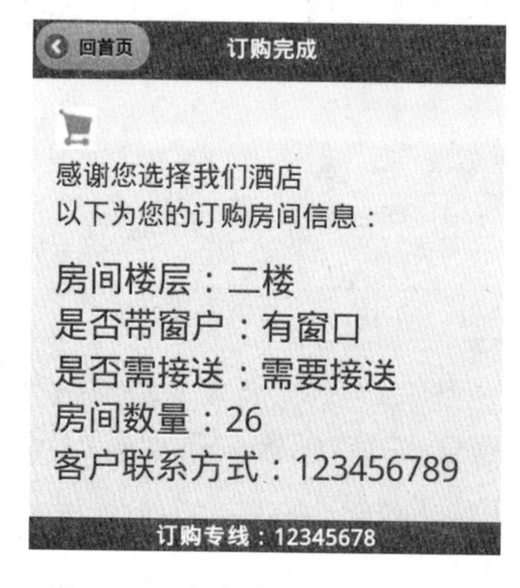

图 18-13　订单完成信息页面效果

接收订单的功能是通过 JavaScript 来完成的，代码如下：

```
<script type="text/javascript">
 var orderitem = "orderitem";
 var flavor = "itemflavor";
var flavor1 = "itemflavor1";
 var num = "num";
 var text1 = "text1";
        $("#second").live('pagecreate', function() {
            $('#addToStorage').click(function() {
                localStorage.orderitem=$("select#selectitem").val();
                  localStorage.flavor=$('input[name="flavoritem"]:checked').val();

localStorage.flavor1=$('input[name="flavoritem1"]:checked').val();
                  localStorage.num=$('#num').val();
                               localStorage.text1=$('#text1').val();
                     $.mobile.changePage($('#third'),{transition: 'slide'});
            });
```

```
        });
        $('#third').live('pageinit', function() {
                var itemflavor = "房间楼层："+ localStorage.orderitem+"<br>
是否带窗户："+localStorage.flavor+"<br>是否需接送："+localStorage.flavor1+"<br>
房间数量："+localStorage.num+"<br>客户联系方式：
"+localStorage.text1;
                $('#message').html(itemflavor);
               //document.getElementById("message").innerHTML= itemflavor
          });
</script>
```

其中$符号代表组件，例如$("#second")表示 id 为 second 的组件。live()函数为文件页面附加事件处理程序，并规定事件发生时执行的函数。例如下面的代码表示当 id 为 second 的页面发生 pagecreate 事件时，就执行相应的函数：

```
$("#second").live('pagecreate', function() {…});
```

当 id 为 second 的页面确认订单时，将会把订单的信息保存到 localStorage。当加载到 id 为 third 的页面时，将 localStorage 存放的内容取出来并显示在 id 为 message 的<div>组件中。代码如下：

```
  $('#third').live('pageinit', function() {
                var itemflavor = "房间楼层："+ localStorage.orderitem+"<br>
是否带窗户："+localStorage.flavor+"<br>是否需接送："+localStorage.flavor1+"<br>
房间数量："+localStorage.num+"<br>客户联系方式：
"+localStorage.text1;
                $('#message').html(itemflavor);
          });
```

其中$('#message').html(itemflavor)的语法作用和下面的代码一样，都是用 itemflavor 字符串替代<div>组件中的内容：

```
document.getElementById("message").innerHTML= itemflavor;
```

18.3.3 连锁分店页面

连锁分店页面为 liansuo.html，主要代码如下：

```
<div data-role="page" data-title="全国连锁酒店" id="first" data-theme="a">
<div data-role="header">
<a href="index.html" data-icon="arrow-l" data-iconpos="left" data-
ajax="false">回首页</a>
<h1>全国连锁酒店</h1>
</div>
<div data-role="content" id="content">
    <ul data-role="listview" data-inset="true">
            <li>
                <a href="#" onclick="getmap('上海连锁酒店')" id=btn>
                  <img src="images/shanghai.png" />
                  <h3>上海连锁酒店</h3>
```

```
                <p>咨询热线: 19912345678</p>
              </a>

            </li>
            <li>
              <a href="#" onclick="getmap('北京连锁酒店')" id=btn>
                <img src="images/beijing.png" />
                <h3>北京连锁酒店</h3>
                <p>咨询热线: 18812345678</p>
              </a>

            </li>
            <li>
              <a href="#" onclick="getmap('厦门连锁酒店')" id=btn>
                <img src="images/xiamen.png" />
                <h3>厦门连锁酒店</h3>
                <p>咨询热线: 16612345678</p>
              </a>

            </li>
        </ul>

</div>
<div data-role="footer" data-position="fixed" style="text-align:center">
  连锁酒店总部热线: 12345678
</div>
</div>
```

连锁分店页面的预览效果如图 18-14 所示。

图 18-14　连锁分店页面效果

其中使用 listview 组件来完成列表的功能。通过链接的方式返回到首页，代码如下：

```
<a href="index.html" data-icon="arrow-l" data-iconpos="left" data-
ajax="false">回首页</a>
```

18.3.4 查看订单页面

查看订单页面为 dingdan.html，显示内容的代码如下：

```
<div data-role="page" data-title="订单列表" id="first" data-theme="a">
<div data-role="header">
<a href="index.html" data-icon="arrow-l" data-iconpos="left" data-
ajax="false">回首页</a><h1>订单列表</h1>
</div>
<div data-role="content" id="content">
<a href="#" data-role="button" data-inline="true" onclick="deleteOrder();">
删除订单</a>
以下为您的订购列表:
<div class="ui-grid-b">
  <div class="ui-block-a ui-bar-a">房间楼层</div>
  <div class="ui-block-b ui-bar-a">是否带窗户</div>
 <div class="ui-block-b ui-bar-a">是否需接送</div>

  <div class="ui-block-a ui-bar-b" id="orderitem"></div>
  <div class="ui-block-b ui-bar-b" id="flavor"></div>
 <div class="ui-block-b ui-bar-b" id="flavor1"></div>
 <div class="ui-block-c ui-bar-a">订购数量</div>
  <div class="ui-block-c ui-bar-a">客户联系方式</div>
 <div class="ui-block-c ui-bar-a"></div>
  <div class="ui-block-c ui-bar-b" id="num"></div>
  <div class="ui-block-c ui-bar-b" id="text1"></div>
</div>
</div>
<div data-role="footer" data-position="fixed" style="text-align:center">
  订购专线: 12345678
</div>
```

查看订单页面的预览效果如图 18-15 所示。

该页面的主要功能是将 localStorage 的数据取出并显示在页面上，主要由以下代码实现：

```
<script type="text/javascript">
$('#first').live('pageinit', function() {
    $('#orderitem').html(localStorage.orderitem);
    $('#flavor').html(localStorage.flavor);
        $('#flavor1').html(localStorage.flavor1);
    $('#num').html(localStorage.num);
        $('#text1').html(localStorage.text1);
});
</script>
```

图 18-15 查看订单页面效果

通过单击页面中的【删除订单】按钮，可以删除订单。通过以下函数实现删除功能：

```
function deleteOrder(){
    localStorage.clear();
```

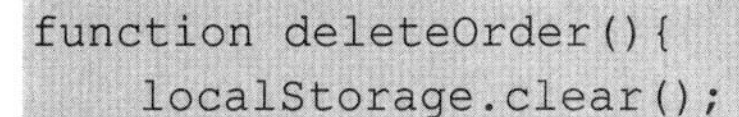

```
    $(".ui-grid-b").html("已取消订单!");
}
```

18.3.5 酒店介绍页面

酒店介绍页面为about.html，该页面的主要代码如下：

```
<div data-role="page" data-title="全国连锁酒店" id="first" data-theme="a">
<div data-role="header">
<a href="index.html" data-icon="arrow-l" data-iconpos="left" data-
ajax="false">回首页</a><h1>千谷连锁酒店</h1>
</div>
<div data-role="content" id="content">

<img src="images/about.png" /><br>
<font style="font-size:20px;">千谷连锁酒店集团定位于全国连锁高级酒店的发展,完善的酒
店预订系统,让您预订酒店客房更加轻松快捷,是您出差、旅游的好选择。</font>

</div>
<div data-role="footer" data-position="fixed" style="text-align:center">
  连锁酒店总部热线: 12345678
</div>
</div>
```

酒店介绍页面的预览效果如图18-16所示。

图18-16　酒店介绍页面效果